T0329707

Environmental Policy

Environmental Policy

An Economic Perspective

Edited by

Thomas Walker
Full Professor, John Molson School of Business,
Concordia University, Montréal, Québec, Canada

Northrop Sprung-Much
PhD Candidate, Department of Finance, John Molson School of Business,
Concordia University, Montréal, Québec, Canada

Sherif Goubran
PhD Candidate, Individualized Program, Concordia University,
Montréal, Québec, Canada

Registered Offices
John Wiley & Sons, Inc., 111 River Street, Hoboken, NJ 07030, USA
John Wiley & Sons Ltd, The Atrium, Southern Gate, Chichester, West Sussex, PO19 8SQ, UK

Editorial Office
9600 Garsington Road, Oxford, OX4 2DQ, UK

For details of our global editorial offices, customer services, and more information about Wiley products visit us at www.wiley.com.

Wiley also publishes its books in a variety of electronic formats and by print-on-demand. Some content that appears in standard print versions of this book may not be available in other formats.

Library of Congress Cataloging-in-Publication Data

Names: Walker, Thomas (Thomas J.) editor. | Sprung-Much, Northrop, 1991– editor. | Goubran, Sherif, 1991– editor.
Title: Environmental policy : an economic perspective / edited by Thomas Walker, Northrop Sprung-Much, Sherif Goubran.
Other titles: Environmental policy (Wiley Blackwell)
Description: Hoboken, NJ : Wiley Blackwell, 2020. | Includes bibliographical references and index.
Identifiers: LCCN 2020019234 (print) | LCCN 2020019235 (ebook) | ISBN 9781119402596 (cloth) | ISBN 9781119402572 (adobe pdf) | ISBN 9781119402558 (epub)
Subjects: LCSH: Environmental policy–Economic aspects. | Global environmental change–Economic aspects. | Climatic changes–Economic aspects. | Sustainable development–Economic aspects. | Renewable energy resources–Economic aspects.
Classification: LCC HC79.E5 E57877 2020 (print) | LCC HC79.E5 (ebook) | DDC 333.7–dc23
LC record available at https://lccn.loc.gov/2020019234
LC ebook record available at https://lccn.loc.gov/2020019235

Cover Design: Wiley
Cover Image: © Sylvain Sonnet / Getty Images

Set in 9.5/12.5pt STIXTwoText by SPi Global, Pondicherry, India
Printed and bound by CPI Group (UK) Ltd, Croydon, CR0 4YY

10 9 8 7 6 5 4 3 2 1

Contents

Preface

Environmental legislation has a profound effect on everyone's lives as it has the power to save lives by ensuring proper health standards, to create and maintain green jobs, and to protect natural habitat all around the globe. The role it will play in the next couple of years will be significant, so in this book, we aim to understand where we are and where we are going as a society in fighting against climate change. Our book does this by examining the new advances made in the environmental legislation field. We as people have the power to influence policies through our role in society, so it is important that we understand these policies. Our wellbeing and that of our planet depend on it.

This book addresses the many different themes involved in environmental policies. It includes topics such as climate change legislation, water pricing and the conservation of water, biodiversity of the marine environment, wildlife ranching, emission trading schemes, green job strategies, market-based instruments, sustainable investing, and lastly food security. The book takes a global perspective as many of the contributions cover different geographical locations. The global perspective allows readers to have a macro understanding of environmental policies, while examining specific local areas also allows the reader to have a micro understanding of specific issues.

The content provided in the book is presented by leading experts and practitioners in related fields. The authors provide an in-depth analysis of the key issues to be addressed and the new developments made through their own research efforts. Each chapter has a different focus which spans across the different topics mentioned above.

The book's main goal is to inform readers on the recent developments made in the field of environmental policy and to provide a granular view of the different subtopics within it. The contents of this book can be used by academic researchers in the field, graduate students who have an interest in the topic and any reader who aims to be informed about the developing environmental legislations affecting them. Our book aims to provide cutting-edge research and to capture the latest trends in the transition toward a more sustainability-oriented legislative system.

About the Editors

Thomas Walker is a full professor of finance at Concordia University, Montreal. He previously served as associate dean, department chair, and director of Concordia's David O'Brien Centre for Sustainable Enterprise. Prior to his academic career, he worked for firms such as Mercedes Benz, KPMG, and Utility Consultants International. He is well published with over 50 journal articles and books.

Northrop Sprung-Much is a research staff member at the David O'Brien Centre for Sustainable Enterprise, and a PhD candidate in finance at the John Molson School of Business at Concordia University. He holds an MBA (focus in finance) from Concordia University. His research interests center around risk management, carbon trading, and environmental economics.

Sherif Goubran is a PhD candidate in the Individualized Program at Concordia University and a Vanier Scholar (SSHRC). His interdisciplinary research is focused on sustainable building practices within the fields of design, building engineering, and finance. His PhD research investigates the alignment between sustainable practices and global sustainability goals. Sherif holds a BSc in Architecture and a MASc in Building Engineering.

List of Contributors

Elia E. Cia Alves
Federal University of Paraíba,
Joao Pessoa, Paraíba, Brazil

Bruno Arcand
University of Quebec, Montreal,
Quebec, Canada

Andrew Brennan
School of Economics, Finance
and Property, Faculty of Business
and Law, Curtin University, Bentley,
Western Australia, Australia

Stephanie Cairns
Smart Prosperity Institute, Ottawa,
Ontario, Canada

Amelia Clarke
School of Environment, Enterprise
and Development (SEED), University
of Waterloo, Waterloo, Ontario,
Canada

Sherif Goubran
Concordia University, Montreal,
Quebec, Canada

Tariro Kamuti
Department of Public Law, University
of Cape Town, Cape Town,
South Africa

David H. Lont
University of Otago, Dunedin,
New Zealand

Astghik Mavisakalyan
Bankwest Curtin Economics Centre,
Faculty of Business and Law, Curtin
University, Bentley, Western Australia,
Australia

Carol Pomare
Ron Joyce Center for Business Studies,
Mount Allison University, Sackville,
New Brunswick, Canada

Muhammad Abdur Rahaman
Climate Change Adaptation, Mitigation,
Experiment & Training (CAMET) Park,
Noakhali, Bangladesh

Mohammad Mahbubur Rahman
Network on Climate Change
in Bangladesh (NCCB), Dhaka,
Bangladesh

Quintin G. Rayer
P1 Investment Management,
Exeter, UK

Camilo Romero
Technical University of Berlin, Berlin,
Germany

Sharanya Basu Roy
University of Derby, Derby, UK

Horatiu A. Rus
Department of Economics and Department
of Political Science, University of Waterloo,
Waterloo, Ontario, Canada

Guneet Sandhu
School of Environment, Enterprise and
Development, University of Waterloo,
Waterloo, Ontario, Canada

Tyler Schwartz
John Molson School of Business,
Concordia University, Montreal,
Quebec, Canada

Eleni Sfakianaki
School of Social Sciences, Hellenic Open
University, Patras, Greece

Andrea Q. Steiner
Federal University of Pernambuco, Recife,
Pernambuco, Brazil

Yashar Tarverdi
Bankwest Curtin Economics Centre,
Faculty of Business and Law, Curtin
University, Bentley, Western Australia,
Australia

Olaf Weber
School of Environment, Enterprise and
Development, University of Waterloo,
Waterloo, Ontario, Canada

Michael O. Wood
School of Environment, Enterprise and
Development, University of Waterloo,
Waterloo, Ontario, Canada

Theodoros Zachariadis
Cyprus University of Technology,
Limassol, Cyprus

Ying Zhou
School of Environment, Enterprise and
Development (SEED), University of
Waterloo, Waterloo, Ontario, Canada

Ivelin M. Zvezdov
AIR Worldwide, VERISK Analytics
Corporation, Boston, MA, USA

Notes on Contributors

Elia E. Cia Alves is an associate professor of international relations in the Department of Political Science of the Federal University of Pernambuco (UFPE), Brazil, and professor at the graduate program in Public Management and International Cooperation. She is also a researcher at the Center for Research in Comparative and International Politics. She holds a PhD in political science from UFPE, where she also conducted postdoctoral studies. She holds a Master's degree in economics from Unicamp, and Bachelor's degrees in international relations from the University of São Paulo and in economics from Unisul. Her research interests include international political economy, environmental policy, policy diffusion, and energy policy.

Bruno Arcand is a candidate for the Master's degree in political science at the Université du Québec à Montréal where he is completing his thesis on Ontario's Green Energy and Green Economy Act. His research explores the relationship between economic development and climate change mitigation to better understand, evaluate, and develop environmental policies. His areas of interest include green industrial policy and the governance of disruptive policies to manage the sustainable transition.

Andrew Brennan is a senior lecturer at Curtin University in the School of Economics, Finance and Property. He has a strong innate desire to teach in and research on economics. His research has focused on critically evaluating measures of environmental and social welfare. He has a special interest in political economy, particularly ecological economics where he has published several papers. He enjoys communicating knowledge in creative and innovate ways so his students can attain the necessary skills to apply economic concepts to everyday events and be able to explain real-world phenomena.

Stephanie Cairns is the Director of Circular Economy, and former Director of the Cities and Communities program at the Smart Prosperity Institute, Canada's leading source of research and policy insights for a stronger, cleaner economy. She has done policy research on energy, environment, and economics for 30 years, authoring major reports and advising government agencies, think-tanks, nonprofit organizations, corporations, and industry associations. She has a BA (Hons) in environmental studies and economics from the University of Toronto, and an MSc from the Institute for Industrial Environmental Economics at Lund University, Sweden.

Amelia Clarke has been working on environment and sustainability issues since 1989, including as President of Sierra Club Canada (2003–2006). She holds a PhD in management (strategy) from McGill University and is now a tenured faculty member in the School of Environment, Enterprise and Development (SEED) at the University of Waterloo. As of May 2018, she is also the associate dean of research for the Faculty of Environment at the University of Waterloo. Her research interests include community sustainable development strategies, local decarburization pathways, corporate social and environmental responsibility, and youth-led social entrepreneurship and innovation.

Tariro Kamuti is a postdoctoral research fellow with the South African National Biodiversity Institute, in conjunction with the Global Risk Governance Program of the Department of Public Law at the University of Cape Town, South Africa. Kamuti holds a joint PhD from Vrije University Amsterdam, Netherlands, and the University of the Free State, South Africa. His research interests include land and environmental policy, governance issues surrounding the sustainable use of natural resources, and the social dimensions of biodiversity conservation. Kamuti is currently researching the rampant persecution of predators in areas surrounding livestock and private wildlife ranches.

David H. Lont is a professor of accounting at the University of Otago and a senior scholar in Management at the University of California, Davis. He has published extensively on financial reporting, auditing, financial markets, and climate risk-related issues. His research has also been featured in the international media. He is an editor for two accounting academic journals: *Abacus* and *Accounting and Finance*. He is a fellow of the Chartered Accountants Australia and New Zealand, CPA Australia, and a past president and fellow of the Accounting and Finance Association of Australia and New Zealand.

Astghik Mavisakalyan is an associate professor at the Bankwest Curtin Economics Centre at Curtin University. Her research is interested in political economy, with recent work focusing on the sources of differences in climate change policy action around the world, as well as the distributional aspects of natural resource abundance. Mavisakalyan's research has been published in a number of reputable journals including the *European Economic Review*, *Energy Economics*, *Journal of Comparative Economics*, *European Journal of Political Economy* and *Southern Economic Journal*, among others.

Carol Pomare is an associate professor of accounting at Mount Allison University. Pomare's market-based accounting research, in collaboration with Dr Lont and Dr Griffin, focuses on the impact of disclosure of greenhouse gas emissions on stock valuation and in event studies. Pomare is grateful for the Financial Accounting Research Grant from the Chartered Professional Accountant (CPA) and the Canadian Academic Accounting Association (CAAA). She is also grateful for the Marjorie Young Bell Faculty Fellowship and travel grants from the Marjorie Young Bell Faculty Fund.

Muhammad Abdur Rahaman has a MSc degree from Chittagong University in geography and environmental studies. In his professional history, he has served national and international development organizations with research and development activities, including the

Planning Commission, Worldwatch Institute, and the International Institute for Vegetable and Ornamental Crops. He has conducted much research on health, climate change, disaster risk reduction, water, agriculture, and gender perspectives. He is currently serving as director for the Climate Change Adaptation, Mitigation Experiment & Training (CAMET) Park. He is also serving different organizations as a freelance expert, and is a scientific expert for international networks on Home Garden for Response and Recovery (HG4RR).

Mohammad Mahbubur Rahman is working as a research and advocacy officer at Network on Climate Change in Bangladesh (NCCB). He has extensive experience in both scientific and social research, and over three years of experience in planning and reporting on climate change, public health, environmental, social, and development interventions. He has worked as a researcher with the University of Victoria, the Refugee and Migratory Movements Research Unit, the Christian Commission For Development In Bangladesh, Bangladesh Centre for Advanced Studies, the Bangladesh Institute of ICT in Development, and Grameen Communications on their different development and research projects. After obtaining a Bachelor's degree in environmental sciences from Jahangirnagar University, Bangladesh, he completed a Master's degree from the same department. He has also published several articles in peer-reviewed international journals.

Quintin G. Rayer holds a PhD in physics from the University of Oxford, is a fellow of the Institute of Physics, a chartered fellow of the Chartered Institute for Securities & Investment (CISI), a chartered wealth manager, and a graduate of the Sustainable Investment Professional Certification (SIPC) program at Concordia University. With experience in actuarial consultancy and wealth management, he founded P1's ethical investment proposition in January 2017.

Camilo Romero holds a degree in civil engineering from the University of Los Andes in Colombia. He started his academic career in history, doing research about the construction controlling profession in Colombia. In 2015, he left Colombia and lived between Egypt and Germany while obtaining a Master's degree in urban development at the Technical University of Berlin. His research interests are politics of urban development, urban political economy, and urban political ecology.

Sharanya Basu Roy is an associate lecturer of economics, business, law and social sciences at the University of Derby, UK. Her field of specialization is primarily environmental and natural resource economics, but she has also been exploring interdisciplinary methodologies, such as the "economic analysis of law." She completed her PhD in environmental law (using interdisciplinary approaches) from University College Cork (UCC), Ireland, and was funded by the Government of Ireland International Education Scholarship and UCC School of Law PhD Scholarship. While at UCC, she was also awarded the President James Slattery Prize and Medal in Law. Basu Roy has worked on various government funded projects like the WetlandLIFE, and has experience working with various government organizations, such as India's National Institute of Science, Technology and Development Studies (NISTADS-CSIR) and the Indo-French Centre for the Promotion of Advanced Research (CEFIPRA).

Horatiu A. Rus is an economist by training and works as an associate professor of economics and political science at the University of Waterloo. His earlier research work straddles the fields of natural resource economics and international trade, while his more recent projects focus on water resources and climate change. In particular, he is interested in the interplay between climate change mitigation and adaptation in international coalitions, as well as in the link between public policy and technological innovation in the field of water provision. Rus teaches courses on the economics and politics of international trade and natural resource economics.

Guneet Sandhu has completed a Master of Environmental Studies degree in sustainability management (water) at the University of Waterloo. Her research focuses on designing water pricing policies and strategies to promote sustainable water management. Sandhu also holds a Master of Engineering (chemical) from Cornell University, USA, and Bachelor of Engineering (chemical) with Honors from Panjab University, India. She has previously worked as a project engineer with AguaClara Reach, Inc. She has helped develop and implement sustainable drinking water treatment technologies for marginalized communities while gaining exposure to the interdisciplinary facets of water management.

Tyler Schwartz currently serves as a research assistant at the University of Concordia in the Department of Sustainable Finance. He recently completed his undergraduate degree at the John Molson School of Business in which he received Honors in Finance. As part of his undergraduate degree, he completed a thesis project in which he wrote a paper focusing on the relationship between data breaches, security prices, and crisis communication. He was also presented with the CUSRA scholarship in 2016, which is awarded to undergraduate students who have an interest in pursuing research activities. His research interests include sustainable finance, machine learning, data breaches, and cognitive science.

Eleni Sfakianaki is an assistant professor at the School of Social Sciences of the Hellenic Open University, in Greece. Her areas of expertise include environmental management, sustainability, and total quality management. Over the last years, she has focused on researching the implementation of environmental methods, frameworks, and policies in business sectors, with a particular focus on the construction industry.

Andrea Q. Steiner is head of the Department of Political Science at the Federal University of Pernambuco (UFPE), where she is also an associate professor. She is part of the graduate program in Political Science and the professional Master's program in Public Policy. She holds a Bachelor's degree in biology from the Rural Federal University of Pernambuco, a Master's degree in biological sciences (zoology) from the Federal University of Paraíba, and a PhD in political science from UFPE, where she also conducted postdoctoral studies. Her research interests include international regimes, environmental governance, marine governance, international environmental policy, and gender and international relations.

Yashar Tarverdi is lecturer at Curtin University in the School of Economics, Finance and Property. Primarily interested in data analytics and applied econometrics, Tarverdi has conducted research on the evaluation of macroeconomic policies concerning the environment, energy, and health. He has an active publication portfolio in highly ranked journals such as *Journal of Comparative Economics*, *Energy Economics*, *Applied Economics*, and *Environmental & Resource Economics*.

Olaf Weber is a professor and university research chair in sustainable finance at the University of Waterloo. His main research interests include sustainable lending and investing, corporate sustainability performance, as well as sustainability regulations and codes of conduct in the financial industry.

Michael O. Wood is the associate director of undergraduate studies and a continuing lecturer at the School of Environment, Enterprise and Development (SEED) of the University of Waterloo. He holds a PhD in strategy and sustainability from the Ivey Business School. His research examines organizational perceptions and responses to sustainability issues through the lens of space, time, scale, and social license to operate within the contexts of the insurance industry, mining, carbon management, Blue Economy, climate change and global security. His research has most recently been published in the *Academy of Management Review*, *British Journal of Management*, and the journal *Resources*.

Theodoros Zachariadis is an associate professor at the Cyprus University of Technology, teaching environmental economics and energy resource management. He has a PhD in mechanical engineering from the Aristotle University of Thessaloniki, Greece. He is a member of the Scientific Committee of the European Environment Agency and an associate editor of the international journal *Energy Economics*. He is also the recipient of the 2009 Research Prize of the Republic of Cyprus. His research interests include energy and environmental economics and policy, especially the assessment of decarbonization policies in the transport sector and long-term energy and environmental modeling.

Ying Zhou is a second-year PhD student in sustainability management at the University of Waterloo. Her Master's research is on the use of market-based instruments (MBIs) for implementing sustainable community plans. She also developed the Sustainability Alignment Methodology (SAM), a tool that aligns MBIs under municipal jurisdiction with environmental goals. Her current PhD research focuses on the deep decarburization of Canadian cities. Specifically, she investigates the local decarburization pathways and governance arrangements for achieving the 2050 Greenhouse Gas reduction target.

Ivelin M. Zvezdov is an assistant vice president in the Product Development group of AIR Worldwide, a natural catastrophe and extreme event modeling company. He holds an MA in economics from St Andrews University, UK, and a MPhil in EU studies from the University of Oxford, UK. Zvezdov is a Certified Portfolio Management Professional (CPMP) from the New York Institute of Finance. His previous career experiences include quantitative financial modeling and risk management for investment banking, securities trading and commodities. He has published several practitioner articles on insurance and reinsurance technical pricing, risk management, and capital reserving.

Copy Editing Team

Julie Brown is currently completing a BFA in art history and studio arts at Concordia University, with a minor in professional writing. Her academic interests chiefly concern visual literacy and radical accessibility through the bias of zines, comics, and graphic novels. Brown was Editor-in-Chief for *Yiara Magazine*'s seventh volume, published in March 2019. She was previously a copy editor for *CUJAH* and the *ASFA Academic Journal*, and has also written for the FOFA, the VAV Gallery, and *Yiara Magazine*. She now works as a research assistant for Concordia's Department of Finance.

Adèle Dumont-Bergeron is an MA student in English literature and creative writing at Concordia University, Montreal. She currently serves as a research assistant for Concordia's Finance Department. She recently completed a research project about plastic taxes and has co-authored an article, soon to be published, on the topic. In her department, Dumont-Bergeron has won the Compton-Lamb Memorial Scholarship and has been awarded a fellowship to pursue her work in literature. She has presented and published her work at several literary conferences. Her interests include sustainability, feminism, and modernity.

Kalima Amber Vico is a research associate at the John Molson School of Business at Concordia University, Montreal. She has served with Concordia's David O'Brien Centre for Sustainable Enterprise and is part of Sustainable Concordia. She is currently finishing her Bachelor of Commerce in finance with a concentration in economics at Concordia. She has participated in and worked on well over 25 research papers and projects within the Finance Department. She has organized the Emerging Risks in Finance conference that is related to this book. Her research interests include diverse topics in economics, psychology, organizational behaviour, finance, and sustainability.

List of Figures

Chapter 8

Chapter 11

Chapter 12

Chapter 15

Chapter 16

List of Tables

Acknowledgments

We acknowledge the financial support provided through the Autorité des marchés financiers and the David O'Brien Centre for Sustainable Enterprise at Concordia University. In addition, we greatly appreciate the research and administrative assistance provided by Kalima Vico and Tyler Schwartz as well as the excellent copy-editing and editorial assistance we received from Adele Dumont-Bergeron and Julie Elizabeth Brown. Finally, we feel greatly indebted to Arlene Segal, Joseph Capano, Stephane Brutus, Anne-Marie Croteau, and Norma Paradis (all at Concordia University) who in various ways supported this project.

1

An Introduction to the Current Landscape

Environmental Policy and the Economy

Tyler Schwartz[1] and Sherif Goubran[2]

[1] *John Molson School of Business, Concordia University, Montreal, Quebec, Canada*
[2] *Concordia University, Montreal, Quebec, Canada*

1.1 Background

Global environmental challenges have caused a range of policy solutions, approaches, and models to emerge. As these challenges are expected to intensify in the near future, environmental policy and its instruments are increasingly becoming a topic of discussion, action, and disagreement in academic, professional, and mass media outlets. This fixation on the topic of policy is well-justified considering the consequences policy can have on all levels of society – global, national, sectoral, organizational, and even personal. Policy has a vital role in reducing environmental damage, incentivizing positive environmental behavior, and guiding practice toward a more sustainable future. While most policies have economic repercussions, environmental policies, and specifically new environmental policy instruments, have exhibited a special and complex relation to the economy (Jordan et al. 2003). The environment can thus be considered an envelope encompassing and sustaining the economic system – much more than just a factor of production (Hawken et al. 1999).

Two main arguments relate to economic and environmental health. The first proposes that continuous growth is inevitably contradictory to improvements in environmental health, citing reasons such as geographical shifts in production, lock-in, the social-political feasibility of environmental agreements and policies, and obstacles to behavioral change, among others (Antal and Van Den Bergh 2016). The second proposes that economic growth can and will eventually lead to environmental improvements (Faure 2012), in what is commonly referred to as the Environmental Kuznets Curve (Kuznets 1995). This claim is usually supported by empirical evidence (Shen and Hashimoto 2004; Kong and Khan 2019). In the second argument, the relationship between regulation and environmental health is not direct. Instead, environmental health depends on external economic parameters: the macro (national economic performance) and the micro (personal income) (Faure 2012). It is important to note the broad agreement between the two views on the need to strengthen institutions and regulatory regimes to improve environmental performance (Faure 2012).

Environmental Policy: An Economic Perspective, First Edition. Edited by Thomas Walker, Northrop Sprung-Much, and Sherif Goubran.
© 2020 John Wiley & Sons Ltd. Published 2020 by John Wiley & Sons Ltd.

Table 1.1 The typology of environmental policy instruments

		Goals to be achieved	
		Specified	Not specified
Method to achieve goals	**Specified**	Command and control	Technology-based regulatory standards
	Not specified	Voluntary agreements	Market-based instruments

Source: Based on Russell and Powell (1996) and elaborated by Jordan et al. (2003).

The debates about the compatibility of economic growth, including "green growth," with the imperative of climate change action create the underlying concerns of environmental policy design: Do we design policies to encourage growth or to slow it down? What are the policy instruments that can help reconcile the health of the environment and the economy?

Environmental policy can be broadly organized in a two-by-two matrix, first proposed by Russell and Powell (1996) and elaborated by Jordan et al. (2003) (reproduced in Table 1.1). This matrix categorizes the available environmental policy instruments based on their inclusion of the following: (i) the specific environmental goals to be achieved, and (ii) the specific means of improving environmental performance. This puts at opposite poles market-based instruments (MBIs) and regulatory approaches (i.e. command and control) – two approaches which occupy a prominent role in debates today.

Faure (2012) provides an overview of the environmental policy instruments and their relevant empirical evidence. He pays specific attention to the interaction between the policy, on the one hand, and the economy and the private actors on the other, and identifies the following key policy instruments.

- *Liability rules*: instruments that aim to hold polluters directly accountable for their impact on the environment. With regard to their success, Faure comments that stricter liabilities could result in higher rates of avoidance and insolvency. He proposes that policy packages should be considered with liabilities to address their unintended consequences (specifically related to the higher insolvency rates they create).
- *Regulation*: minimum performance standards (based on absolute environmental goals or on effluent or emission levels). Faure indicates that these instruments usually present a risk since industry actors (especially the biggest polluters) will usually interfere in their development. An example of such interference is the "grandfather clauses" which exclude existing firms or products from regulations. Additionally, and as Nobel Prize winners Buchanan and Tullock proposed, many firms prefer standards and regulations because they create additional barriers to market entry.
- *Taxation*: instruments designed to increase the cost of activities that have known impacts on the environment and could be levied on energy, transportation, or other activities. Considered as a command and control instrument, taxes present the same risk of interference seen in other regulatory approaches (Carl and Fedor 2016).

- *MBIs*: frameworks set by regulators to induce market adjustment. This system has been considered successful in a number of cases – such as the SO_2 trading scheme (Stavins 1998; Murray and Rivers 2015) where cost savings were achieved along with environmental goals. However, Faure highlights that such mechanisms can only be successful in locations where institutional and administrative structures are well established. Additionally, the process of setting emission or effluent allocations to existing firms tends could lead to overallocations.
- *Voluntary agreements*: energy or emission reduction agreements, often between public and private sector actors. While Faure does not directly comment on voluntary agreements, their non-legally binding nature makes them usually less effective on the environmental front (Bizer 1999), and only subgroups of economic actors will end up participating in, or abiding by, such agreements (Dawson and Segerson 2008).

In this complex landscape of instruments, environmental policies are seen to present a perfect case of wicked problems (Rittel and Webber 1973), where the actual framing of the problem becomes a complex and contested task (Coyne 2005). As Faure highlights, various combinations of instruments are usually required in order to ensure their success – for instance, regulatory, enforcement, and governance policies are needed to ensure the success of MBIs. In other words, there is no single instrument that could be successful for all environmental problems in all regions. It is also important to consider that no single actor – public or private – has all the required knowledge or capacity to solve the complex environmental challenges we face (Jordan et al. 2003). Instead, policymakers must ensure that they recognize the barriers to the success of various instruments before policy implementation. These barriers include lack of economic expertise, cultural antipathy, opposition from stakeholders (both for requiring more or less stringent approaches), legal constraints, and possible market imbalances and disturbances (Jordan et al. 2003).

While policies, environmental policies included, are usually developed and studied by lawyers, political scientists, and policymakers, their effectiveness is mainly studied by economists and other researchers from the social sciences (Faure 2012). This creates a number of limitations: (i) the results are often published in diverse journals and venues which are not readily reviewed by policymakers; (ii) many of the studies are focused on single and specific instruments; and (iii) the studies tend to be country- or region-specific (Faure 2012).

In the face of these limitations, this collection aims to provide an integrated reference which features contributions from social scientists, economics, political scientists, and practitioners to expose and explore the economic repercussions and limitations of environmental policies. Through the policy recommendations proposed in these chapters, the volume also intends to bridge some of the current gaps in the full policy development cycle.

1.2 Overview of Content

This volume aims to address how different aspects of environmental policies are both shaping and impacting the world around us. It describes the role that environmental policies are playing in the fight against climate change, and how they can be more effectively developed to better combat this issue. Both global and local perspectives are taken throughout the publication as many chapters use case studies from specific locations to examine how environmental policies function in different areas of the world. The book has four sections: (I) *An Overview*, (II) *Governing and Protecting Natural Resources*, (III) *Energy, Emissions, and the Economy,* and lastly, (IV) *Financing Environmental Transition*. Each of these sections is summarized below.

1.2.1   Section I: An Overview

The first section provides an overview of the existing conditions operating within the domain of environmental policy and the resulting economic incentives. The section covers environmental legislation from a global perspective while also using case studies to illustrate points made within the chapters.

The section begins with Chapter 2, *Responses to Climate Change: Individual Preferences and Policy Actions around the World*, which is an overview of existing global climate change policies. Brennan, Mavisakalyan, and Tarverdi pose the question of whether proenvironmental individual preferences are consistent with climate change policies at the macro level. A case study of Sweden and Turkey demonstrates how heavy-handed policies are often connected with proenvironmental individual preferences, yet proenvironmental attitudes do not always translate into stringent policies.

Chapter 3, *Legislation or Economic Instruments for a Successful Environmental Policy? Reflections after "Dieselgate,"* asks whether stricter legislation can be an effective tool in addressing environmental challenges. By conducting a case study of the Volkswagen emission scandal "Dieselgate," Zachariadis compares whether stricter legislation results in improved environmental conditions, social equity, legal and political obstacles, and behavioral barriers. The resulting conclusion is that a hybrid approach works best to address environmental concerns, as it offers the flexibility of economic instruments as well as the robustness of environmental legislation.

Chapter 4, *Environmental Legislation and Small and Medium-Sized Enterprises (SMEs): A Literature Review*, looks at environmental legislation through the perspective of SMEs. Sfakianaki demonstrates that SMEs have fewer resources to evaluate and understand environmental legislation and are therefore less able to adapt to their surrounding environment and address environmental challenges. It is also shown that environmental legislation is a driver toward environmental compliance of SMEs, due to the fear these organizations have of being penalized for noncompliance. The main conclusion the authors reach is that environmental legislation is both a driver and a deterrent for SMEs to implement greener operations.

1.2.2   Section II: Governing and Protecting Natural Resources

The second section of the publication takes a more applied approach to environmental policies by looking at their impact on natural resources such as water, marine life, and the wilderness.

The section begins with Chapter 5, *Bulk Water Pricing Policies and Strategies for Sustainable Water Management: The Case of Ontario, Canada*, which looks at bulk water pricing in Ontario as an instrument for managing demand and encouraging use-efficiency and conservation. Wood, Rus, Weber, and Sandhu provide a tangible framework for developing water extraction charges that can foster sustainable water management and trigger Ontario's transition to a more water-efficient economy.

Similarly, Chapter 6, *The Role of Water Pricing Policies in Steering Urban Development: The Case of the Bogotá Region*, explores the role that water pricing policy has played in shaping discursive representations of water use. Romero looks closely at the conceptualization of urban political ecology (UPE) as a theoretical framework and additionally provides evidence to show the importance of integrating water pricing and trading policies within urban development policies.

Chapter 7, *Effective Environmental and Regulatory Quality: A National Case Study of China*, analyzes the regulatory quality of pollution control policy by conducting a regression analysis of selected environmental policies and water pollution measures. Basu Roy's main goal is to quantify the impact of the major tools of pollution control policies in China and help identify the key elements of the policy and governance frameworks that have had the greatest impact.

Chapter 8, *The EU Legal and Regulatory Framework for Measuring Damage Risks to the Biodiversity of the Marine Environment*, reviews the frameworks within which it is possible to estimate the financial risks and costs placed upon the marine environment by European environmental legislation. Zvezdov's research aims to provide a robust economic valuation of natural resource conservation within the offshore marine environment.

The concluding chapter in the section, Chapter 9, *Redefining Nature and Wilderness Through Private Wildlife Ranching: An Economic Perspective of Environmental Policy in South(ern) Africa*, offers a critique of wildlife policy in southern Africa by tracing the development of wildlife ranching and its trajectory in shifting the conservation narrative. Kamuti argues that the observed trends in wildlife ranching development contributed to the redefinition of nature and wilderness as part of a broad economic policy. It further contends that there is a need to embrace the idea of inclusive conservation which focuses on placing indigenous communities at the forefront of conservation efforts by empowering them to take charge of natural resources.

1.2.3   Section III: Energy, Emissions, and the Economy

This section focuses on energy-related environmental policies that concern emissions and pollution.

The first chapter in the section, Chapter 10, *Climate Change Regulations and Accounting Practices: Optimization for Emission-Intensive Publicly Traded Firms*, looks at environmental legislation from the perspective of emission-intensive companies. Lont and Pompare argue for an increase in transparency and the improvement of the information quality required and provided by legislators, in order to avoid off-balance sheet liabilities which may negatively impact investors.

In Chapter 11, *The Economic Aspects of the Adoption of Renewable Energy Policies in Developing Countries: An Overview of the Brazilian Wind Power Sector*, Alves and Steiner address the economic triggers that have led to the diffusion of renewable energy policies worldwide. A case study of the Brazilian wind sector is used to illustrate this idea; it concludes that six economic triggers are responsible for the diffusion: income, energy prices, financing, trade, foreign direct investment (FDI), and lobbying.

A similar idea is evaluated in Chapter 12, *Ontario's Energy Transition: A Successful Case of a Green Jobs Strategy?*, which looks at the case of Ontario's Green Energy and Green Economy Act (GEGEA) as part of a green jobs strategy, and examines the challenges that an energy transition poses to employment. Arcand takes a critical look at whether the GEGEA's reported job growth and creation is a false perception. It concludes that it remains difficult to properly assess job creation in this area due to many factors, which are critically discussed in the chapter.

The final chapter in Section III, Chapter 13, *Ethical and Sustainable Investing and the Need for Carbon Neutrality*, looks at recent technological developments, in particular the Internet of Things technologies, and makes the case for their use by businesses to achieve net zero carbon emissions (NZCE). Rayer looks at the role ethical investors must play in this transition toward NZCE by focusing on their selection of companies that have adopted carbon-neutral processes.

1.2.4   Section IV: Financing the Environmental Transition

The final section, *Financing Environmental Transition*, takes a financial approach to environmental policies. The three chapters in the section respectively cover MBIs, fund allocation in the food sector, and the investment process of ethical and sustainable fund managers.

The first chapter in the section, Chapter 14, *Building Sustainable Communities through Market-Based Instruments*, discusses the role of MBIs in the implementation of Local Agenda 21s (LA21s), which are strategic plans that integrate ecological, social, and economic areas of research. Zhou, Clarke and Cairns argue that the use of MBIs has the potential to bridge the gap between plan and implementation which would help address local water challenges. It presents over 15 MBIs across four different water subtopics and provides an improved understanding of MBIs for implementing LA21s.

Chapter 15, *Climate Justice in Food Security: Experience from Climate Finance in Bangladesh*, evaluates climate change policies impacting food security by conducting a case study of Bangladesh. Rahaman and Rahman show that there are significant gaps in fund allocation from the government's four major funding mechanisms, including the Bangladesh Climate Change Resilience Fund, the Strategic Program for Climate Resilience, the Bangladesh Climate Change Trust, and nongovernmental efforts. The chapter concludes that the projects designed by the Bangladesh government to address food security are beyond the scope of climate justice.

The book concludes with Chapter 16, *A Survey of UK-Based Ethical and Sustainable Fund Managers' Investment Processes Addressing Plastics in the Environment*, where Rayer looks at the commitment of fund managers in addressing environmental issues, in particular the use of plastics. This study conducted a survey of 12 UK-based fund management firms that were deemed to have superior ethical investment policies. The results suggest that while efforts are being made to address the use of plastics, the commitment by fund managers to address it may be somewhat weak.

References

Antal, M. and Van Den Bergh, J.C.J.M. (2016). Green growth and climate change: conceptual and empirical considerations. *Climate Policy* 16 (2): 165–177.

Bizer, K. (1999). Voluntary agreements: cost-effective or a smokescreen for failure? *Environmental Economics and Policy Studies* 2 (2): 147–165.

Carl, J. and Fedor, D. (2016). Tracking global carbon revenues: a survey of carbon taxes versus cap-and-trade in the real world. *Energy Policy* 96: 50–77.

Coyne, R. (2005). Wicked problems revisited. *Design Studies* 26 (1): 5–17.

Dawson, N.L. and Segerson, K. (2008). Voluntary agreements with industries: participation incentives with industry-wide targets voluntary agreements with industries: participation incentives with industry-wide targets Na Li Dawson and Kathleen Segerson. *Land Economics* 84 (1): 97–114.

Faure, M.G. (2012). Instruments for environmental governance: what works? In: *Environmental Governance and Sustainability* (eds. P. Martin, L. Zhiping, Q. Tianbao, et al.), 3–23. Cheltenham: Edward Elgar Publishing.

Hawken, P., Lovins, A., and Lovins, H. (1999). The next industrial revolution. In: *Natural Capitalism*, 1–21. Boston, MA: Little, Brown & Co.

Jordan, A., Wurzel, R.K.W., and Zito, A.R. (2003). "New" instruments of environmental governance: patterns and pathways of change. *Environmental Politics* 12 (1): 1–24.

Kong, Y.S. and Khan, R. (2019). To examine environmental pollution by economic growth and their impact in an environmental Kuznets curve (EKC) among developed and developing countries. *PLoS One* 14 (3): 1–23.

Kuznets, S. (1995). Economic growth and income inequality. *American Economic Review* 45 (1): 1–28.

Murray, B. and Rivers, N. (2015). British Columbia's revenue-neutral carbon tax: a review of the latest "grand experiment" in environmental policy. *Energy Policy* 86: 674–683.

Rittel, H.W.J. and Webber, M.M. (1973). Dilemmas in a general theory of planning. *Policy Sciences* 4 (December 1969): 155–169.

Russell, C.S. and Powell, P.T. (1996). *Choosing Environmental Policy Tools and Practical Considerations (No. ENV-102)*. Washington, DC: Inter-American Development Bank.

Shen, J. and Hashimoto, Y. (2004). *Environmental Kuznets Curve on Country Level: Evidence from China* (Discussion Papers in Economics and Business No. 04–09). Osaka: Toyonaka.

Stavins, R.N. (1998). What can we learn from the grand policy experiment? Lessons from SO2 allowance trading. *Journal of Economic Perspectives* 12: 69–88.

Section I

An Overview

2

Responses to Climate Change

Individual Preferences and Policy Actions around the World

Andrew Brennan[1], Astghik Mavisakalyan[2], and Yashar Tarverdi[2]

[1] *School of Economics, Finance and Property, Faculty of Business and Law, Curtin University, Bentley, Western Australia, Australia*
[2] *Bankwest Curtin Economics Centre, Faculty of Business and Law, Curtin University, Bentley, Western Australia, Australia*

2.1 Introduction

Human-induced climate change is a global environmental problem. Due primarily to the burning of fossil fuels, emissions build up as the scale of human enterprise grows and we continue to industrialize in an ever more global economy. Economic expansion is a waste-generating process restricted to the finite global ecosystem. Today, global climate change remains one of the greatest perils in humanity's history. With potentially irreversible catastrophic impacts on ecosystems, societies, and economies, addressing climate change has become one of the main agenda items in political debates and international summits.

The atmospheric concentration of carbon dioxide (CO_2) has been constant at around 280 parts per million by volume (ppm) for many centuries; however, since the Industrial Revolution in the early 1800s, it has risen by almost a half to 410 ppm in 2017. According to the 2014 IPCC Fifth Assessment Report, the average temperature of the planet's surface rose by 0.89 °C from 1901 to 2012. The force of inertia means that current emissions could potentially define future stocks. Long time-lags between today's mitigatory actions of reducing CO_2 and tomorrow's outcomes are thus built into the system because of cumulative (and fairly irreversible) processes from the building up of greenhouse gases. How much carbon and other greenhouse gases we emit in the future determines the amount of warming we will see.

There are a large range of "carbon budget" estimates – the total amount of CO_2 emissions compatible with a given global average warming. For example, according to Millar et al.'s modeling of future anthropogenic (human-induced) warming, we may have up to 20 years before we use up the carbon budget to stay below 1.5 °C (2017). However, there is uncertainty in such budget estimates, as much depends on the complex response of global temperatures to carbon emissions – for instance, the thawing permafrost is an impact factor not currently included in the estimates. Despite the current range of carbon budget estimates, they all show a consistent picture: that limiting warming to 2 °C or less remains a challenging goal for policy (Met Office 2017).

The natural environment is a vital asset, both for production and sustainable economic welfare. Effective climate protection policies must center around achieving a balance between the forces of profit and the climate-environment – where a number of institutional factors are key drivers influencing this balance. Policy action is not determined by the political and economic standing of countries alone. The preferences of a population can often feed into the government's decision to amend policies in response to climate change.[1] As Rodrigues argues, individual motivations and preferences cannot be understood without reference to the institutional context that partly shapes and defines them (2004, p. 192). It is thus an empirical question as to how individuals and institutional factors influence countries' performance in climate change mitigation.

This chapter concerns policies designed to mitigate climate change, as opposed to policies for adapting to the impacts of climate change. Studies on climate change have mostly focused on performance outcomes rather than policy measures (e.g., Böhringer and Jochem 2007; Cole and Fredriksson 2009; Kellenberg 2009; Greenstone and Hanna 2014). This could be due in part to the small amount of existing policy indicators for measuring climate change regulations, as most of the indicators referenced in literature on the matter address a wider concept of the environment. Nevertheless, distinguishing such conceptual differences in practice is not a trivial task, as the interrelationships between elements of the environmental system are still puzzling to many scientists.

For this purpose, the analysis in section 2.2 focuses on one climate change policy indicator in particular: the Climate Laws, Institutions, and Measures Index (CLIMI) from Steves et al. (2013). The CLIMI measures a country's adopted policies that address climate change through mitigation. Components of the CLIMI incorporate the following dimensions, subject to various weightings: the extent of international cooperation (such as ratification of the Kyoto Protocol, or the use of Joint Implementation or Clean Development Mechanism); the strength of a state's domestic climate framework (such as a cross-sectoral climate change legislation, a carbon emissions target, or a dedicated climate change institution); and sectoral fiscal or regulatory measures or targets (such as energy supplies and renewables). Section 2.2 provides a broad overview of climate change policies around the world using the CLIMI. We demonstrate the differences in dedicated policies and measures to address climate change, drawing comparisons across different regions of the world and by level of economic development.

Accepting climate change policy indicators at face value can give a skewed representation of their strength and potential effectiveness. In Section 2.3, our analysis focuses on individual perceptions and values over the significance of climate change and their link to climate change policies across countries. This is significant because of the heterogeneous roles of the individual in society; an individual is simultaneously a product of society and a producer of change in that culture. As John B. Davis writes, individuals "are acted upon by society" but "are also themselves agents who act upon and change society" (2003, p. 111). This individual is also self-reflective because they possess a capacity to act upon and influence social structures, including language. Individuals may develop and change substantially, both in their character and in their fundamental tastes or perceptions; their actions can then in principle affect and change society. An individual in the affluent West, for instance, can change for the better the practice of market activity through their interactions (e.g., through the demand for Fairtrade commodities and energy efficiency developments).

Particularly, this section analyzes the measures of climate change policies in relation to data on individual preferences in the World Values Survey (WVS) – nationally representative surveys conducted since 1981 in almost 100 countries using a common questionnaire. Among other things, the respondents in several waves of WVS are asked to provide information on their perceptions of the significance of climate change as well as on various environmental values and actions.

The overall results of this chapter are discussed in section 2.4. In this section we also explore, as a case study, the differences between Sweden and Turkey in structural and institutional influences on environmental policies. In democracies, parties and individual politicians in the government have reason to consider the views of their constituents, perhaps even more so in a high-income country such as Sweden. Educated and well-informed individuals might thus be able to influence the environmental policy of a democratic state. The more responsive the democracy and the level of human and social capital, the more the preferences of the electorate seem to matter; the reverse is also true for countries such as Turkey. A number of factors determine the adoption of policies that improve the environment: the political participation of the population, the voting behavior of citizens, and the electoral preference for long-term over short-term programs, which foster a good balance between the forces of environmental protection and economic development. Specifically, section 2.4 focuses on the important roles of social capital and trust as the core factors that determine the adoption of policies to improve the natural environment.

In summary, there are three major sections of this chapter. In Section 2.2, we discuss the prevalence of climate change policies, as captured by the CLIMI, around the world, including variations by various regions of the world and by different levels of economic and human development. Section 2.3 aims to provide a descriptive assessment of individual values and perceptions of the significance of climate change, and to see whether or how they are linked with observed climate change policies across countries. Section 2.4 provides a discussion of our key findings and a case study between Sweden and Turkey examining the complex relationships between social and environmental factors. Section 2.5 briefly concludes the chapter.

2.2 Climate Change Policies Across Countries

2.2.1 Measures of Climate Change Policies

There have been several attempts to measure the climate change policies across countries (see Surminski and Williamson 2012 for a review). Nevertheless, several shortcomings of existing measures exist. First, these policies are often restricted to a relatively small group of countries. For example, the Climate Policy Index (CPI) developed by Kunkel, Jacob, and Busch in 2006 is available for 24 countries only, and is also restricted to their national climate policies. There are some measures that are more comprehensive in coverage and include international alongside national dimensions of countries' climate change policies. One prominent example is the climate policy component of the Climate Change Performance Index (CCPI) from Germanwatch, available for 58 countries (Burck et al. 2014). However, this measure is based on subjective assessments by experts on the commitments of these countries to climate policies and regulations.

Table 2.1 Policy areas of the Climate Laws, Institutions and Measures Index (CLIMI)

International cooperation (0.1)	Domestic climate framework (0.4)	Significant sectoral fiscal or regulatory measures or targets (0.4)	Additional cross-sectoral fiscal or regulatory measures (0.1)
Kyoto ratification (0.5) Joint Implementation or Clean Development Mechanism host (0.5)	cross-sectoral climate change legislation (0.33) carbon emissions target (0.33) dedicated climate change institution (0.33)	energy supplies/ renewables (0.3) transport (0.13) buildings (0.07) agriculture (0.13) forestry (0.17) industry (0.2)	cross-sectoral policy measures (1)

Notes: Figures in parenthesis represent the weights used to reflect the contribution of each of the components and areas to climate change mitigation.

In view of these considerations, our analysis is based on a third measure of countries' climate change policies, the CLIMI, constructed by Steves et al. (2013). The usefulness of this measure is that it allows for the systematic global comparison of climate change policies across a large group of countries. Furthermore, based on the 2005–2010 annual national communication to the United Nations Framework Convention on Climate Change (UNFCCC), it is an objective and accurate measure of actual climate change commitments taken up by countries.

The CLIMI is based on 12 components grouped into four key policy areas, with weights used to reflect the contribution of each of the components and areas to climate change mitigation. As seen in Table 2.1, which details the construction of the index, the CLIMI places greater weight on a country's domestic laws and regulations.

The CLIMI accords value scores between 0 and 1; higher values represent stricter policies, signaling more cooperative political behavior regarding contributions to the global environmental public good. Figure 2.1 depicts the CLIMI scores of countries around the world. Tonga has the lowest CLIMI score in the sample at 0.011, while the UK has the highest score at 0.801.

The list of countries with the top CLIMI scores is largely dominated by high-income countries in Europe. Conversely, countries with the lowest CLIMI scores are often developing countries in Africa, Central America, and the Middle East. There is some heterogeneity in the involvement of domestic versus international efforts to address climate change. In some countries, for example Fiji, both domestic and international efforts play a comparable role in addressing climate change. In others, like the UK, Belgium, and Greece, there is more emphasis on domestic climate change policy framework.

2.2.2   Climate Change Policies in Different Regions of the World

There are rather significant differences in the CLIMI scores across different parts of the world. Observations in Figures 2.1 and 2.2 demonstrate that the countries in Europe and Central Asia have the most stringent climate change policies in the world, with their CLIMI

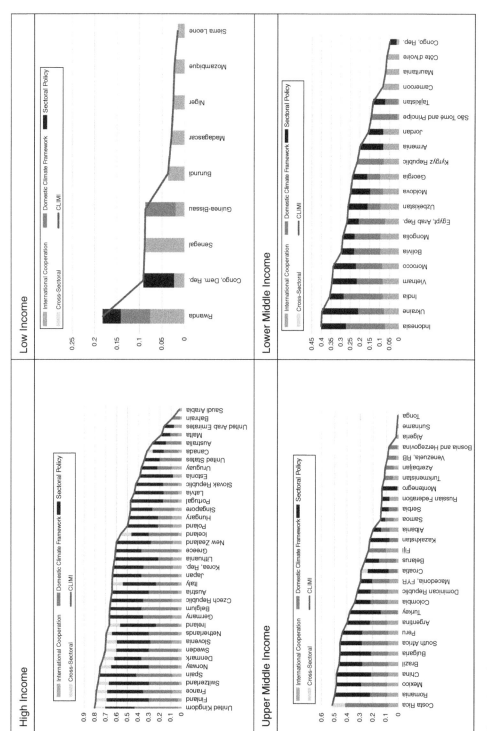

Figure 2.1 Climate Laws, Institutions and Measures Index (CLIMI) and its components across countries. *Source:* Authors' calculations and visualization based on raw data taken from Steves et al. (2013).

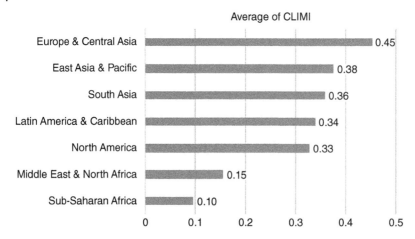

Figure 2.2 Climate change policies across geographic regions. *Source:* Authors' calculations and visualization based on raw data taken from Steves et al. (2013).

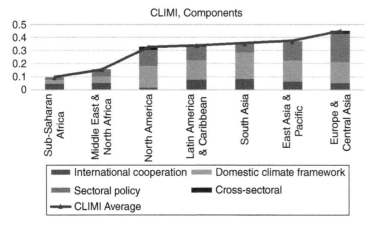

Figure 2.3 Dimensions of climate change policies across geographic regions. *Source:* Authors' calculations and visualization based on the raw data taken from Steves et al. (2013).

scores averaging 0.45. Countries in sub-Saharan Africa are ranked lowest, with an average CLIMI score of 0.10. Middle Eastern and North African countries are not doing much better; the average CLIMI score for this region is 0.15. Other parts of the world fall between these two extremes, with CLIMI scores averaging from 0.33 to 0.38.

We can also note differences in the relative weight of the various dimensions of climate change policies across regions. Figure 2.3 shows that in North America, the emphasis of climate change policy framework is on domestic policies, accounting for over half of the contribution toward its overall CLIMI score. In sub-Saharan Africa, on the other hand, nearly half of the contributions toward the overall index come from the international climate policy commitments of its countries.

2.2.3   Economic Development and Climate Change Policies

Furthermore, the CLIMI helps us discern whether the stringency of climate change policies increases with economic development. In Figure 2.4, we map the CLIMI scores of countries and indicate their level of development, as measured by their gross domestic product (GDP) per capita. In many cases, we observe stringent climate change policies in more developed countries, especially within northern Europe. The UK, Finland, and Norway, for instance, are among the countries with the highest CLIMI scores in the world. At the other extreme, we have less economically developed countries, such as Sierra Leone, Algeria, and Mozambique, with some of the lowest CLIMI scores in the world.

Observing the CLIMI averages for countries grouped by level of economic development, as seen in Figure 2.5, reveals these patterns more vividly. In high-income countries, the average CLIMI score is 0.54. This is in contrast with the 0.06 average CLIMI score observed in low-income countries. Middle-income countries have CLIMI scores averaging between 0.23 and 0.27.

In Figure 2.6, we consider a specific form of commitment to address climate change – ratification of the Kyoto Protocol from the UN Framework Convention on Climate Change. We consider the proportion of countries that have taken up this commitment, grouped by their level of economic development. Consistent with the patterns observed above, climate change policy commitments increase with economic development. In 2010, 37 high-income countries had ratified the Kyoto protocol, compared to only nine countries in the low-income group. Encouragingly, however, we have seen the number of countries committing to the protocol increase rapidly over the preceding decade in developing countries, particularly those in the

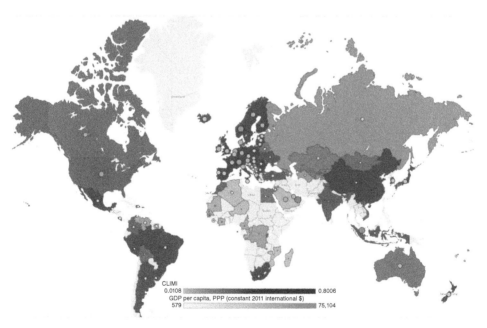

Figure 2.4   Climate change policies and GDP per capita. *Source:* Authors' calculations and visualization based on raw data taken from Steves et al. (2013).

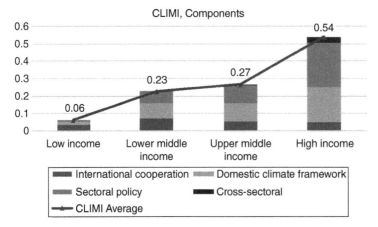

Figure 2.5 Climate change policies and GDP per capita by income class. *Source:* Authors' calculations and visualization based on raw data taken from Steves et al. (2013) and World Bank (2018).

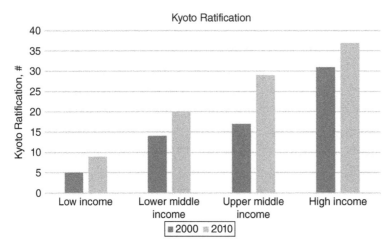

Figure 2.6 Kyoto ratification and GDP per capita. *Source:* Authors' calculations and visualization based on raw data taken from Steves et al. (2013) and World Bank (2018).

middle-income group. In upper-middle income countries, for example, the number of countries which had ratified the Kyoto protocol increased from 17 in 2000 to 29 in 2010. In lower-middle income countries, this number increased from 14 in 2000 to 20 in 2010.

2.3 Environmental Preferences of Individuals

2.3.1 Measures of Individual Environmental Preferences

Our individual-level analysis of environmental preferences is based on data from the WVS, a collection of nationally representative, individual-level surveys conducted in nearly 100 countries (almost 90% of the world's population). The surveys started in 1981–1984 and have been conducted six times since, with the latest wave covering the years 2010–2014. However, survey questions have changed across this time period.

We consider three measures of proenvironment attitudes of individuals. First, we analyze the respondents' prioritization of protecting the environment versus ensuring economic growth and creating jobs. In particular, the respondents to the WVS are asked: "Here are two statements people sometimes make when discussing the environment and economic growth. Which of them comes closer to your own point of view?" The response options are: "Protecting the environment should be given priority, even if it causes slower economic growth and some loss of jobs" or "Economic growth and creating jobs should be the top priority, even if the environment suffers to some extent."

Furthermore, we look at individuals' willingness to participate financially in addressing environmental issues. This is captured by two measures. First, the respondents are asked about their willingness to give part of their income for the environment. Second, there is information on whether the respondents would specifically support an increase in taxes if the extra money is used to prevent environmental pollution.

2.3.2 Socioeconomic Standing and Environmental Preferences

Individual preferences for environment policies vary by socioeconomic status. In Table 2.2, we present the proportions of those who agree that protecting the environment is more important than ensuring economic growth and creating jobs. This information is grouped by the income and level of education of individuals – WVS provides information on the income scale based on the individual reports of household incomes. As Table 2.2 demonstrates, the share of those who prioritize environmental protection increases with socioeconomic standing. Sixty-three percent of those in the fifth – therefore highest – quintile of income distribution agree that protecting the environment is more important, compared to around 53% among those in the first quintile. Similarly, among those with tertiary education, nearly 66% prioritize environmental protection over ensuring economic growth and job creation, while only around half of those with primary educational attainment do the same.

Table 2.2 Individual preferences: protecting the environment is more important than economic growth

		Education level:			
		Primary	Secondary	Tertiary	Total
Income quintile:	1st quintile	49.8%	54.9%	55.4%	53.2%
	2nd quintile	49.7%	55.0%	59.4%	55.4%
	3rd quintile	51.8%	54.4%	59.6%	56.5%
	4th quintile	53.4%	54.0%	60.9%	58.2%
	5th quintile	51.2%	55.0%	65.8%	63.0%
	Total	50.7%	54.6%	60.0%	

Source: Authors' calculations based on raw data taken from World Values Surveys – Years: 2005–2010.

Table 2.3 Individual preferences: willingness to sacrifice a part of income for environment

		Education level:			
		Primary	**Secondary**	**Tertiary**	**Total**
Income quintile:	1st quintile	55.6%	60.9%	67.1%	60.9%
	2nd quintile	57.9%	62.1%	68.4%	63.6%
	3rd quintile	68.2%	68.2%	71.0%	69.6%
	4th quintile	72.3%	69.0%	72.5%	71.6%
	5th quintile	71.0%	70.5%	74.9%	74.0%
	Total	61.6%	65.4%	70.6%	

Source: Authors' calculations based on raw data taken from World Values Surveys – Years: 2005–2010.

Table 2.4 Individual preferences: willingness to support increase in tax to prevent pollution

		Education level:			
		Primary	**Secondary**	**Tertiary**	**Total**
Income quintile:	1st quintile	48.7%	51.9%	55.3%	51.8%
	2nd quintile	49.9%	52.8%	57.8%	54.1%
	3rd quintile	58.5%	59.7%	61.1%	60.2%
	4th quintile	67.2%	63.0%	64.8%	64.7%
	5th quintile	60.2%	60.6%	67.2%	65.7%
	Total	53.8%	57.0%	61.0%	

Source: Authors' calculations based on the raw data taken from World Values Surveys – Years: 2005–2010.

Looking at individuals' willingness to pay for environmental action in Tables 2.3 and 2.4 yields similar patterns. Seventy percent of individuals in the fifth quintile of income distribution are happy to give part of their income for the environment. Only 61% among the poorest do. The share of those willing to give part of their income for environmental action ranges from around 62% among individuals with primary educational attainment to around 71% among tertiary degree holders.

A lower share of individuals specifically opt for a tax increase to prevent pollution compared to those expressing more general willingness to give part of their income. However, here too we see similar patterns by socioeconomic characteristics of individuals. Around 66% of individuals in the fifth quintile of income distribution agree on a tax increase to prevent pollution. Only 55% of individuals in the first quintile of income distribution do so. Among tertiary degree holders, the share of those opting for a tax increase to prevent pollution is 61%, while among primary degree holders it is under 54% only.

In sum, individual preferences for environmental action vary by socioeconomic status, as defined by income and education level.

2.3.3 Individual Preferences and Policy Actions

We also consider whether individual preferences for environmental protection are linked to the environmental policies observed at the macro level. In a standard median voter model, political decisions reflect the preferences of the electorate (Downs 1957). Hence, in places with more proenvironmental attitudes, we can expect to observe more stringent climate change policies. Alternatively, consistent with a "citizen-candidates" model, the politicians themselves may have proenvironmental preferences and implement policies consistent with those preferences once elected (Osborne and Slivinski 1996; Besley and Coate 1997).

In Figures 2.7 and 2.8, we map CLIMI scores of countries and indicate individual willingness to pay for environmental action based on WVS responses. In many cases, we see a direct link between the proenvironmental preferences of individuals and the environmental policies adopted by countries. Scandinavian countries present some of the most significant examples; Norway and Sweden both have a high share of proenvironmental individuals, as well as being among the countries with the most stringent environmental policies. However, the proenvironmental attitudes of individuals does not always translate into stringent environmental policies. Turkey is a notable case; with over 80% of the country's population willing to give a part of their income for an environmental cause, the country's CLIMI score is nevertheless among the lowest in the sample used in this analysis. Jordan is another case, with over 70% of population willing to give a part of their income for environment, but a CLIMI score of only 0.156. The quality of institutions facilitating the link between individual preferences and policies adopted by countries' leaders is likely to be at the core of such differences across countries.

Figure 2.7 CLIMI and willingness to sacrifice a part of income for the environment. *Source:* Authors' calculations and visualization based on raw data taken from World Values Surveys – Years: 2005–2010 and Steves et al. (2013).

Figure 2.8 CLIMI and willingness to support increase in tax to prevent pollution. *Source:* Authors' calculations and visualization based on raw data taken from World Values Surveys – Years: 2005–2010 and Steves et al. (2013).

2.4 Discussion

2.4.1 Overall Findings

Our contribution has some expected and confounding results. Stronger climate change mitigation policies at the macro level appear to be consistent with proenvironmental individual preferences at the micro level when comparing high-income versus low-income economies. At the macro level, the average CLIMI score for low-income economies was a mere 0.06 in comparison to the average score of 0.54 for high-income economies. At the micro level, our results show that the citizens of countries with high levels of national wealth, as measured by GDP per capita, tend to have more individual concern for (and likely involvement in) environmental action than citizens of countries with lower levels of wealth. On average, proenvironmental attitudes, described by intention or willingness to make sacrifices for the environment, are positively correlated across nations with higher levels of wealth.

These findings are similar to previous research. For example, Pisano and Lubell (2017) found that postindustrialized (high-income) societies that have more dense communication structures and mass education demonstrate more environmental engagement and activism than developing (low-income) countries. As the level of development and position of power of a country increase, there is a stronger correlation between environmental attitudes and climate action.

We also find that an individual's income and education level have a positive effect on proenvironmental attitudes. This is also a consistent finding within other literature (e.g., see Aklin et al. 2013). Income changes an individual's preferences and underpins the government's incentive to decrease negative environmental externalities, including pollution and waste (Dasgupta et al. 2002). The environmental Kuznets curve hypothesis posits that environmental degradation follows an inverted U-shaped trajectory as a country's income grows (Galeotti et al. 2006). Education matters in emerging countries too. For example, in investigating the case of the Protected Area of Junyun Mountain in China, Liu et al. (2010) found empirical evidence that education levels strongly predict proenvironmental attitudes.

However, we need to be mindful of the uncertainty and level of complexity when interpreting these data. There is not always a direct translation from having a high level of environmental *concern* (as expressed by proenvironmental attitudes) to proenvironmental *behavior*. Some studies demonstrate that environmental concerns do not predict proenvironmental behavior. For instance, in theory, we might expect that people with proenvironmental values and concerns about the risks of climate change would also be more likely to give up voluntary air travel, or at least to fly less far, so as to lessen their personal contribution to fuel emissions. Alcock et al. (2017) found that, in the United Kingdom, there was no association between individuals' concern for climate change and their propensity to reduce discretionary air travel (i.e., take fewer nonwork-related flights or reduce the distances flown by those who do so). Mitigating climate change problems requires behavioral changes by individuals all over the world, including the well-educated and affluent.

The results of our study reveal the complex connectivity between the proenvironmental concern of individuals and climate policy action. Environmental behavior depends on the interaction between individual and institutional factors. Even if individuals within less developed (e.g., lower-middle to upper-middle income) countries express higher levels of proenvironmental attitudes, they might find it difficult to find appropriate resources to help them act on their intentions.

It is sensible to expect that the concern and behavior association is weaker in societies where economic resources, information about environmental issues, environmental facilities, and green consumption options are lacking. This implies that there are structural or institutional barriers between individual preferences for environmental concern and climate policy action. Barriers to individual action reinforce inequality in climate change outcomes across countries with different levels of development; however, these barriers vary across societies, so it is challenging to systematically identify the specific structural-institutional barriers for individuals across different regions, income groups, and so on. We believe that an important direction for future research is to develop a more specified framework for conceptualizing and measuring structural-institutional barriers, and thereby examine their effects.

In our study, we found that a high proportion of the population in the increasingly industrialized state of Turkey appear to have solid proenvironmental attitudes, yet these do not translate into strong environmental policies. On the other hand, the high-income nations of Norway and Sweden both feature a high share of proenvironmental individuals, as well as some of the most rigorous environmental policies and an important use of green energy technologies. They are also known for progressive social policies, economic equality, and participatory democratic institutions (see O'Hara 2012). There are thus much higher structural-institutional barriers in Turkey than for Scandinavia, and this would partly explain (in the case of Turkey) the divergence between individual preferences for environmental concern and climate policy action.

We posit that one of the key institutional barriers to examine would be a lack of social capital. This concept can be understood as the quality of social relationships in society. There is an emerging literature that links social capital to individuals' concern for protection of the natural environment. Defining social capital in the form of intellectual human capital, or social cognitive abilities, Obydenkova and Salahodjaev find that both democratic institutions and the cognitive abilities of the population have a positive impact on environmentalism (2017, p. 185). The authors find that a 10-point increase in a country's social cognitive capital is associated with a nearly 16-point increase in its CLIMI, which is quite significant. Conversely, defining *a*social capital in the form of distrust – a general negative view of human nature and distrust of people and institutions – Tam and Chan find that proenvironmental concern and behavior are weaker in societies characterized by higher levels of distrust (2017, p. 219). Therefore, lack of trust (i.e., distrust) is another important barrier to consider in the research problem. In section 2.4.2, we investigate the relationship between the environment and social capital, including trust, using Sweden and Turkey as case studies to better understand the key structural-institutional barriers.

2.4.2 Case Study: Sweden and Turkey

We use three different measures with sufficient data over the past 20 years to compare the trends between Sweden and Turkey. For a suitable indicator of effective environment policies, we use the OECD's Environmental Policy Stringency Index. This index is defined as an "economy-wide" indicator that includes market-based and nonmarket-based instruments, such as emissions limits for greenhouse gases or government expenditure on R&D for renewable technologies (Botta and Koźluk 2014). It is a useful composite index, as it allows for intercountry assessments over time and considers different kinds of measurements: abatement costs, public expenditures, policy assessments and emissions. For example, in terms of taxes on CO_2 emissions, a higher price implies higher stringency.

To measure generalized trust, we draw on various World Value Surveys, where surveyed citizens are asked the following question: "Would you say that most people can be trusted, or that you can't be too careful?" Generalized trust refers to a general positive outlook on human nature or an expectation about other people's benevolence (Nannestad 2008). Trust values are taken from Wave 3 (1995–1999), Wave 4 (2000–2004), Wave 5 (2005–2009), and Wave 6 (2010–2014) (World Bank 2018). To measure social capital, we draw on the Legatum Prosperity Index's "social capital pillar," which measures the strength of personal and social relationships, social norms, and civic participation in a given country (Legatum Institute 2018). Where available, data on the Environmental Policy Stringency Index, social capital, and generalized trust for Sweden and Turkey from 1998 to 2018 are shown in Table 2.5.

As seen in Table 2.5, Sweden surpasses Turkey in all three measures. Compared to Sweden, Turkey has a lower level of rigor in environmental policy, although the index shows some improvement over time. Generalized trust is vastly different too: very high in Sweden, very low in Turkey. Among the 149 countries ranked, Turkey has a far lower level

Table 2.5 Environmental policy, social capital, and trust in Sweden and Turkey, 1998–2018

Indicator	Country	1998	2003	2008	2013	2018
Environmental Policy Stringency Index[a] (higher the better)	Sweden	1.25	2.43	2.92	3.10 (2012)	—
	Turkey	0.50	0.69	1.50	1.83 (2012)	—
Generalized trust[b] ("most people can be trusted")	Sweden	56.6%	63.7%	65.2%	60.1%	—
	Turkey	6.5%	18.6%	4.8%	11.6%	—
Social capital[c] (lower the better, rank among 149 countries)	Sweden	—	—	18	17	22
	Turkey	—	—	104	121	100

Source: Data adapted from [a]Botta and Koźluk (2014), [b]World Bank (2018) and [c]Legatum Institute (2018).
— = data not available.

of social capital compared to Sweden, reflected in, for example, the lower number of people who say they help strangers, volunteer or donate.

People and relationships matter. Societies are better places to live when people trust and support one another and have extensive communities and social networks. As a result, social trust is one of the most important components of social capital. It is influenced by social interactions, and includes norms of reciprocity and networks (Putnam 1993). Trusting attitudes are also rooted in and shaped by policies and political institutions (Bergh and Öhrvall 2018, p. 1147).

Trust is vital for the acceptance of increased state intervention. To accept policies, individuals must both trust that public institutions can actually manage and implement policies in a noncorrupt and efficient way, and trust that fellow citizens and business actors are following policy guidelines (Harring 2018, pp. 1–2). There are thus different forms of trust: horizontal trust – trust among citizens – and vertical trust – trust between citizens and government actors. Pitlik and Kouba argue that both concepts matter in forming environmental preferences (2015, p. 359).

Sweden is among a small number of countries which are pacesetters in environmental policymaking, with significant implementation success (Jänicke 2005). In 1991, it was among the first countries in the world to introduce a carbon tax on fossil fuels, making it more expensive to use fossil fuels. Subsidies and investment schemes urging households to change from fossil fuels to other energy systems, such as biofuels, hydropower, and nuclear power, are also a part of Sweden's innovative set of policies (Schmidt et al. 2019, p. 429). However, the introduction of environmental taxes alone cannot explain Sweden's triumphs in greening its economy.

Horizontal trust shapes people's perception and expectation that other people would act likewise – for instance, comply with environmental policies – to achieve a collective good for their society. As Marbuah argues, a high level of trust in Sweden among citizens suggests that they believe other citizens have proenvironmental preferences (2019, p. 469).

People in turn might be willing to serve the collective good of their communities (Tam and Chan 2018, p. 183). Environmental policy support in general is also linked to vertical trust, both political and institutional. The reasoning is that if people trust the institutions that implement environmental policies, then they are more likely to support these policies. Hammar and Jagers show that trust in the institutions implementing the policies is important for the acceptance of CO_2 taxes (2006). Therefore, high levels of both horizontal and vertical trust in Sweden and other Nordic countries explain the support for environmental taxes (Harring 2016, p. 586; Marbuah 2019, p. 455). The credibility and legitimacy of environmental policies in Sweden, and the success of their implementation, depend on the extent to which the people trust each other and responsible institutions.

The Swedish government is more proactive and extensive in its climate change mitigation than most other countries. Sweden has various national policies and strategies that aim to lower its unsustainable consumption. It has established all-encompassing national environmental quality objectives, with the central component being the "generational goal." According to the Swedish Environmental Protection Agency, "the overall goal of Swedish environmental policy is to hand over to the next generation a society in which the major environmental problems in Sweden have been solved, without increasing environmental and health problems outside Sweden's borders" (2012, p. 3). Sweden is also a signatory of the Agenda 2030 and Sustainable Development Goals, with sustainable consumption and production listed as Goal 12.

To meet the generational goal is a major challenge but with this objective, Sweden is at the forefront of countries recognizing the need to address their displaced emissions. It is difficult to decrease or even halt the greenhouse gas emissions that happen abroad due to Swedish consumption, in particular that of manufactured goods. In recent years, Swedish climate policy measures have aligned to the generational goal by implementing policies to target emission intensity improvements in Swedish production, and enhancing information flows to consumers about the environmental performance of various products (Isenhour and Feng 2016, p. 326). In summary, the evidence suggests that there is a convergence between individual preferences for environmental concern and climate policy action in Sweden due to high levels of trust and social capital formation.

In Turkey, however, matters are completely different. It is rational for an individual to *not* participate in collective action, simply if they do not trust that others will act, giving rise to what is called a "social trap." Smith and Mayer argue that this "trap will remain shut until these deep-rooted patterns of distrust are resolved, an enduring problem in itself" (2018, p. 143). The authors conclude that in nations where both social and institutional trust are apparent, individual behaviors to address climate change are more likely (Smith and Mayer 2018, p. 149). In Turkey, there is a high level of distrust in news, which Yanatma explains as an indicator of news media in the country and a very polarized society (2017). In contrast to countries such as Sweden with high levels of trust, societies such as Turkey are unable to engage in collective action because of high levels of distrust.

Moreover, as Turkey is an energy-dependent country, its policymakers face different challenges compared to Sweden. At present, Turkey imports around 75% of its primary energy supply, which is mostly composed of Russian gas and Iranian oil and natural gas (Erşen and Çelikpala 2019, p. 128). Turkey is dependent on fossil fuel imports, as it also has the second-highest energy consumption growth, after China. Turkey is also a net producer of coal – primarily lignite – and harbors considerable domestic reserves (Jones et al. 2017). In this case, low levels of trust and import dependence on fossil fuels are obvious structural barriers. However, the situation requires a more in-depth critical analysis of the socio-institutions at play.

Over the past decade and a half, the economy of Turkey has undergone major economic growth in construction, real estate, and infrastructure. Along with construction and transportation, mining, carbon-based energies, and renewables such as hydropower have been among the fastest growing sectors in Turkey. These sectors have been liberalized since the early 2000s, and had ready access to the cheap credit that was being propelled into the Turkish financial market. State-led real estate creation and speculation occurred, and it is evident that local participation, environmental justice, and sustainability have not been priorities since the 2002 election of the Justice and Development Party (AKP) government. Their top-down approach to growth has had a negative effect on Turkey's ability to bridge social capital links between different groups.

During the 2000s and early 2010s, economic growth was dominated by speculative finance at the expense of social and environmental considerations, in both rural and urban areas, including peripheral towns. This especially affected the poor and middle class urban populations due to massive transfers of wealth. A notable example of speculation-driven investments incited by the AKP elite would be the commercial redevelopment of the most centrally located public park in Turkey (Erensü and Karaman 2017, p. 21). As a whole, the growth fueled by speculative finance "led to Turkey's vast countryside being opened up to an unprecedented scramble to set up extractive industries and infrastructural projects" (Erensü and Karaman 2017, p. 28).

The AKP government used its power to transform the countryside without consultation with or concern for rural individuals. It demonstrated a clear preference for private investors, as the government deregulated environmental directives such as the Environmental Impact Assessment procedures 15 times, and refused to ratify the Framework Convention and the Kyoto Protocol (Erensü 2018, pp. 149–154; Erensü and Alemdaroğlu 2018, p. 25). In a 2016 postcoup context, the Turkish government is keen to consume coal to strive for more economic growth and energy security – a clearly unhelpful climate change mitigation strategy. Discourse in the Colombia-Turkey coal chain therefore lends itself to a lack of local participation in the decision-making process, repression of social movements, and weak environmental regulations (Cardoso 2018, pp. 54–55; Cardoso and Turhan 2018, p. 406).

Moreover, as Tansel's research shows, the AKP-led urban transformation and housing projects, from 2003 to 2017, are marked by a nonparticipatory approach to urban "renewal" (2019a, p. 329). In short, local resistance and voices were stifled, and private companies' and multinationals' interests were bolstered. Turkey's low level of interest in "bridging" social capital suggests that the majority of social ties are between exclusive groups within the AKP and the private sector, resulting in a society orientated to the needs of only a select few.

Critical scholars – Sinan Erensü and Cemal Tansel, for instance – argue that the AKP regime has both positive and negative aspects, judiciously blending and deploying rationalities of reform *and* repression. Starting from the mid-1990s, Turkish governments took the initiative and signed several oil and gas pipeline projects, linking the rich natural resources of the Caucasus and the Middle East to European consumers. By 2017, there were 10 major pipeline projects passing through in Turkey, either in operation or under construction. With these pipelines, Turkey hoped to retain its geostrategic relevance by becoming an "energy hub" in the region. But the present mode of Turkish authoritarianism used to bring about reform has led to numerous problems: the erosion of the rule of law; the lack of democratic governance; the rising land use disputes; the systematic displacement of the working class away from working-class jobs; the use of legal and extralegal intimidation to discipline dissenting citizens; the destruction of cultural heritage embedded in the party's urban policy; and the steady transformation and weakening of the regulatory and administrative bodies responsible for environmental protection (Erensü and Alemdaroğlu 2018, pp. 20, 24; Tansel 2019b, p. 12). The political economic milieu in Turkey has not fostered peaceful class relations or vertical trust, and as a result, the stock of social capital is worn out.

Since the 2010s, there have been uneven patterns of development and industrialization across Turkey, with a widening gap between high-income and low-income regions. High-pollution industries have been relocating from the more developed areas of the country to the less developed central and eastern regions (Acar and Yeldan 2018, pp. 97, 105). For Turkey to make any significant progress in tackling future CO_2 emissions, it would require fairer income distribution, as policies that reduce income inequality will help improve environmental quality (Uzar and Eyuboglu 2019, p. 156). This will create an important balance in the distribution of political power in Turkey, in addition to restoring trust and bridging social capital. A glimmer of hope for mitigating environmental problems, as found in the results of this chapter, is in the decent proportion of individuals in Turkey expressing solid proenvironmental concern. However, these attitudes will only translate to proenvironmental actions when there is a high level of generalized trust and a restoration in the stock of social capital as a whole.

2.5   Conclusion

There are multiple factors to consider at the cross-country level of data, and there is likely a diversity of subnational regions in terms of political institutions, democracy, social capital, public policies, and environmentalism. Indeed, Turkey is simply a case in point, and yet the rising industrialized states are a very diverse group in terms of their emissions, economic activities, regional relations, and energy possibilities. Some of these states are now taking active steps to reduce their emissions, invest in renewable energy, and develop low-carbon solutions. They provide low-cost manufacturing for renewable energy sources – especially solar photovoltaics, as in the case of China – but a considerable number of new patents for clean energy are coming out of emerging economies like China and South Korea, countries that feature above-average CLIMI scores (Lachapelle et al. 2017).

In summary, policy actions to mitigate climate change vary geographically according to the position of a state in the global economy and its particular institutional forms of economic-ecological intervention. From a political economy perspective, climate change is best understood as a global struggle between fossil fuel interests – corporate lobbyists and the states that support them – and the rest of humanity over the future trajectory of economic development. A key argument by Paterson and P-Laberge is that these interests are supported by the deeply embedded social and practical uses of energies that affect the environment (2018, p. 4). These practices reflect the way that the economy is lived in daily life – driving, flying, cooking, heating, cooling, and so on – and constitute, in a sense, the social and cultural life of greenhouse gas emissions. It is in this context that we need to think more critically about the design of international and domestic agreements, as shown in the CLIMI, and the population's deep concern for protecting the natural environment.

Note

1 "Preferences are reasons for behavior, that is, attributes of individuals that (along with their beliefs and capacities) account for the actions they take in a given situation" (Bowles 1998, p. 78).

References

Acar, S. and Yeldan, A.E. (2018). Investigating patterns of carbon convergence in an uneven economy: the case of Turkey. *Structural Change and Economic Dynamics* 28: 96–106.

Aklin, M., Patrick, B., Harish, S.P., and Urpelainen, J. (2013). Understanding environmental policy preferences: new evidence from Brazil. *Ecological Economics* 94: 28–36.

Alcock, I., White, M.P., Taylor, T. et al. (2017). 'Green' on the ground but not in the air: pro-environmental attitudes are related to household behaviours but not discretionary air travel. *Global Environmental Change* 42: 136–147.

Bergh, A. and Öhrvall, R. (2018). A sticky trait: social trust among Swedish expatriates in countries with varying institutional quality. *Journal of Comparative Economics* 46: 1146–1157.

Besley, T. and Coate, S. (1997). An economic model of representative democracy. *Quarterly Journal of Economics* 112: 85.

Böhringer, C. and Jochem, P.E.P. (2007). Measuring the immeasurable: a survey of sustainability indices. *Ecological Economics* 63 (1): 1–8.

Botta, E. and Koźluk, T. (2014). *Measuring Environmental Policy Stringency in OECD Countries: A Composite Index Approach*. OECD Economics Department Working Papers, no. 1177. Paris: OECD.

Bowles, S. (1998). Endogenous preferences – the cultural consequences of markets and other economic institutions. *Journal of Economic Literature* 36 (March): 75–111.

Burck, J., Marten, F., and Bals, C. (2014). The Climate Change Performance Index Results 2014. Bonn: Germanwatch and CAN.

Cardoso, A. (2018). Valuation languages along the coal chain from Colombia to the Netherlands and to Turkey. *Ecological Economics* 146: 44–59.

Cardoso, A. and Turhan, E. (2018). Examining new geographies of coal: dissenting energyscapes in Colombia and Turkey. *Applied Energy* 224: 398–408.

Cole, M.A. and Fredriksson, P.G. (2009). Institutionalized pollution havens. *Ecological Economics* 68 (4): 1239–1256.

Dasgupta, S., Laplante, B., Wang, H., and Wheeler, D. (2002). Confronting the environmental Kuznets curve. *Journal of Economic Perspectives* 16 (1): 147–168.

Davis, J.B. (2003). *The Theory of the Individual in Economics*. London: Routledge.

Downs, A. (1957). *An Economic Theory of Democracy*. New York: Harper and Row.

Erensü, S. (2018). Powering neoliberalization: energy and politics in the making of a new Turkey. *Energy Research & Social Science* 41: 148–157.

Erensü, S. and Alemdaroğlu, A. (2018). Dialectics of reform and repression: unpacking Turkey's authoritarian "turn". *Review of Middle East Studies* 52 (1): 16–28.

Erensü, S. and Karaman, O. (2017). The work of a few trees: Gezi, politics and space. *International Journal of Urban and Regional Research* 41 (1): 19–36.

Erşen, E. and Çelikpala, M. (2019). Turkey and the changing energy geopolitics of Eurasia. *Energy Policy* 128: 584–592.

Galeotti, M., Lanza, A., and Pauli, F. (2006). Reassessing the environmental Kuznets curve for CO2 emissions: a robustness exercise. *Ecological Economics* 57 (1): 152–163.

Greenstone, M. and Hanna, R. (2014). Environmental regulations, air and water pollution, and infant mortality in India. *American Economic Review* 104 (10): 3038–3072.

Hammar, H. and Jagers, S.C. (2006). Can trust in politicians explain individuals' support for climate policy? The case of CO_2 tax. *Climate Policy* 5 (6): 613–625.

Harring, N. (2016). Reward or punish? Understanding preferences toward economic or regulatory instruments in a cross-national perspective. *Political Studies* 64 (3): 573–592.

Harring, N. (2018). Trust and state intervention: results from a Swedish survey on environmental policy support. *Environmental Science and Policy* 82: 1–8.

IPCC (2014). Climate Change 2014: Synthesis Report. Contribution of Working Groups I, II and III to the Fifth Assessment Report of the Intergovernmental Panel on Climate Change. Geneva: IPCC.

Isenhour, C. and Feng, K. (2016). Decoupling and displaced emissions: on Swedish consumers, Chinese producers and policy to address the climate impact of consumption. *Journal of Cleaner Production* 134: 320–329.

Jänicke, M. (2005). Trend-setters in environmental policy: the character and role of Pioneer countries. *European Environment* 15: 129–142.

Jones, C.R., Kaklamanou, D., and Lazuras, L. (2017). Public perceptions of energy security in Greece and Turkey: exploring the relevance of pro-environmental and pro-cultural orientations. *Energy Research & Social Science* 28: 17–28.

Kellenberg, D.K. (2009). An empirical investigation of the pollution haven effect with strategic environment and trade policy. *Journal of International Economics* 78 (2): 242–255.

Kunkel, N., Jacob, K., and Busch, P.-O. (2006). *Climate Policies: (the Feasibility of) a Statistical Analysis of Their Determinants*. Berlin: Human Dimensions of Global Environmental Change.

Lachapelle, E., MacNeil, R., and Paterson, M. (2017). The political economy of decarbonisation: from green energy 'race' to green 'division of labour'. *New Political Economy* 22 (3): 311–327.

Legatum Institute (2018). The Legatum Prosperity Index 2018. https://www.prosperity.com/download_file/view/3578/1692

Liu, J., Ouyang, Z., and Miao, H. (2010). Environmental attitudes of stakeholders and their perceptions regarding protected area–community conflicts: a case study in China. *Journal of Environmental Management* 91 (11): 2254–2262.

Marbuah, G. (2019). Is willingness to contribute for environmental protection in Sweden affected by social capital? *Environmental Economics and Policy Studies* 21: 451–475.

Met Office (2017). How can we limit warming? www.metoffice.gov.uk/binaries/content/assets/metofficegovuk/pdf/weather/learn-about/climate/cop/how-can-we-limit-warming-v4.pdf

Millar, R.J., Fuglestvedt, J.S., Friedlingstein, P. et al. (2017). Emission budgets and pathways consistent with limiting warming to 1.5°C. *Nature Geoscience* 10: 741–747.

Nannestad, P. (2008). What have we learned about generalized trust, if anything? *Annual Review of Political Science* 11: 413–436.

O'Hara, P.A. (2012). Short-, long-, and secular-wave growth in the world political economy: periodicity, amplitude, and phases for 8 regions, 108 nations, 1940–2010. *International Journal of Political Economy* 41 (1 Spring): 3–46.

Obydenkova, A.V. and Salahodjaev, R. (2017). Climate change policies: the role of democracy and social cognitive capital. *Environmental Research* 157: 182–189.

Osborne, M.J. and Slivinski, A. (1996). A model of political competition with citizen candidates. *Quarterly Journal of Economics* 111: 65–96.

Paterson, M. and P-Laberge, X. (2018). Political economies of climate change. *WIREs Climate Change* 9: e506.

Pisano, I. and Lubell, M. (2017). Environmental behavior in cross-national perspective: a multilevel analysis of 30 countries. *Environment and Behavior* 49 (1): 31–58.

Pitlik, H. and Kouba, L. (2015). Does social distrust always lead to a stronger support for government intervention? *Public Choice* 163 (3): 355–377.

Putnam, R. (1993). *Making Democracy Work: Civic Traditions in Modern Italy*. Princeton, NJ: Princeton University Press.

Rodrigues, J. (2004). Endogenous preference and embeddedness: a reappraisal of Karl Polanyi. *Journal of Economic Issues* 38 (1): 189–200.

Schmidt, S., Södersten, C.-J., Wiebe, K. et al. (2019). Understanding GHG emissions from Swedish consumption – current challenges in reaching the generational goal. *Journal of Cleaner Production* 212: 428–437.

Smith, E.K. and Mayer, A. (2018). A social trap for the climate? Collective action, trust and climate change risk perception in 35 countries. *Global Environmental Change* 49: 140–153.

Steves, F., Teytelboym, A., and Treisman, D. (2013). *Political Economy of Climate Change Mitigation Policy*. Working Paper. European Bank of Reconstruction and Development and University of Oxford, London and Oxford.

Surminski, S. and Williamson, A. (2012). *Policy indexes? What do they tell us and what are their applications? The case of climate policy and business planning in emerging markets*. Working Paper 101. Centre for Climate Change Economics and Policy, London.

Swedish Environmental Protection Agency (2012). *Sweden's Environmental Objectives – An Introduction*. Stockholm: Swedish Environmental Protection Agency. www.swedishepa.se/Documents/publikationer6400/978-91-620-8620-6.pdf?pid=6759

Tam, K.-P. and Chan, H.-W. (2017). Environmental concern has a weaker association with pro-environmental behavior in some societies than others: a cross-cultural psychology perspective. *Journal of Environmental Psychology* 53: 213–223.

Tam, K.-P. and Chan, H.-W. (2018). Generalized trust narrows the gap between environmental concern and proenvironmental behavior: multilevel evidence. *Global Environmental Change* 48: 182–194.

Tansel, C.B. (2019a). Reproducing authoritarian neoliberalism in Turkey: urban governance and state restructuring in the shadow of executive centralization. *Globalizations* 16 (3): 320–335.

Tansel, C.B. (2019b). The shape of 'rising powers' to come? The antinomies of growth and neoliberal development in Turkey. *New Political Economy* 12: 1–22.

Uzar, U. and Eyuboglu, K. (2019). The nexus between income inequality and CO_2 emissions in Turkey. *Journal of Cleaner Production* 227: 149–157.

World Bank (2018). *World Development Indicators*. http://databank.worldbank.org/data/reports.aspx?source=world-development-indicators

Yanatma, S. (2017). *Reuters Institute Digital News Report 2017: Turkey Supplementary Report*. Oxford: Reuters Institute for the Study of Journalism.

3

Legislation or Economic Instruments for a Successful Environmental Policy?

Reflections After "Dieselgate"

Theodoros Zachariadis

Cyprus University of Technology, Limassol, Cyprus

3.1 Introduction

In September 2015 the US Environmental Protection Agency issued a "notice of violation of the Clean Air Act," thereby announcing that it had started investigations against Volkswagen. The allegation was that the company intentionally installed illegal software which allowed some diesel-powered models to activate pollutant emission controls only under regulatory test conditions; thus the models were able to pass stringent laboratory emission tests required by legislation, although in real-world operation these emission controls were deactivated by the software. As a result, emissions of nitrogen oxides (NOx) from these models were up to 40 times higher under on-road operation than under standardized testing procedures. It turned out that hundreds of thousands of diesel-fueled Volkswagen vehicles circulating in the US (and several million worldwide) were equipped with such "defeat devices" (Ewing 2015).

In the months and years that followed, further investigations led to several indictments of Volkswagen and its top managers; up to summer 2018, this case had cost the firm more than 25 billion euros in fines, vehicle buybacks and compensations – most of which occurred in the US. It also led to investigations in other automobile firms of the same group, causing the arrest of the Chief Executive Officer of Audi in June 2018. Lawsuits from individual car owners and investors followed and other car makers have also undergone investigations worldwide.[1]

The Volkswagen emission scandal, also known as "Dieselgate," poses very interesting and timely questions about the appropriate way to tackle environmental policy problems. Is a more stringent regulatory policy sufficiently effective to improve air quality and mitigate climate change, or are there more effective ways through economic incentives? This chapter discusses these issues by using North American and European evidence (both theoretical and empirical) from recent environmental policy research. It formulates the advantages and disadvantages of different approaches to environmental policy. Although this chapter focuses on road transportation, it can be extrapolated, as the discussion on environmental policy and its challenges can be applied to other sectors.

Environmental Policy: An Economic Perspective, First Edition. Edited by Thomas Walker, Northrop Sprung-Much, and Sherif Goubran.

A preview of the chapter's main conclusion: More emphasis should be given to economic incentives for phasing out carbon-intensive and highly polluting fuels and technologies. However, in a complex and imperfect world, a mix of regulatory and economic measures remains the preferred policy response to energy and environmental problems. To enable this, there needs to be an improvement in the communication between economists and technology experts. Economists and other social scientists need to realize that there will never be perfectly optimal prices and perfectly functional institutions, while technology experts need to understand that no policy is free of costs.

This chapter briefly reviews the regulatory and technological challenge that led to the Dieselgate scandal; describes the policy response of authorities; reports similar challenges in other energy-consuming sectors; outlines the pros and cons of legislative and economic measures for environmental policy; and summarizes the main conclusions.

3.2 The Policy Challenge for Cleaner Vehicles and Fuels

Transportation accounts for one-third of global final energy consumption and one-fifth of global primary energy demand.[2] Three-quarters of transport's energy needs are for road transport modes – cars, trucks, buses, trains, and motorcycles. According to the International Energy Agency, in the coming decades, this sector will be the main driver in global growth of fossil energy use and carbon emissions (IEA 2017). Moreover, the sector is the main cause of air quality problems compromising human health, despite impressive technological progress in fuel quality, engine technologies, and exhaust gas after-treatment. As of now, in a world with growing car numbers, technological improvements have not been sufficient to eliminate these adverse effects of transportation (OECD 2015).

To address these problems, industrialized nations have enacted environmental legislation, mainly in the form of emission standards for different air pollutants and all vehicle types. This environmental legislation, that was started in the 1970s, attempts to address all kinds of pollution that road vehicles cause. In principle, it involves determining the maximum values of pollutant emissions which a vehicle or engine must not exceed when it is tested under predefined and controlled laboratory conditions, in order for the vehicle or engine to be approved to enter the market.

Transportation experts worldwide have recognized that these vehicle emission tests are carried out with outdated test procedures, which ultimately allows many vehicle models to be easily compliant with the emission standards. This easy compliance happens because:

- tests are conducted with driving cycles that were devised two or three decades ago and are not representative of today's actual driving conditions (where there is more traffic congestion on the roads) and do not match the capabilities of today's vehicles, which are more powerful than those of some decades ago. This leads to reduced load and acceleration requirements from modern vehicles, which leads to relatively low officially reported emission levels.
- test procedures enable car makers to exploit "flexibilities" with regard to tires used, ambient test temperature, use of auxiliary devices such as air conditioning, etc. so as to yield artificially low emission results (EEA 2016).

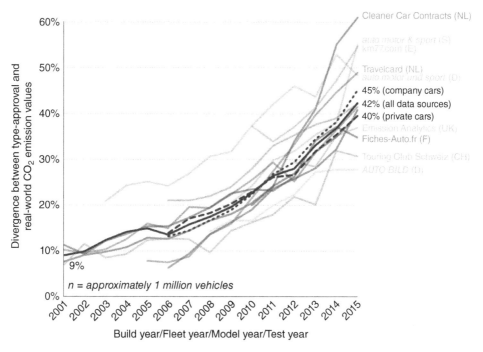

Figure 3.1 Divergence between standardized-test and real-world CO_2 emissions of new cars in Europe according to Tietge et al. (2017), with data from different sources.

The resulting artificially low emission levels have been highlighted in several studies. Tietge et al. (2017), for example, found that on-road carbon dioxide (CO_2) emissions of cars in 2015 in Europe were more than 40% higher than their formal test emissions; this gap was less than 10% in the early 2000s. This trend is illustrated in Figure 3.1; it should be noted that Tietge et al. (2017) found such gaps to be higher in Japan than in Europe, and lower in the United States. In the case of NOx, a major air pollutant related to serious human health impacts, this gap has been reported to be substantially higher, in particular for diesel-powered cars (Barrett et al. 2015; Franco et al. 2014; Weiss et al. 2012). As shown in Figure 3.2, measurements in Europe reveal an increasing NOx gap which has reached a factor of five on average – although, as mentioned in the introductory section, individual diesel vehicles tested in the US had up to 40 times higher NOx emissions. According to post-Dieselgate analysis, even the EU's recently adopted stringent emissions type-approval procedure for passenger cars, which includes a new real-driving emissions test conducted with on-board portable measurement systems, cannot lead to the elimination of the NOx gap (Miller and Franco 2016).

The discrepancy between officially reported and actual on-road emissions perfor-mance – and, most importantly, the increasing trend of this gap over time – renders much of the current environmental policy ineffective. Because of this rising gap between reported and real performance, progress in air quality improvement and climate change mitigation is much slower than one would anticipate based on legislative requirements and official emis-sion reports. CO_2 is the main greenhouse gas contributing to global warming and NOx-related

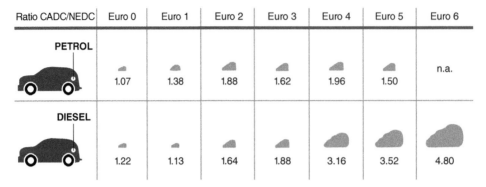

Ratio CADC/NEDC	Euro 0	Euro 1	Euro 2	Euro 3	Euro 4	Euro 5	Euro 6
PETROL	1.07	1.38	1.88	1.62	1.96	1.50	n.a.
DIESEL	1.22	1.13	1.64	1.88	3.16	3.52	4.80

Figure 3.2 Ratio of NOx emissions measured under real-world driving conditions ("CADC" driving cycle) versus the European standard NEDC cycle. There is clear evidence of a growing gap between on-road and standardized-test emissions in more recent vehicle models. "Euro 0" corresponds to models that entered the EU market around 1990, whereas "Euro 6" are models that entered after September 2014. The gap is growing much faster in diesel-powered cars, which reflects the technical challenges of emission abatement in this type of engines. *Source:* EEA (2016).

pollution is among the most serious health issues worldwide, with transport being the major contributor to urban NOx emissions. Inability to curb emissions compromises national and global efforts to combat air pollution, climate change, and rising health concerns.

There is a clear technical rationale behind why this gap exists. In internal combustion engines – the engines that have been powering motor vehicles since the nineteenth century – there is a trade-off between NOx emissions and fuel efficiency (and hence CO_2 emissions); optimal fuel efficiency and minimum CO_2 emissions are achieved with fast and high-temperature combustion in the engine, and this is done by having a high-pressure fuel injection. Especially in diesel engines, these conditions also reduce the production of carcinogenic particulate matter (PM), but lead to higher NOx emissions. Conversely, NOx control techniques such as exhaust gas recirculation reduce combustion temperature, lead to lower fuel efficiency, and potentially higher PM emissions.

With increasingly stringent emission standards for air pollutants (such as NOx and PM) and greenhouse gases (mainly CO_2), it becomes ever more difficult for internal combustion engines to meet all legislative mandates at the same time. Apart from the fact that internal combustion engines cannot become more efficient because there are thermodynamic limits to their achievable thermal efficiency, stringency in emissions requires additional emission control systems which pose an additional cost to automakers. These conditions increase the incentives for auto companies to apply legal and illegal strategies to circumvent some of the requirements. It is therefore not surprising that the defeat devices revealed during Dieselgate affected precisely NOx emissions.

Although generally prohibited, modern software makes it feasible for vehicles to detect an emission test and modulate engine operation or emission control accordingly. Under controlled laboratory conditions, the nonpowered axle of the vehicle (i.e., the wheels that are not connected to the engine) is stationary; tests are regularly conducted under temperatures of 22–28 °C; the steering wheel does not move; air conditioning is off; and the vehicle is preconditioned with given profiles before the test starts. It is easy for a modern vehicle's control system to sense these conditions and program engine operation in a specific manner. One could assume, then, that the combination of today's electronic capabilities with the existence of stringent environmental legislation and imperfect monitoring of automakers' compliance made Dieselgate an accident waiting to happen. In fact, technology and "flexibilities" of legislation allow cheating in the opposite direction as well. European authorities have reportedly found that very recent emission tests have led to inflated emission measurements, because car makers may have exploited regulatory loopholes in order to report artificially high emissions – which would help make future legislation more lenient (McGee 2018).

3.3   The Reaction of Policymakers to Dieselgate – Business as Usual

Reacting to Dieselgate, policymakers vowed to accelerate the ongoing transition to a more stringent regulatory regime for vehicle emissions, especially in Europe. Indeed, the European Commission – the EU's executive body – decided with EU leaders that from 2019 onwards, the CO_2 emission levels of cars sold in the EU will be reported on the basis of emission tests conducted in the new WLTP (Worldwide Harmonized Light Vehicle Test Procedure) driving cycle. The WLTP is supposed to replace the New European Driving Cycle (NEDC) test, as the WLTP is deemed more representative of today's car capabilities and driving conditions. As a complement to this new test, it was decided that emissions of pollutants will be measured more reliably by portable emission measuring systems attached to a number of cars while driving in real conditions on the road. Finally, the type of approval procedure is supposed to become more robust by strengthening the independence of authorities designated for the testing and inspection of vehicle compliance with EU legislation and by introducing a market surveillance system to control the conformity of cars already in circulation (European Commission 2018).

National governments have responded too, mainly by conducting their own investigations and filing lawsuits against Volkswagen and other auto manufacturers.[3] As it became increasingly evident that diesel cars emit much higher amounts of NOx than officially reported, many local authorities started considering banning diesel cars in urban areas. London, Paris, Madrid, Hamburg, Stuttgart, and other cities announced such plans for the years to come.[4,5] London has a congestion charge in place, a £11.50 daily fee paid for each vehicle within the charging zone between 07.00 and 18.00, Monday to Friday[6]. An independent study recommended augmenting the congestion charge by including a NOx emissions component in the calculation of this charge. Based on that logic, diesel would have to pay a fee higher than £11.50, this fee increasing further the older the vehicle is (Drummond and Ekins 2016). Moreover, some of the major diesel car manufacturers have agreed to cooperate on real-world testing and emission reductions (Brand 2016).

With the exception of the NOx surcharge for London mentioned above (which was suggested but not implemented by policymakers up to the time of this writing), all other policy responses have been targeted toward strengthening aspects of the "command-and-control" approach where a public authority legally imposes a minimum performance or technology standard, and tries to enforce compliance of all market actors with this standard.

Strengthening command and control means making the standards more stringent and at the same time implementing more intensive and complicated (and hence more costly) compliance monitoring policies.

But are stricter and more expensive command-and-control policies the only possible response to Dieselgate? Are there alternative policy options that would be worth examining?

3.4   Dieselgate-Type Issues in Other Energy-Using Sectors

Motor vehicles are not the only products that suffer from demanding regulatory policies and have integrated smart software. Energy-using appliances are also subject to standardized test procedures which lead to their classification with an Energy Label in Europe[7] and similar labeling systems around the world such as the Energy Star in the US,[8] indicating how well each product performs in terms of energy efficiency. Suspecting that Dieselgate-type problems may occur in such appliances as well, a group of European nongovernmental organizations conducted energy consumption tests both in standardized conditions and in conditions that are regarded as more representative of real-life use. They tested three types of appliances: TV sets, refrigerators, and dishwashers. Their findings resemble those of car emissions: according to standards, dishwashers are only tested on a very efficient but infrequently used wash program; TVs are tested with a video clip from 2007 that does not reflect typical home viewing or increasingly common TV technologies, whereas tests with a modern video clip in a higher quality format led to considerably higher electricity consumption; and refrigerators are tested without opening the doors and without any load in the fresh food compartments. The same group mentions a similar problem with standardized tests on vacuum cleaners: their energy labels are based on a test made with a completely empty bag, while under real-world conditions they have at least half-full bags which leads them to consume more energy (CLASP et al. 2017). In a similar investigation in the US, the Natural Resources Defense Council found evidence that TV sets change behavior when identifying that they are being tested according to standardized procedures prescribed by the US Department of Energy (NRDC 2016).

This challenge is usually described in the economics literature as "Goodhart's Law," from a statement made by the economist Charles Goodhart for monetary policy, which (paraphrased) is often cited as: "When a measure becomes a target, it ceases to be a good measure" (Reynaert and Salle 2016). Dieselgate and the above examples from energy-consuming appliances are manifestations of this phenomenon, which call for a careful examination of policy options.

3.5   Comparing Command-and-Control and Economic Measures in Environmental Policy

This section offers a comparison between the two major types of instruments used in environmental policy – legislation and economic incentives. The focus is on transport-related policies, but the insights can be generalizable to other environmental challenges. The reader may initially have the impression of a bias against command-and-control approaches, but this is mainly due to the fact that Dieselgate has exposed the weaknesses of this approach. A balanced discussion of policy options follows in section 3.5.3.

3.5.1   Features of Command-and-Control Policies

Mandatory regulations in environmental policymaking are largely preferred by authorities because they are both environmentally effective and acceptable to citizens as their costs to society are indirect. The usual criticism directed to authorities is not whether environmental legislation is necessary but that this legislation is not stringent enough or not adequately implemented (ECA 2018). Without question, much of the improvement in air and water quality in industrialized countries during the last decades has been due to the enforcement of mandatory environmental standards (EEA 2019; EPA 2019). Apart from policymakers and the public, command-and-control measures are often well received by the industry, as they provide regulatory certainty for the short and medium term because legislation includes mandatory standards to be fully implemented some years later. Legislation can also lead to large environmental benefits faster than voluntary or economic measures (Harrington and Morgenstern 2004).

Natural scientists and engineers are also typical supporters of implementing environmental legislation and tightening it over the years. They claim that, if affordable technological solutions are available for reducing emissions or fuel consumption, they should be legally enforced in order to promote technical progress and achieve energy and environmental objectives.

In this argument, environmental effectiveness is important but cost-effectiveness (i.e., achieving an environmental target with the lowest cost to society) is not sufficiently questioned. In fact, as discussed previously, legal or illegal avoidance of regulatory requirements often arises where compliance with a mandatory environmental standard is needed, especially if compliance costs are high and enforcement is difficult. In the case of vehicle emission standards, T&E (2018) reports recent evidence about manipulation of tests and exploitation of "flexibilities" by automakers in Europe but as mentioned in section 4, several additional examples of legal avoidance have been found by experts.

A large part of the research on the economic efficiency of command and control has been conducted in the US and is related to the analysis of Corporate Average Fuel Economy (CAFE) standards that have been implemented since the 1970s. Predictably, most of this work has been conducted by economists who often dismiss command-and-control policies as economically inefficient. Anderson and Sallee (2016) provide an overview of this research. The bottom line is that CAFE standards – and environmental regulations in general – cause costs to society which are not immediately observable. This makes command-and-control policies attractive to the public but not less costly. Reasons for this are provided in the following paragraphs.

First, legislation usually contains many exemptions and derogations. Sallee and Slemrod (2012) report examples of CAFE legislation whose design allows the industry to take advantage of loopholes so that it complies with regulatory commitments by applying less effort than would otherwise be required. Such loopholes are more evident in regulations for flexible fuel vehicles (which can run for example both on gasoline and on a biofuel blend). For example, Anderson and Sallee (2011) describe the case where CAFE standards included a loophole that treated flexible fuel vehicles as if they ran on ethanol 50% of the time, although in practice these vehicles rarely burned ethanol. Moreover, regulations often treat electric cars as zero emission vehicles irrespective of the amount of pollutant and carbon emissions generated during the production of electricity they consume. Hugosson and Algers (2012) reported a similar case from Sweden: the regulatory definition of "clean vehicles" in national legislation allowed flexible fuel cars to gain a large share in the Swedish market, but in practice these vehicles were often driven on gasoline alone and therefore offered no environmental benefit compared to conventional gasoline cars. Another loophole of regulatory approaches in transport applies to electric vehicles which often enjoy credits or exemptions without real-world checks of whether they can be considered "clean cars"; their environmental impact largely depends on the fuel mix of the electricity grid from which they are charged (Jacobsen et al. 2016).

Second, there are side-effects on the car market which can compromise the effectiveness of environmental legislation. For example, an emissions or fuel economy standard does not provide an incentive to citizens to own fewer cars. To the extent that legislation makes new cars more costly because they enforce stringent emission control technologies, they may also encourage many owners of older cars to use them for a longer time and scrap them later than usual, which leads to an increased number of less efficient and more polluting vehicles on the road (Jacobsen and van Benthem 2015). Moreover, to the extent that legislation leads to the adoption of fuel-saving technologies, it reduces the fuel costs of new cars and encourages drivers to use their cars more, thereby reducing the net environmental benefit – the so-called rebound effect (Hymel and Small 2015). Economists have developed methods to assess the additional costs borne by society because of these effects (see for example the review by Anderson and Sallee 2016). Since these costs are sometimes indirect and may not occur during the implementation of a regulatory policy but only in the longer term, they tend to be neglected in the policy debate.

Moreover, environmental legislation may also have unintended consequences on social equity. Davis and Knittel (2016) examined the distributional implications of CAFE standards; for example, to what extent standards affect rich and poor households in the US. Their analysis showed that CAFE is mildly regressive, which means that high-income households bear less cost of this legislation as a fraction of their income than low-income households. This indicates that it is difficult to argue for legislated fuel economy standards out of concerns for social equity.

Another feature of legislated standards is that they are usually blunt policy instruments. They impose the same obligation (e.g., in grams of pollutant per mile driven or per kWh of electricity generated) irrespective of whether the polluting activity takes place in a densely populated area or in the countryside, in a street canyon where pollutants are trapped and cause serious health problems, or in an isolated region where health impacts are negligible. This further compromises cost-effectiveness.

As we have seen with Dieselgate, cheating is also a problem to take into consideration. Reynaert and Salle (2016) analyzed the effects of such cheating behavior on consumer costs and social welfare. They found that when companies manipulate emissions data, they benefit both themselves and consumers, because the costs of environmental protection are lower than what they should be but this is done to the detriment of social welfare because environmental challenges are not properly addressed.

To summarize, command-and-control policies can be *effective* but not necessarily *cost-effective*. Imperfect targeting of a mandatory standard and the existence of loopholes compromise the cost-effectiveness of such an approach. A mandatory standard may have been ill-conceived by bureaucrats and turn out to be too stringent (and hence costly to the industry), or too lenient (thus favoring some polluters, which may even have an influence on the definition and stringency of the standard). A regulation may also be costly to enforce because it requires resources (in the form of staff, equipment, etc.) to ensure enforcement compliance – this is evident in the case of the European policy response to Dieselgate, since the recently introduced more stringent regulations require sophisticated equipment and continuous monitoring. And, as shown by the evidence mentioned above, it may have unintended consequences both in environmental effectiveness and in social equity.

3.5.2 The Alternative of Economic Measures

Economists do not share the view of engineers and most policymakers toward ever stricter environmental regulations and enforcement mechanisms. For the reasons mentioned above, they consider command-and-control approaches to be economically inefficient, and therefore they promote the use of economic measures through appropriate pricing of polluting behavior. In this case, the economic approach is to impose an economic disincentive to the polluting or resource-wasting activity (in the form of a tax/levy that is proportional to the amount of pollution caused) and let the market adjust. For example, some firms may invest in emission abatement technology, whereas others may prefer to pay the additional tax if this is cheaper for them. Economic measures are not confined to outright taxes; they include subsidies and emissions trading. According to economic theory, added environmental fees correct market failures, so that prices of all products and services reflect the actual costs borne by society for each activity. By pricing products at their actual cost, market equilibrium is achieved. Furthermore, reducing harmful activities as much as necessary ensures cost-effectiveness, ultimately improving social welfare (Parry et al. 2014).

The price mechanism is transparent, so that all producers and consumers understand the costs of their actions and adjust their behavior accordingly. Moreover, it provides a continuous incentive to reduce energy consumption or emissions because every unit of activity is taxed. Conversely, regulatory mandates impose a limit (e.g., emission or fuel economy standard, minimum technology diffusion rates, etc.); firms must meet this limit but have no further incentives to overshoot the mandatory target.

Command and control is justified by economists mainly as a "paternalistic" policy, that is, a policy which is chosen by the government to promote the welfare of society by limiting the autonomy of citizens to choose. This may be the case where consumers are imperfectly informed or inattentive when purchasing energy-using goods, that is, they are unaware of what is economically best for them. Government-imposed standards can address such imperfections by limiting the options available to consumers in the market, thereby protecting them from buying, for example, light bulbs, electric appliances or cars with an energy efficiency that is "too low." However, as Allcott (2014) has shown, the size of such biases is not sufficient to justify many paternalistic energy policies.

Seen from a certain perspective, Dieselgate seems to confirm economists' concerns. To avoid test cheating and high implementation costs, a government might impose – instead of environmental legislation – a levy on car drivers that is proportional to the distance traveled per car. Today's technology allows tracking the distance driven by each car and automatically charging each driver with the amount of money that corresponds to their car's mileage. As demonstrated in several case studies of the application of congestion charges around the world,[9] it is also possible to impose different levies by time of day or by location, thereby charging higher rates when and where most harm is caused to society (e.g., driving within a city during peak traffic hours). Higher costs from driving larger distances and at certain times would discourage car travel – and in fact, most of the economic damage caused by motor vehicles to society, including air pollution, is due to the number of miles traveled (Parry et al. 2007). A reduction in car travel in most urban areas, although causing some inconvenience to parts of society especially if public transportation is not improved at the same time, has a clearly positive effect on overall social welfare because several social costs associated with congestion, accidents, noise, air pollution, and fuel use decline. A distance-based charge affects more strongly those drivers who travel most, and therefore introduces a penalty that is proportional to the social cost caused by each car; this achieves what economists call "efficiency" and is also in line with the "polluter pays" principle, which is a cornerstone of environmental policymaking.

The economic arguments are more convincing and straightforward for climate policy, that is, in the case of carbon dioxide emissions, which are proportional to the amount of fuel consumed. In fact, the vast majority of economists' work that leads to favoring taxes over standards has analyzed energy conservation and carbon mitigation policies (Anderson and Sallee 2016). In those studies, the economic instrument is simple and effective; for example, a carbon tax (amounting to a tax per liter of fuel depending on the carbon content of each fuel) will encourage people both to purchase low-emission vehicles and to drive less with their cars, in order to reduce their fuel bill. This means that a proper economic measure such as carbon pricing promotes a wide range of behavioral responses that can improve the environment, whereas an emissions standard encourages only the purchase of low-emission cars but has numerous negative side-effects, as mentioned in the previous section. Numerous economic studies, summarized by Anderson and Sallee (2016), conclude that fuel economy standards clearly involve higher social costs and hence are inferior to a fuel or carbon tax.

However, economic analysis has been less vocal when dealing with air or water pollution challenges. In such cases, despite some success stories (Harrington and Morgenstern 2004), economic measures alone may not be sufficient and should probably be complemented by command-and-control policies. In the case of distance-based charges described above, a regulatory standard for the sulfur content of fuels and the emissions of pollutants such as NOx or particles would still be necessary. It is nonetheless worth observing the economic argument here: standards are necessary but should not become so stringent that they encourage the use of defeat devices. When combined with economic incentives in mandating the areas where pollution is more serious, they can deliver cost-effective pollution abatement.

3.5.3 Legislation or Economic Instruments? Probably Both

The review of sections 3.5.1 and 3.5.2, based both on economic theory and empirical evidence, indicates that regulatory approaches are effective but probably not cost-effective, and that economic approaches may have many more advantages compared to regulations, especially for tackling specific environmental problems. These considerations could make one conclude that we should rather phase out command-and-control policies in favor of flexible market-based instruments. However, such an argument overlooks some real-world policy considerations, away from the world of textbooks. These are explained below and point toward a recommendation for blending elements from both approaches and arriving at a hybrid policy design in order to address anthropogenic climate change, air and water pollution, and energy security cost-effectively.

First, legislation is often the only realistic solution in a political environment that is unfavorable to tax increases, "a trade-off between lower political costs and higher economic costs" (JTRC 2008). Most economic measures are faced with the barrier of political acceptance: they offer a clear price signal but at the same time impose a tangible economic burden, which makes them unattractive and hence politically difficult to enforce. Recycling the revenues of such a tax in the economy, for example by returning all proceeds of environmental taxation to citizens, can make such a policy far more acceptable (Klenert et al. 2018). Environmental fiscal reforms, which involve increases in taxation of polluting, resource-depleting activities and recycling their revenues in the economy (e.g., through direct payments to citizens or reductions in labor taxes), have been proposed in the past.[10] Although their economic performance is often disputed, such reforms have been tested in several countries and have led to interesting and positive economic outcomes in terms of economic growth, employment, and environmental improvement.[11]

Fast and effective action to mitigate climate change and pollution requires politically feasible measures rather than economically superior ideas that are impossible to implement. The combination of regulatory standards with tax reforms that favor environmental charges may be a realistic second-best solution.

Second, not all market-based policies are cheap to implement. A carbon tax may be a straightforward measure, essentially with no transaction costs; however, emissions trading or congestion charging involves substantial administrative costs both for installing the corresponding systems and monitoring their operation and enforcement efficacy. Thus, an important weakness of regulatory mandates can also be seen in pricing mechanisms.

Third, consumers and firms are far from perfectly informed and rational. Therefore, price signals may not lead to optimal behavior adjustments, in which case a paternalistic regulatory standard may indeed be preferable because consumers and firms do not have the full information necessary to make economically optimal decisions. Additionally, we cannot forget that some of the economic instruments also are subject to cheating. Several taxation schemes for vehicles or household equipment calculate tax levels on the basis of an officially reported energy consumption or emissions figure. These figures are the same as those used for legislated standards, hence the same caveats apply to such market-based policies.

In fact, several hybrid policy approaches have occurred in recent years. CAFE standards, for example, involve trading of credits obtained by a company that overperforms; this adds flexibility to the regulatory approach which resembles an economic instrument (Kiso 2019). Moreover, "feebate" systems have been introduced in several industrialized countries: a car that emits carbon below a threshold receives a rebate, which lowers its retail price, while a car exceeding the threshold has to pay an additional fee that may be proportional to its excess emissions. Feebates essentially convert CO_2 emission standards to flexible price-based mechanisms that promote the purchase of low-emitting vehicles, thereby combining a regulatory with a market-based system.[12]

3.6 Conclusions

In their review of the EU Emissions Trading System, Convery and Redmond (2007, p. 88) started by declaring that "Every profession has its *idée fixe*, its core response and solution to problems ... The economist sees prices as the solution."

The Dieselgate case provoked concerns across different fields, and they all attempted to find solutions to avoid another similar scenario. Technology enthusiasts (usually engineers) called for stricter enforcement systems which would enable the full implementation of emission abatement technologies; political scientists underlined the need for properly working institutions and control authorities; organization theorists focused on the way decisions are made within a large corporation which can lead to cheating; educators always declare that environmental awareness is key; and economists typically support proper pricing of all activities.

There are obviously elements of truth in all approaches. Our review, keeping in mind both economic theory and empirical evidence, has shown the importance of a proper mix of regulatory and economic approaches. It seems to be clear that economic incentives for phasing out carbon-intensive and highly polluting fuels and technologies deserve more attention. Environmental tax reforms (e.g., carbon taxation or emissions trading) are universally accepted as necessary ingredients of sustainability policies as they can be implemented to showcase the actual cost borne by society (IMF 2014; OECD 2016). Market-based instruments can generally be more flexible than uniform regulations, and hence can complement environmental legislation in order to improve cost-effectiveness. For example, time-variant tariffs for electricity or water, or distance-based vehicle charges that differ based on location and peak or off-peak hours may be combined with CO_2, NOx or PM emission standards. The rationale for economic incentives is stronger in climate policy, whereas their advantages are not as clear when tackling air and water pollution; in the latter case synergies with legislation should be exploited.

At any rate, in a complex and imperfect world, a mix of regulatory and economic measures remains the preferred hybrid policy response to global energy and environmental problems. This is in line with the recommendation of Ostrom (2009) to adopt "polycentric" approaches to address environmental challenges. The appropriate blend of this mixture (e.g., how strict the standards and how high the emission charges) has to be assessed in each case by weighing the social costs and benefits of each policy option. Proper monitoring of real-world environmental behavior will also be an indispensable element of both regulatory and economic approaches to environmental protection.

Such solutions can be facilitated by improved communication between disciplines, so that experts temper their *idées fixes* and accommodate to the ideas of other fields. Technology experts, for example, need to understand that no policy is free of costs, and that even command and control can cause substantial costs to society beyond the obvious administrative costs, while economists should remember that pure and optimal pricing exists only in a perfectly rational and fully informed society, which has yet to appear.

Notes

1 "Five things to know about VW's 'dieselgate' scandal": https://phys.org/news/2018-06-vw-dieselgate-scandal.html (last accessed on 16 August 2018).
2 The difference between final and primary energy demand is that the latter includes final energy plus all fuel consumption used for the transformation of energy from one source to another – mainly power plants.
3 See www.bbc.com/news/business-34352243 (last accessed on 28 October 2019).
4 See, e.g., www.reuters.com/article/us-germany-emissions-factbox/factbox-german-cities-ban-older-diesel-cars-idUSKCN1NK28L (last accessed on 28 October 2019).
5 https://e360.yale.edu/digest/diesel-vehicles-face-a-grim-future-in-europes-cities (last accessed on 28 October 2019).
6 See https://tfl.gov.uk/modes/driving/congestion-charge (last accessed on 28 October 2019).
7 https://ec.europa.eu/info/energy-climate-change-environment/standards-tools-and-labels/products-labelling-rules-and-requirements/energy-label-and-ecodesign_en
8 www.energystar.gov.
9 See, e.g., the case of Sweden: https://transportstyrelsen.se/en/road/Congestion-taxes-in-Stockholm-and-Goteborg (last accessed on 28 October 2019).
10 See the recent comprehensive review of Freire-González (2018).
11 See, e.g., studies of the Green Fiscal Policy Network (www.greenfiscalpolicy.org) and the Green Growth Knowledge Platform (www.greengrowthknowledge.org).
12 See www.theicct.org/spotlight/feebate-systems for a presentation of feebates.
13 All webpages mentioned in the list of references were last accessed on 28 October 2019.

References[13]

Allcott, H. (2014). Paternalism and energy efficiency: An overview. Working Paper 20363. National Bureau of Economic Research, Cambridge, MA.

Anderson, S. and Sallee, J. (2011). Using loopholes to reveal the marginal cost of regulation: the case of fuel-economy standards. *American Economic Review* 101: 1375–1409.

Anderson, S. and Sallee, J. (2016). Designing policies to make cars greener: a review of the literature. *Annual Review of Resource Economics* 8: 157–180.

Barrett, S.R.H., Speth, R.L., Eastham, S.D. et al. (2015). Impact of the Volkswagen emissions control defeat device on US public health. *Environmental Research Letters* 10: 114005.

Brand, C. (2016). Beyond dieselgate: implications of unaccounted and future air pollutant emissions and energy use for cars in the United Kingdom. *Energy Policy* 97: 1–12.

CLASP, ECOS, EEB, and Topten.EU (2017). Closing the 'reality gap' – ensuring a fair energy label for consumers. http://eeb.org/wp-content/uploads/2017/06/Reality-Gap-report.pdf

Convery, F. and Redmond, L. (2007). Market and price developments in the European Union emissions trading scheme. *Review of Environmental Economics and Policy* 1: 88–111.

Davis, L.W. and Knittel, C.R. (2016). Are Fuel Economy Standards Regressive? NBER Working Paper 22925. Cambridge, MA: National Bureau of Economic Research.

Drummond, P. and Ekins, P. (2016). Tackling air pollution from diesel cars through tax: options for the UK. Brussels: Green Budget Europe. https://green-budget.eu/wp-content/

uploads/Tackling-air-pollution-from-diesel-cars-through-tax-options-for-the-UK-Full-Report-June-2016-Web.pdf

ECA (2018). Air pollution: Our health still insufficiently protected. European Court of Auditors, Luxembourg. www.eca.europa.eu/Lists/ECADocuments/SR18_23/SR_AIR_QUALITY_EN.pdf.

EEA (European Environment Agency) (2016). Explaining road transport emissions – a non-technical guide. Copenhagen: EEA. doi:https://doi.org/10.2800/71804.

EEA (European Environment Agency) (2019). Air quality in Europe – 2019 report. Copenhagen: EEA. doi:https://doi.org/10.2800/822355.

EPA (US Environmental Protection Agency) (2019). Our Nation's Air: Status and Trends through 2018. https://gispub.epa.gov/air/trendsreport/2019.

European Commission (2018). Testing of emissions from cars. Press release. http://europa.eu/rapid/press-release_MEMO-18-3646_en.htm

Ewing, J. (2015). Volkswagen Says 11 Million Cars Worldwide Are Affected in Diesel Deception. *New York Times*, 23 September.

Franco, V., Sanchez, F.P., German, J., and Mock, P. (2014). Real-world exhaust emissions from modern diesel cars. Berlin: International Council on Clean Transportation. https://www.theicct.org/publications/real-world-exhaust-emissions-modern-diesel-cars

Freire-González, J. (2018). Environmental taxation and the double dividend hypothesis in CGE modelling literature: a critical review. *Journal of Policy Modelling* 40: 194–223.

Harrington, W. and Morgenstern, R.D. (2004). Economic Incentives versus Command and Control. www.rff.org/research/publications/economic-incentives-versus-command-and-control-whats-best-approach-solving

Hugosson, M.B. and Algers, S. (2012). Accelerated introduction of 'clean' cars in Sweden. In: *Cars and Carbon: Automobiles and European Climate Policy in a Global Context* (ed. I.:.T. Zachariadis). New York: Springer.

Hymel, K.M. and Small, K.A. (2015). The rebound effect for automobile travel: asymmetric response to price changes and novel features of the 2000s. *Energy Economics* 49: 93–103.

IEA (International Energy Agency) (2017). *World Energy Outlook 2017*. Paris: IEA.

IMF (International Monetary Fund) (2014). *Getting Energy Prices Right*. Washington, DC: IMF.

Jacobsen, M.R. and van Benthem, A.A. (2015). Vehicle scrappage and gasoline policy. *American Economic Review* 105: 1312–1338.

Jacobsen, M.R., Knittel, C.R., Sallee, J.M., and van Benthem, A.A. (2016). Sufficient statistics for imperfect externality-correcting policies. Working Paper 22063. Cambridge, MA: National Bureau of Economic Research.

JTRC (Joint Transport Research Centre of the International Transport Forum) (2008). The cost and effectiveness of policies to reduce vehicle emissions. Discussion Paper No. 2008-9. Paris: International Transport Forum.

Kiso, T. (2019). Evaluating new policy instruments of the corporate average fuel economy standards: footprint, credit transferring, and credit trading. *Environmental and Resource Economics* 72: 445–476.

Klenert, D., Mattauch, L., Combet, E. et al. (2018). Making carbon pricing work for citizens. *Nature Climate Change* 8: 669–677.

McGee, P. (2018). EU finds evidence carmakers are manipulating results. *Financial Times*, 24 July.

Miller, J. and Franco, V (2016). Impact of improved regulation of real-world NOx emissions from diesel passenger cars in the EU, 2015–2030. Berlin: International Council on Clean Transportation. www.theicct.org/publications/impact-improved-regulation-real-world-nox-emissions-diesel-passenger-cars-eu-2015%E2%88%922030.

NRDC (Natural Resources Defense Council) (2016). The secret costs of manufacturers exploiting loopholes in the government's TV energy test. Report R-16-09-B. www.nrdc.org/resources/secret-costs-manufacturers-exploiting-loopholes-governments-tv-energy-test

OECD (2015). *The Metropolitan Century: Understanding Urbanisation and its Consequences.* Paris: OECD https://doi.org/10.1787/9789264228733-en.

OECD (2016). *Effective Carbon Rates – Pricing CO2 through Taxes and Emissions Trading Systems.* Paris: OECD https://doi.org/10.1787/9789264260115-en.

Ostrom, E. (2009). A Polycentric Approach for Coping with Climate Change. Background Paper to the 2010 World Development Report, Policy Research Working Paper 5095. Washington, DC: World Bank.

Parry, I.W.H., Walls, M., and Harrington, W. (2007). Automobile externalities and policies. *Journal of Economic Literature* 45: 374–400.

Parry, I.W.H., Evans, D., and Oates, W.E. (2014). Are energy efficiency standards justified? *Journal of Environmental Economics and Management* 67: 104–125.

Reynaert, M. and Salle, J.M. (2016). Corrective policy and Goodhart's law: The case of carbon emissions from automobiles. NBER Working Paper 22911. Cambridge, MA: National Bureau of Economic Research.

Sallee, J. and Slemrod, J. (2012). Car notches: strategic automaker responses to fuel economy policy. *Journal of Public Economics* 96: 981–999.

T&E (Transport & Environment) (2018). CO2 Emissions from Cars: The Facts. Brussels: T&E. www.transportenvironment.org/publications/co2-emissions-cars-facts

Tietge, U., Diaz, S., Yang, Z, and Mock, P. (2017). From laboratory to road – International. International Council on Clean Transportation, Berlin. www.theicct.org/publications/laboratory-road-intl.

Weiss, M., Bonnel, P., Kühlwein, J. et al. (2012). Will Euro 6 reduce the NOx emissions of new diesel cars? Insights from on-road tests with portable emissions measurement systems (PEMS). *Atmospheric Environment* 62: 657–665.

4

Environmental Legislation and Small and Medium-Sized Enterprises (SMEs)

A Literature Review

Eleni Sfakianaki

School of Social Sciences, Hellenic Open University, Patras, Greece

4.1 Introduction

Over the last 40 years, environmental issues have attracted increasing attention. The Brundtland Report in 1987 describes the concept of sustainable development as an approach that would meet the needs of the present and the immediate future without compromising the ability of later generations to meet their own needs (Brundtland 1987). A set of pillars defines this idea of sustainability: attention to economic, social, and environmental concerns (Edum-Fotwe and Price 2009), which in turn lead to the assessment of environmental impacts (Sfakianaki and Stovin 2002). Sustainable development and related concepts greatly influenced the Earth Summit that took place in 1992 in Rio de Janeiro, Brazil, as well as the third UN Conference on Environment and Development in 2002 in Johannesburg, South Africa. In many countries, environmental legislation gradually attracted more interest. As Wilson et al. (2012) show, more than 80 000 pages of European legislation and 500 EU Directives have been produced on this topic in the last couple of decades.

The business sector has not remained unaffected by the enforcement of environmental legislation and policy. Indeed, environmental concerns may have had a significant effect on the sustainability and profitability of business organizations, while business operations have also tended to pollute and consequently impact the environment. According to Ferenhof et al. (2014), as customers gain greater consciousness of environmental impacts, they demand that companies, irrespective of their size, exhibit more responsiveness to environmental concerns. Small and medium-sized enterprises (SMEs) are a very significant business target group with an influential role in national economies. In the context of the global economy and social development, SMEs are key players as can be seen in Europe where SMEs generate 66% of employment and create 58% of added value (Muller et al. 2018). In the UK specifically, Wilson et al. (2012) state that SMEs numerically account for 99.7% of the 4.7 UK million businesses, 47.5% of employment, and 48.7% of turnover, concluding that their existence is vital to the UK economy. Greece is also noteworthy in this

Environmental Policy: An Economic Perspective, First Edition. Edited by Thomas Walker, Northrop Sprung-Much, and Sherif Goubran.

aspect, with an industrial make-up consisting of approximately 650 000 SMEs, which among them generate 18% of the country's gross domestic product (GDP) (EU 2016; PwC 2015).

Although, on a global scale, SMEs do substantially contribute to the economy and have an impact on the environment, the effects remain understudied (Diabat et al. 2014; Michelsen and Fet 2010). Indeed, SMEs are frequently unaware of their environmental impact and do not have the expertise or the knowledge required to respond to environmental challenges. Although each SME may not have a significant environmental impact, because of their large number, the collective impact may be enormous. In Europe, for example, SMEs are considered responsible for approximately 64% of industrial contamination (DG Enterprise 2010). According to CEC (2002), the majority of SMEs are "vulnerably compliant," meaning that they are unaware of existing legislation and environmental legislation in particular. According to Hillary (2004), due to the heterogeneity of SMEs and the scarcity of larger-scale strategic action, practices have not been seriously generalized.

Environmental organizations such as the Environment Agency tend to concentrate their attention on larger businesses because they are perceived to be more prominent polluters. However, the majority of environmental incidents are caused by SMEs (Williamson et al. 2006; Wilson et al. 2011). In the same context, Blundel et al. (2013) also argue that the promotion of environmental policies traditionally concentrates on larger firms, although attention is gradually shifting to focus on the role of SMEs and how they can improve their performance.

Following the above considerations, this study evaluates the literature on environmental legislation and its enforcement in SMEs. Using the research method of a systematic literature review (SLR), this study explores and synthesizes relevant research to provide an understanding of the application of environmental legislation to SMEs overall. Gaps in knowledge have been identified and the literature organized. Three research questions are posed.

- *RQ1*. Authors: How frequently do authors publish in the specific research field? Is there an emerging trend of new authors in this field? Where are authors located geographically, and how are they dispersed? Is there collaboration among them?
- *RQ2*. Publications: Which publications mainly restrict themselves to the specific field? What are the broader publication trends for this period of time? How many citations do publications usually receive?
- *RQ3*. Content: What keywords do authors use in this field? What observations can be made about them?

The remainder of this report has the following structure. The methodology used to select the pertinent publications is presented in the subsequent section, followed by the findings of the research. After that, a discussion is presented, which is followed by the conclusion and subsequent potential directions for future research.

4.2 Methodology

The present work follows the SLR methodology proposed by Tranfield et al. (2003). An SLR methodology offers combined insights regarding the areas of review through the synthesis of empirical and theoretical work, following a replicable and transparent process to examine published work using search criteria; such an approach is appropriate for data volumes of smaller size (Tranfield et al. 2003). As Engert et al. (2016) find, an SLR examines current research in a more thorough way and delivers a summary of the key issues and themes animating a specific research area. It furthermore documents, as Meredith (1993) demonstrates, the analyses of a specific field in combination with the theoretical frameworks. Thus, in a specific research context, strengths and weaknesses of approaches can be identified and discussed. According to Denyer and Tranfield (2009), such a methodology is especially suitable for reviewing issues such as the "what" and "how" in a literature inquiry and for identifying opportunities for further research once the studies related to such issues are organized and examined.

The present review follows the SLR guidelines of Tranfield et al. (2003), which outline six stages: (i) definition of the problem, (ii) search strategy, (iii) exclusion criteria, (iv) data gathering, (v) analysis, and (vi) reporting. The first three stages are described below, and the description of stages 4, 5, and 6 follows in the analysis and findings section.

4.2.1 Problem Definition

The limited number of research papers published over the past decade on environmental legislation and SMEs made this review a challenging prospect, but also created an opportunity to document progress and the growth of knowledge on environmental legislation related to SMEs. The methodology used includes analyses of authors, publications, and contents, enabling the examination of different aspects of the existing scholarly literature. In this context, the SLR approach adopted illustrates and reports on the literature currently available. Additionally, such an approach provides a useful tool for organizing future development, hence setting priorities for the given field.

4.2.2 Search Strategy

To ensure the quality of the data sources and to follow the method of Wetzstein et al. (2016), two experts with extensive knowledge and experience in databases and performing SLRs were sought to acquire a wide-ranging view of the enforcement of environmental legislation on SMEs. The exploratory stage of the present work led to the selection of three major research organizations: Wiley, Emerald, and Elsevier. These publishers have both sufficient size and significance for the purposes of this study, and they all enjoy full acceptance in the academic community. Furthermore, all three permit electronic search queries, which are deemed the most effective means of a literature review (Levy and Ellis 2006).

This review is restricted to peer-reviewed articles published in English. The timeliness of the target set was ensured by limiting the publication period to the last decade (Alwan et al. 2017), specifically from 2008 to 2018. The search strategy included the search term "environmental legislation" and the three following synonymous terms: "small and medium enterprises," "SME," and "SMEs." For each database, the term "environmental legislation" was used with each of the other three keywords, one at a time. A hit was returned only if the search terms appeared in the keyword, title, or abstract. This research protocol was examined and adapted several times before the final list of publications was completed. Separate searches were executed for all databases to increase the quality of the overall search quality. Following the snowball principle, the total number of publications was subsequently expanded, using the references of the articles identified (Coenen et al. 2017).

4.2.3 Exclusion Criteria

The complete database searches found a total of 192 publications, after the removal of duplicates. This group of publications was screened through several stages. Although the search terms used are clear enough in the target field, they can have other meanings in other research contexts (for example, SME can also stand for statutory medical examinations). Articles with no direct relation to the topic were thus excluded. Finally, publications related to the topic of the study but not directly focused on the search subjects were omitted once the full texts were examined.

After the exclusion criteria were incorporated, the publications selected for further examination in this review included a total of 33 articles, representing 17% of the initial set of returns (after the removal of duplicates). This set was examined for a number of characteristics, including the number of authors, the geography of authors, the journals of publication, and the number of articles published per year. Following this, the study synthesizes the identified literature to document how this research field is developing and to respond to the research questions set in the introduction related to authors, publications, and content.

4.3 Analysis of Findings

4.3.1 Authors

To form a response to RQ1, the following characteristics of the authors were examined: numbers, geographic locations and dispersion, and partnerships. The number of authors in the publication set was calculated first. In the 33 publications, there were 93 unique authors, of which 10 were repeated: Christopher D.H. Wilson, Ian David Williams, Simon Kemp, Mathiyazhagan Kaliyan, Devika Kannan, A.N. Haq, Kannan Govindan, Muhammad D. Abdulrahman, Y. Liu, and Nachiappan Subramanian, each of whom had two publications. On some occasions, authors published together more than once (for example, Christopher D.H. Wilson, Ian David Williams, and Simon Kemp published together in 2011 and 2012). The dispersion of contributors indicates that no main, central core of contributors has yet been established in the field, nor any emerging trends (Martin 2012).

Next, the frequency of new authors appearing in the publication set was determined. Figure 4.1 shows the frequency for the given period of time. It is clear that the number of new authors did not increase with each passing year but, on the contrary, decreased significantly for some years with a couple of years even having no new authors (2009, 2016). Thus, no definite conclusions can be drawn for the period 2008–2018, even with the increase observed in 2018.

Next, we investigated the origins of the authors. The 93 authors represented 20 countries. The best-represented countries were Spain (24%, 22 authors), the UK (15%, 14 authors), China (14%, 13 authors), and India (4.5%, seven authors). The Netherlands, Brazil, Australia, Portugal, and the USA also appeared, with six, five, three, three, and three authors respectively, and a few other countries appeared with fewer representations. The

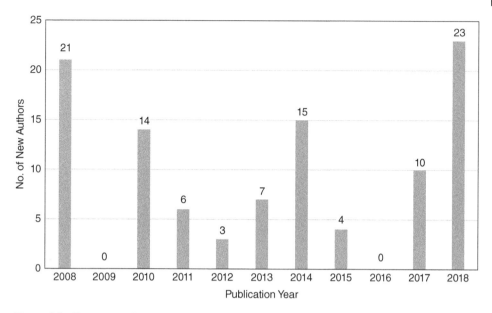

Figure 4.1 Frequency of appearance of new authors.

top six countries account for 67 of the 93 authors in the study. This relatively limited range covers three continents, which indicates an emerging global trend. However, more extensive contributions from other countries and continents could add extra value to the field.

Lastly, the number of authors per paper and cross-border collaborations were examined. Of the 33 publications, only three had a single author, and two or more authors were responsible for the rest. Multicountry collaborations were observed in 11 publications, representing 33% of the total publications. This result is a noteworthy percentage, and it may indicate that the study of environmental legislation and SMEs has left the emerging research stage (Borrego and Bernhard 2011). Collaboration work was observed between Spain and Portugal; China and Australia; Brazil, the UK, and Denmark; India, Denmark, and China; Sweden and Japan; the USA, China, and the Netherlands; Qatar and India; Turkey and the USA; Venezuela and Spain; and China and the USA.

4.3.2 Publications

As a response to RQ2, a review of publication outlets was conducted. In total, 10 publication outlets were identified for the set of 33 publications, as illustrated in Table 4.1.

The findings of Table 4.1 demonstrate that there is one central contributor, *Journal of Cleaner Production*, which published 18 of the total 33. The remaining outlets retained very low numbers of publications in this subject. It can be concluded that work in this research area is not widely distributed across scientific journals. The distribution of publications for different journals is limited to one, two, or in one case three publications per outlet. This distribution means that there is only one publication outlet with a dominant presence, and this presence is substantial. Given the relative importance of the *Journal of Cleaner Production*, it may be perceived that the research area is showing an evolving maturity (Maloni et al. 2012).

Table 4.1 Academic journals and number of publications per journal in the publication set

Journal	Count
Journal of Cleaner Production	18
Business Strategy and the Environment	3
International Journal of Production Economics	2
Journal of Environmental Management	2
Journal of World Business	2
Renewable and Sustainable Energy Reviews	2
Bioresource Technology	1
International Journal of Hospitality Management	1
Resources, Conservation, and Recycling	1
Transportation Research Part E	1

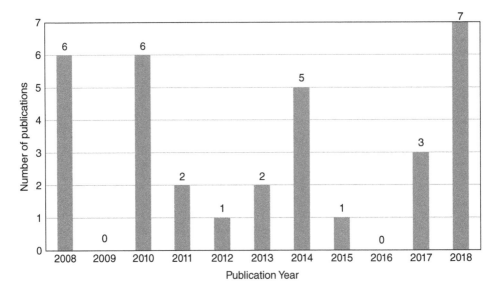

Figure 4.2 Number of publications per year (2008–2018).

Figure 4.2 shows the fluctuation of publications for the examination period, and this study attempts to evaluate trends in the results. Nevertheless, no particular patterns can be observed as the number of publications fluctuates to low averages from zero to three a year for the majority of years and from five to seven for four years (2008, 2010, 2014, 2018). The rate of publication reveals no increasing trends; on the contrary, it shows a rather low level of interest in the subject. The increase noted in 2018 requires further analysis to conclude whether it signifies an increasing trend in the publication rate.

Following Ferenhof et al. (2014), Table 4.2 presents the set of 33 publications according to the number of times each was cited. The most frequently cited publication was "An ISM approach for the barrier analysis in implementing green supply chain management"

Table 4.2 Publication set, number of citations

Year	Author	Title	Citations
2013	Mathiyazhagan et al.	An ISM approach for the barrier analysis in implementing Barriers analysis for green supply chain management implementation in Indian industries using analytic hierarchy process green supply chain management	358
2014	Govindan et al.	Barriers analysis for green supply chain management implementation in Indian industries using analytic hierarchy process	331
2010	Revell et al.	Small businesses and the environment: turning over a new leaf?	315
2008	Shi et al.	Barriers to the implementation of cleaner production in Chinese SMEs: government, industry, and expert stakeholders' perspectives	282
2008	Chan	Barriers to EMS in the hotel industry	258
2008	Zhang et al.	Why do firms engage in environmental management? An empirical study in China	235
2014	Abdulrahman et al.	Critical barriers in implementing reverse logistics in the Chinese manufacturing sectors	199
2011	Murillo-Luna et al.	Barriers to the adoption of proactive environmental strategies	196
2010	Zorpas	Environmental management systems as sustainable tools in the way of life for the SMEs and VSMEs	147
2013	Agan et al.	Drivers of environmental processes and their impact on performance: a study of Turkish SMEs	145
2010	Martín-Tapia et al.	Environmental strategy and exports in medium, small, and micro-enterprises	137
2008	Balzarova and Castka	Underlying mechanisms in the maintenance of ISO 14001 environmental management system	102
2010	Heras and Arana	Alternative models for environmental management in SMEs: the case of Ekoscan vs. ISO 14001	93
2010	Fernández-Viñé et al.	Eco-efficiency in the SMEs of Venezuela. Current status and future perspectives	65
2012	Wilson et al.	An evaluation of the impact and effectiveness of environmental legislation in small and medium-sized enterprises: experiences from the UK	61
2014	Martín-Peña et al.	Analysis of benefits and difficulties associated with firms' Environmental Management Systems: the case of the Spanish automotive industry	61

(*Continued*)

Table 4.2 (Continued)

Year	Author	Title	Citations
2014	Granly and Welo	EMS and sustainability: experiences with ISO 14001 and Eco-Lighthouse in Norwegian metal processing SMEs	56
2010	Brust and Heyes	Environmental management intentions: an empirical investigation of Argentina's polluting firms	55
2008	Cloquell-Ballester et al.	Environmental education for small- and medium-sized enterprises: methodology and e-learning experience in the Valencian region	42
2011	Wilson et al.	Compliance with producer responsibility legislation: experiences from UK small and medium-sized enterprises	37
2014	Subramanian et al.	Integration of logistics and cloud computing service providers: cost and green benefits in the Chinese context	34
2018	Gandhi et al.	Ranking of drivers for integrated lean-green manufacturing for Indian manufacturing SMEs	27
2017	Neto et al.	Framework to overcome barriers in the implementation of cleaner production in small and medium-sized enterprises: multiple case studies in Brazil	24
2018	Ormazabal et al.	Circular economy in Spanish SMEs: challenges and opportunities	22
2015	Thollander et al.	A review of industrial energy and climate policies in Japan and Sweden with emphasis toward SMEs	21
2017	Graafland and Smid	Reconsidering the relevance of social license pressure and government regulation for environmental performance of European SMEs	19
2018	Aboelmaged	The drivers of sustainable manufacturing practices in Egyptian SMEs and their impact on competitive capabilities: a PLS-SEM model	18
2018	Xia et al.	Conceptualizing the state of the art of corporate social responsibility (CSR) in the construction industry and its nexus to sustainable development	10
2017	Diana et al.	Putting environmental technologies into the mainstream: adoption of environmental technologies by medium-sized manufacturing firms in Brazil	10
2018	Osseweijer et al.	A comparative review of building integrated photovoltaics ecosystems in selected European countries	4
2018	Seth et al.	Green manufacturing drivers and their relationships for small and medium (SME) and large industries	3

Table 4.2 (Continued)

Year	Author	Title	Citations
2008	Ackroyd et al.	A critical appraisal of the UK's largest rural waste minimisation project: business excellence through resource efficiency (betre) rural in East Sussex, England	2
2018	Alvarez-García et al.	The influence of motivations and barriers in the benefits. An empirical study of EMAS certified business in Spain	2

(Mathiyazhagan et al. 2013), with 358 citations, the second most cited was "Barriers analysis for green supply chain management implementation in Indian industries using analytic hierarchy process" (Govindan et al. 2014), with 331 citations, followed by "Small businesses and the environment: turning over a new leaf?" (Revell et al. 2010), with 315 citations. Of the 33 publications, three had more than 300 citations, three had between 200 and 300, six had between 100 and 200, six had between 50 and 100 citations, and all publications had at least one citation. Citations data were extracted from Google Scholar on April 23, 2019.

4.3.3 Keyword Analysis

Following Narayanamurthy and Gurumurthy (2016) and in response to RQ3, an analysis of the keywords of the selected articles in environmental legislation and SMEs was created to evaluate how the authors positioned their publications. This analysis was used as a means of classifying knowledge and for analyzing the content of the publication set identifying the research direction. Keywords for all the publications were collected and grouped into seven different clusters, as indicated in Table 4.3. The topic of environmental legislation and SMEs attracted interest in relation to many different topics from different contexts, signifying that there are many issues explored in the articles. Thus, the research conducted could be described as heterogeneous. This analysis intended to identify the main issues addressed in the most recent publications.

As anticipated, environmental keywords topped the list, appearing 63 times in 31 of the publications, a group that was called cluster 1 (CL1). This cluster included keywords such as environment, sustainability, and sustainable development. The second cluster (CL2) included keywords related to SMEs; these 24 keywords appeared in 20 papers. In all, 23 keywords appeared in cluster 3 (CL3) in 15 papers, including keywords such as framework, factor analysis, and multi-criteria decision analysis, followed by CL4 and CL5, which both had the same count of 16, appearing in nine and eight publications, respectively. CL4 included the keywords building, integrated photovoltaics, logistics services, and manufacturing, and CL5 contained keywords like certification, environmental management system (EMS), ISO14001, and EMAS. The keywords found in the first two clusters (CL1 and CL2) examined the topics of environment and SMEs, which related to the criteria for the selected

Table 4.3 Analysis of keywords

Clusters of keywords	Keywords	Count	Citations
CL1	Environment; Environmental performance; Sustainable development goals; Sustainable operations; Sustainable development goals; Sustainability; Sustainable development goals; Natural environment; Environmental management; Eco-efficiency; Environmental strategies; Environmental Proactivity; Environmental behavior; Sustainable transition; Sustainable management systems; External and internal barriers to environmental progress; Corporate social responsibility; Environmental technologies; Sustainable manufacturing; Green Supply Chain Management; Lean-green manufacturing; Green SCM implementation; Cleaner production; Green manufacturing; Environmental training; Corporate environmental management; Barriers; Drivers; Barrier analysis; Stakeholder; Challenges; Motivations; Enablers; Benefits; Built environment; Natural-resource-based view; Circular economy; Circular economy implementation; Industrial symbiosis; Ecodesign; Energy efficiency; Environmental survey; Operational performance; Waste minimisation; Pollution	63	Abdulrahman et al. (2014), Aboelmaged (2018), Ackroyd et al. (2008), Álvarez-Garcia et al. (2018), Balzarova and Castka (2008), Brust and Liston-Heyes (2010), Chan (2008), Cloquell-Ballester et al. (2008), Diana et al. (2017), Fernández-Viñé et al. (2010), Gandhi et al. (2018), Govindan et al. (2014), Graafland and Hugo (2017), Granly and Welo (2014), Heras and Arana (2010), Martín-Peña et al. (2014), Martín-Tapia et al. (2010), Mathiyazhagan et al. (2013), Murillo-Luna et al. (2011), Neto et al. (2017), Ormazabal et al. (2018), Osseweijer et al. (2018), Revell et al. (2010), Seth et al. (2018), Shi et al. (2008), Thollander et al. (2015), Wilson et al. (2012), Xia et al. (2018), Zhang et al. (2008), Zorpas (2010)
CL2	Small- and medium-sized enterprises (SMEs); Small and medium-sized enterprises; Small and medium enterprises; Small to medium enterprises; SME; SMEs; Small and medium manufacturing enterprises; Internationalization; Micro businesses; Indian manufacturing SMEs; Size; Exports	24	Aboelmaged (2018), Ackroyd et al. (2008), Agan et al. (2013), Cloquell-Ballester et al. (2008), Diana et al. (2017), Fernández-Viñé et al. (2010), Gandhi et al. (2018), Graafland and Hugo (2017), Granly and Welo (2014), Heras and Arana (2010), Mathiyazhagan et al. (2013), Martín-Tapia et al. (2010), Neto et al. (2017), Ormazabal et al. (2018), Revell et al. (2010), Shi et al. (2008), Thollander et al. (2015), Wilson et al. (2012), Wilson et al. (2011), Zorpas (2010)

CL3	Analytic hierarchy process; Critical success factors; Procurement effectiveness; MCDM; Multi Criteria Decision Analysis; Determinant factors; Framework; Interpretive Structural Modeling; Fuzzy TOPSIS; Fuzzy SAW; Interpretive structural modeling; Ekoscan; Factor analysis; ANOVA; Borda method; TOE framework	23	Aboelmaged (2018), Álvarez-Garcia et al. (2018), Cloquell-Ballester et al. (2008), Diana et al. (2017), Gandhi et al. (2018), Govindan et al. (2014), Granly and Welo (2014), Heras and Arana (2010), Mathiyazhagan et al. (2013), Osseweijer et al. (2018), Seth et al. (2018), Shi et al. (2008), Xia et al. (2018), Zhang et al. (2008)
CL4	Building integrated photovoltaics, BIPV; Photovoltaics; Construction organizations; Construction industry; Hotel industry; Cloud computing; Energy end-use; Technology adoption; Logistics services; Industry; Manufacturing; Maintenance; Reverse logistics; Industrial firms	16	Abdulrahman et al. (2014), Balzarova and Castka (2008), Chan (2008), Osseweijer et al. (2018), Murillo-Luna et al. (2011), Martín-Peña et al. (2014), Subramanian et al. (2014), Thollander et al. (2015), Xia et al. (2018)
CL5	Environmental Management System; EMS; Certification; Environmental management systems; Management systems; ISO 14001; Eco-management and audit scheme; EMAS; EMAS Benefits	16	Álvarez-Garcia et al. (2018), Balzarova and Castka (2008), Chan (2008), Fernández-Viñé et al. (2010), Granly and Welo (2014), Heras and Arana (2010), Martín-Peña et al. (2014), Zorpas (2010)
CL6	Compliance; environmental legislation; environmental policy; regulation; policy instrument; government regulation; enforcement; environmental law; social license; auditing	12	Brust and Liston-Heyes (2010), Graafland and Hugo (2017), Revell et al. (2010), Thollander et al. (2015), Wilson et al. (2012), Wilson et al. (2011), Zhang et al. (2008)
CL7	Producer responsibility; rural; e-learning; China; esthetics; Egypt; Indian; intended behavior; Valencian region discourse; mindsets; competitive capabilities; prioritization; Argentina	17	Abdulrahman et al. (2014), Aboelmaged (2018), Ackroyd et al. (2008), Agan et al. (2013), Brust and Liston-Heyes (2010), Cloquell-Ballester et al. (2008), Osseweijer et al. (2018), Shi et al. (2008), Subramanian et al. (2014), Wilson et al. (2011), Zhang et al. (2008)

publications and had a dominant presence. The next two clusters (CL3, CL4) related to the methodology followed and the specific sector or field selected.

Conversely, CL5 covered a repeated topic, that of environmental certification. Thus, a possible way that environmental legislation can be enforced is through certification, which can provide a more structured and possibly more accessible approach. Interestingly, the cluster relating most closely to legislation (CL6) came last on the list. Words from this cluster, a total of 12, appeared in seven papers. As no exemptions were made in the collection of keywords, another cluster (CL7) was also found wherein all the keywords that were deemed unrelated or as having an indirect impact on the topic of this study were classified. These 17 keywords, including subjects such as producer responsibility, rural, e-learning, and esthetics, appeared in 11 publications.

These findings from the keyword analysis are in line with observations made during the full-text screening of the selected publications. Although a certain maturity has been reached in the literature of the environment and SMEs, environmental legislation is not the main topic of concern for most publications reviewed. Environmental legislation, regulations, or policy issues are discussed in the context of what can be done to prevent penalties or in the context of the lack of awareness of legislative context and the appropriate enforcement. Indeed, legislation is often mentioned in the literature as a significant driver for companies, as noncompliance may lead to penalties and fines (Agan et al. 2013).

4.4 Discussion

This section investigates the publication set selected from the literature to offer a qualitative overview and reflection of the field of environmental legislation and SMEs. According to the European Commission (2003), SMEs usually have limited resources, both human and financial, and they have difficulty in drawing upon new capital (Bos-Brouwers 2010). These limitations are coupled with the widespread perception among the majority of SMEs that environmental responsibility produces costs and does not return profits to the business (McKeiver and Gadenne 2005). Due to their limited size, the lack of human resources in SMEs leads to the multifunctional use of employees (Granly and Welo 2014), leaving little time for tasks that go beyond the predetermined round of daily activities. In this context, it is quite often challenging to find the time to obtain quality advice or information within the enterprise or ability to find it outside (Hillary 1999; Pimenova and Van der Vorst 2004). Indeed, Wilson et al. (2011) argue that SMEs experience problems following environmental legislation because there may be several legislative acts related to the environment that apply to them without their ability to differentiate their workloads, as is possible in larger companies with greater resources.

In addition, SMEs are usually ignorant of their impact on the environment (Burke and Gaughran 2006; Govindan et al. 2014), the pertinent legislation, and the advantages of following sound environmental practices (European Commission 2007; Wilson et al. 2011; Zorpas 2010). Indeed, Hillary (2000) argues that SMEs do not have the financial or human resources required to deal with the demands of environmental legislation, and they are usually unaware of their impact on the environment, which is a view shared by Wilson et al. (2011). Furthermore, due to their size, SMEs rarely have the resources, time, or knowledge to interpret general guidelines into sector-specific actions, indicating the need for more specific guidelines (Xia et al. 2018).

On one hand, the literature has identified a number of barriers to good environmental practices in SMEs, including the unclear legislative frameworks (Diana et al. 2017; Hillary 2004) and scarce information (Murillo-Luna et al. 2010), which lead to weak or no enforcement at all (Aboelmaged 2018; Mathiyazhagan et al. 2013; Revell et al. 2010; Shi et al. 2008). Equally, a large number of academic authors, working in different contexts and with various backgrounds, have emphasized the difficulties that SMEs face in accessing and understanding environmental legislation (Chan 2008; Martín-Peña et al. 2014). Perron (2005) identifies several categories of barriers that prevent SMEs from adopting green initiatives, among which is the lack of awareness of environmental legislation and requirements (Koefoed and Buckley 2008; Mathiyazhagan et al. 2013).

On the other hand, Williamson et al. (2006) demonstrate in their review of the literature and the research conducted with owners and managers of SMEs in the manufacturing sector that legislation is a key driver of SME environmental initiatives. In a similarly optimistic perspective, Masurel (2007) argues that regulation may not only be the motive to persuade companies to comply with environmental legislation, but also the stimulus for them to behave in a proactive manner, improving their environmental performance beyond compliance. However, such discussions have been criticized because environmental legislation may be excessively restrictive or improperly implemented, both of which may have impacts with a greater meaning for SMEs than for larger companies (Graafland and Hugo 2017).

Furthermore, the enforcement of environmental legislation on SMEs may lead to further consequences. Specifically, companies of this size are usually able to invest financially for environmental reasons only when doing so increases compliance with environmental regulations; they generally cannot afford to invest in clean technologies (Mitchell 2006). Similarly, Neto et al. (2017) argue that cleaner implementation of production in SMEs is prevented by the shortage of sufficient cash flow to invest, finding that SMEs often only have enough resources to comply with regulations and cannot invest further. Günther et al. (2011) show that environmental legislation and regulations can restrain innovation by prescribing only the best available techniques and setting unreasonable deadlines. Consequently, the cost of compliance, due to limited cash flow, will be taken from other sources, such as clean technology and innovation.

SMEs often show a preference for rules related to consenting to noncompliance, losing interest in the norms that could benefit them, evading environmental regulations associated with incentives, opening markets, or increasing access to subsidies, to name just a few (Fernández-Viñé et al. 2010). Indeed, a significant barrier faced by SMEs is that they often do not acknowledge the benefits that could be derived by the enforcement of environmental legislation (van Hemel and Cramer 2002). Ackroyd et al. (2008) claim that businesses with reduced environmental impacts are evidently "more sustainable, profitable, valuable and competitive." Interestingly, according to the Environment Agency (2003), the smallest businesses may be the most motivated by public environmental concern, relative to the pressures imposed by legislation. SMEs should, therefore, consider environmental legislation not only to avoid penalties, but also because this enforcement can offer them a good corporate image, marketing advantages, and environmental protection.

According to Epstein and Roy (2000), although the majority of SMEs have a short-term orientation, investments in sustainability practices generally pay off in the medium to long term (Granly and Welo 2014). It is furthermore argued that SMEs, in addition to the weaknesses that appear in comparison of them with their larger counterparts, have features that are particularly beneficial for the implementation of certain strategies, such as a less bureaucratic nature, a quicker response to change, and adequate internal communication channels (Loucks et al. 2010). Revell et al. (2010) report improvement in the practices and awareness of SMEs with reference to the environment in the UK, reporting that a significant part of their respondents' sample acknowledge their environmental footprint and work toward improving their environmental responses.

Environmental legislation has also prompted SMEs to implement EMS (Seiffert 2008), including standards such as ISO14000 and EMAS (Skouloudis et al. 2013). Such standards require a specific management approach, following measures at the company's strategic and operational level that allow it to comply with environmental legislation and to reduce its environmental impact (Álvarez-Garcia et al. 2018). According to Heras and Arana (2010), compliance with environmental legislation is the reason why companies adopt an EMS, although of course environmental legislation may be a barrier to implementing one (Balzarova and Castka 2008). As Martín-Peña et al. (2014) explain, the benefit provided from such an approach to environmental management is that it helps firms adapt and comply with regulations, decreasing liability, fines, and penalties for noncompliance. Zorpas (2010) is of the same opinion, claiming that through the implementation of an EMS, an organization can facilitate improved awareness of legislative requirements and develop plans for compliance in response.

Globally, the economic contribution and environmental impact of SMEs are substantial (Diabat et al. 2014). Hence, SMEs must adopt environmentally friendly practices, a point that is not debatable. These conclusions here are in line with the findings of the previous section. Both analyses indicate that SMEs show low levels of awareness and comprehension of environmental issues, and they are slow to adopt environmental regulations. It can thus be determined that environmental legislation is seen as both a driver and a barrier for SMEs, an opportunity and a threat. It is, therefore, a matter of perspective and a matter of attitude that decides what type of environmental performance an SME exhibits. Broadly speaking, SMEs show two approaches: either they employ environmental practices in a proactive manner and exploit new opportunities to improve their marketing profile and increase market share, or in a reactive manner, they attempt to merely comply with legislation. Indeed, Worthington and Patton (2005) conclude that although SMEs realize the potential benefits and improvement of performance when voluntarily adopting good environmental practices, their environmental attitude is generally limited to compliance with legislation, the most critical driver for improving the environmental performance of an SME.

Finally, it should be noted that despite the efforts that SMEs may or may not make, developers of environmental policy and legislation must pay closer attention to SMEs. They must take into account the key characteristics of such companies and adapt regulations accordingly with the aim of providing them with the proper frameworks for enforcement. Policymakers and environmental regulators should ascertain the level of compliance of existing legislation and reexamine these modifications to increase motivation and ensure greater compliance with future legislation. The successful adoption and implementation of environmental regulation will help SMEs improve their overall effectiveness and performance while reducing their environmental impacts. More research should be dedicated to studying SMEs, not only because they are key players in the global economy and social development, but also because the aggregated environmental impact of their actions is considerable (Jia et al. 2018).

4.5 Conclusion

The present study employs an SLR to identify and synthesize available research on the topic of environmental legislation and SMEs (Coombes and Nicholson 2013; Tranfield et al. 2003). This approach exploits aspects not exhaustively discussed by scholars, and the study presented an integrated scene of the latest 10 years of literature analyzing the subjects' behaviors for 2008–2018 (de Oliveira et al. 2018).

The enforcement of environmental legislation is undoubtedly necessary, for both large and small corporations. The results of this assessment of publications on environmental legislation and SMEs are of concern to business professionals and research scholars. The present study follows an SLR protocol that enabled the identification of 33 publications, which were subsequently analyzed to find patterns in authorship, venue, and content. In this respect, this study advances the specific research topic and makes contributions to the literature. Besides making itself available as an SLR model, which may be of value to other researchers conducting literature reviews related to environmental legislation, the findings of the analyses offer an evaluation of the examined material.

This review gives no clear indications of an increasing trend in this subject area. Few authors appear repeatedly within the data set, which shows no signs of an established critical mass. The authors are geographically dispersed worldwide, and the level of multi-country collaboration is quite mature. However, the research topic appears to be overall in an early to emerging stage. The analysis of publication outlets also leads in no clear direction, showing a low-level variance in the frequency of recurrence of venue, without any indication of an increasing trend. However, one central publication outlet appears, which may be interpreted as a sign of maturity. Overall, it seems safe to assume that, as before, this is an emerging area of research interest. Citation analysis shows that all publications identified were cited, with several being cited to a significant degree. Keyword analysis shows that, although there is valuable literature on environmental legislation and SMEs, legislation is not usually the main topic of concern for the majority of the publication set. Environmental legislation was found to be considered more in terms of preventing penalties and fines or of lack of awareness rather than as presenting opportunities for development and marketing exposure.

The qualitative analysis presented in the discussion section identifies the main themes and trends presented in the literature, and its results are broadly in line with the findings of the keyword analysis. In summary, it is demonstrated that SMEs have few resources and are less able to dedicate time to understand and respond to legislation, which itself has been criticized as unclear and rigid, as well as not being sector-specific, becoming a barrier to implementation. From a different perspective, it is also considered a driver toward the prevention of penalties; in all cases, the majority of SMEs seem to be unaware of the benefits that could emerge from their responses to legislation.

4.5.1 Research Limitations and Future Research

It is not in question that environmental legislation must be enforced for SMEs. SMEs are significant contributors to the economy that can have significant environmental impacts, although they may appear insubstantial, considered on a case-by-case basis (Kasim and Ismail 2012). The present work conducted an SLR, with the aim of recording and enriching the comprehension of environmental legislation and its implementation for SMEs. Notwithstanding this, this study has, as all reviews do, limitations that require examination. Although the author has put in considerable effort, the analyses presented here reflect only certain aspects of the publications reviewed, resulting in an examination that lacks depth in a certain way. The identified publication set used a number of search terms for a specific period in articles published by a certain number of publishers. Hence, all findings were restricted by these predetermined conditions to a limited sphere of research work. It can be expected, therefore, that some developments of the knowledge of environmental legislation and SMEs have been omitted from the review. However, the author believes that the publication set identified, within the constraints above, can be considered a valuable contribution to the literature on environmental legislation and SMEs.

In the future, additional research should revisit the study protocol used here and expand it further, using more publishers, additional search terms, and, to a lesser extent, a more extended examination period. Empirical work may also be undertaken to better understand and reflect on the relationship that develops between SMEs and environmental legislation. Thus, future research could bring further valuable contributions to knowledge in the given field.

References

Abdulrahman, M.D., Gunasekaran, A., and Subramanian, N. (2014). Critical barriers in implementing reverse logistics in the Chinese manufacturing sectors. *International Journal of Production Economics* 147: 460–471.

Aboelmaged, M. (2018). The drivers of sustainable manufacturing practices in Egyptian SMEs and their impact on competitive capabilities: a PLS-SEM model. *Journal of Cleaner Production* 175: 207–221.

Ackroyd, J., Jespersena, S., Doylea, A., and Phillips, P.S. (2008). A critical appraisal of the UK's largest rural waste minimisation project: business excellence through resource efficiency (betre) rural in East Sussex, England. *Resources, Conservation and Recycling* 52: 896–908.

Agan, A., Acar, M.F., and Borodin, A. (2013). Drivers of environmental processes and their impact on performance: a study of Turkish SMEs. *Journal of Cleaner Production* 51: 23–33.

Álvarez-Garcia, J., del Río-Rama, M.d.l.C., and Saraiva, M. (2018). The influence of motivations and barriers in the benefits. An empirical study of EMAS certified business in Spain. *Journal of Cleaner Production* 185: 62–74.

Alwan, Z., Jones, P., and Holgate, P. (2017). Strategic sustainable development in the UK construction industry, through the framework for strategic sustainable development, using building information modelling. *Journal of Cleaner Production* 140 (1): 349–358.

Balzarova, M.A. and Castka, P. (2008). Underlying mechanisms in the maintenance of ISO 14001 environmental management system. *Journal of Cleaner Production* 16: 1949–1957.

Blundel, R., Monaghan, A., and Thomas, C. (2013). SMEs and environmental responsibility: a policy perspective. *Business Ethics: A European Review* 22 (3): 246–262.

Borrego, M. and Bernhard, J. (2011). The emergence of engineering education research as an internationally connected field of inquiry. *Journal of Engineering Education* 100 (1): 14–47.

Bos-Brouwers, H.E.J. (2010). Corporate sustainability and innovation in SMEs: evidence of themes and activities in practice. *Business Strategy and the Environment* 19 (7): 417–435.

Brundtland Commission (1987). *Report of the World Commission on Environment and Development: Our Common Future*. Oxford: Oxford University Press.

Brust, D.A.V. and Liston-Heyes, C. (2010). Environmental management intentions: an empirical investigation of Argentina's polluting firms. *Journal of Environmental Management* 91 (5): 1111–1122.

Burke, S. and Gaughran, W.F. (2006). Intelligent environmental management for SMEs in manufacturing. *Robotics and Computer-Integrated Manufacturing* 22 (5–6): 566–575.

CEC (Commission of the European Communities) (2002). *European SMEs and Social and Environmental Responsibility*, Observatory of European SMEs, 4. Brussels: EU publications.

Chan, E.S.W. (2008). Barriers to EMS in the hotel industry. *International Journal of Hospitality Management* 27 (2): 187–196.

Cloquell-Ballester, V.-A., Monterde-Díaz, R., Cloquell-Ballester, V.-A., and Torres-Sibille, A.d.C. (2008). Environmental education for small- and medium-sized enterprises: methodology and e-learning experience in the Valencian region. *Journal of Environmental Management* 87: 507–520.

Coenen, L., Hansen, T., McCormick, K., and Palgan, Y.V. (2017). Technological innovation systems for biorefineries: a review of the literature. *Biofuels, Bioproducts and Biorefining* 11 (3): 534–548.

Coombes, P.H. and Nicholson, J.D. (2013). Business models and their relationship with marketing: a systematic literature review. *Industrial Marketing Management* 42 (5): 656–664.

de Oliveira, U.R., Espindola, L.S., da Silva, I.R. et al. (2018). A systematic literature review on green supply chain management. Research implications and future perspectives. *Journal of Cleaner Production* 187: 537–561.

Denyer, D. and Tranfield, D. (2009). Producing a systematic review. In: *The Sage Handbook of Organizational Research Methods* (eds. D. Buchanan and A. Bryman), 671–689. London: Sage.

DG Enterprise (2010). SMEs and the Environment in the European Union. https://op.europa.eu/en/publication-detail/-/publication/aa507ab8-1a2a-4bf1-86de-5a60d14a3977

Diabat, A., Kannan, D., and Mathiyazhagan, K. (2014). Analysis of enablers for implementation of sustainable supply chain management – a textile case. *Journal of Cleaner Production* 83: 391–403.

Diana, G.C., Jabbour, C.J.C., Jabbour, A.B.L.d.S., and Kannan, D. (2017). Putting environmental technologies into the mainstream: adoption of environmental technologies by medium-sized manufacturing firms in Brazil. *Journal of Cleaner Production* 142: 4011–4018.

EC (European Commission) (2016). SBA Fact Sheet Greece. www.ggb.gr/sites/default/files/basic-page-files/Greece%202016%20SBA%20Fact%20Sheet.pdf

Edum-Fotwe, F.T. and Price, A.D.F. (2009). A social ontology for appraising sustainability of construction projects and developments. *International Journal of Project Management* 27 (4): 313–322.

Engert, S., Rauter, R., and Baumgartner, R.J. (2016). Exploring the integration of corporate sustainability into strategic management: a literature review. *Journal of Cleaner Production* 112: 2833–2850.

Environment Agency (2003). SME-nvironment 2003. A survey to assess environmental behavior among smaller UK businesses. www.netregs.org.uk/media/1080/sme_2003_uk_1409449.pdf

Epstein, M.J. and Roy, M.J. (2000). Strategic evaluation of environmental projects in SMEs. *Environmental Quality Management* 9 (3): 37–47.

European Commission (2003). Commission recommendation of 6 May 2003 concerning the definition of micro, small and medium-sized enterprises. *Official Journal of the European Union* 124: 36.

European Commission (2007). Small, Clean and Competitive – a Programme to Help Small and Medium-sized Enterprises Comply with Environmental Legislation. Executive Summary of the Impact Assessment, COM/2007/0379 final. Brussels: EC.

Ferenhof, H.A., Vignochi, L., Selig, P.M. et al. (2014). Environmental management systems in small and medium-sized enterprises: an analysis and systematic review. *Journal of Cleaner Production* 74: 44–53.

Fernández-Viñé, M.B., Gómez-Navarro, T., and Capuz-Rizo, S.F. (2010). Eco-efficiency in the SMEs of Venezuela. Current status and future perspectives. *Journal of Cleaner Production* 18: 736–746.

Gandhi, N.S., Thanki, S.J., and Thakkar, J.J. (2018). Ranking of drivers for integrated lean-green manufacturing for Indian manufacturing SMEs. *Journal of Cleaner Production* 175: 207–221.

Govindan, K., Mathiyazhagan, K., Kannan, D., and Haq, A.N. (2014). Barriers analysis for green supply chain management implementation in Indian industries using analytic hierarchy process. *International Journal of Production Economics* 147: 555–568.

Graafland, J. and Hugo, H. (2017). Reconsidering the relevance of social license pressure and government regulation for environmental performance of European SMEs. *Journal of Cleaner Production* 141: 967–977.

Granly, B.M. and Welo, T. (2014). EMS and sustainability: experiences with ISO 14001 and Eco-Lighthouse in Norwegian metal processing SMEs. *Journal of Cleaner Production* 64: 194–204.

Günther, E., Hoppe, H., and Endikrat, J. (2011). Corporate financial performance and corporate environmental performance: a perfect match? *Journal of Environmental Law and Policy* 3: 279–296.

van Hemel, C. and Cramer, J. (2002). Barriers and stimuli for ecodesign in SMEs. *Journal of Cleaner Production* 10: 439–453.

Heras, I. and Arana, G. (2010). Alternative models for environmental management in SMEs: the case of Ekoscan vs. ISO 14001. *Journal of Cleaner Production* 18: 726–735.

Hillary, R. (1999). *Evaluation of Study Reports on the Challenges, Opportunities and Drivers for Small and Medium-Sized Enterprises in the Adoption of Environmental Management Systems.* London: Department of Trade and Industry.

Hillary, R. (ed.) (2000). Introduction. In: *Small and Medium-Sized Enterprises and the Environment: Business Imperatives.* Sheffield: Greenleaf Publishing.

Hillary, R. (2004). Environmental management systems and the smaller enterprise. *Journal of Cleaner Production* 12 (6): 561–569.

Jia, F., Zuluaga-Cardona, L., Bailey, A., and Rueda, X. (2018). Sustainable supply chain management in developing countries: an analysis of the literature. *Journal of Cleaner Production* 189: 263–278.

Kasim, A. and Ismail, A. (2012). Environmentally friendly practices among restaurants: drivers and barriers to change. *Journal of Sustainable Tourism* 20 (4): 551–570.

Koefoed, M. and Buckley, C. (2008). Clean technology transfer: a case study from the South African metal finishing industry, 2000–2005. *Journal of Cleaner Production* 16S1: S78–S84.

Levy, Y. and Ellis, T.J. (2006). A systems approach to conduct an effective literature review in support of information systems research. *Informing Science: International Journal of an Emerging Transdiscipline* 9: 181–212.

Loucks, E.S., Martens, M.L., and Cho, C.H. (2010). Engaging small- and medium-sized businesses in sustainability. *Sustainability Accounting, Management and Policy Journal* 1 (2): 178–200.

Maloni, M.J., Carter, C.R., and Kaufmann, L. (2012). Author affiliation in supply chain management and logistics journals: 2008–2010. *International Journal of Physical Distribution and Logistics Management* 42 (1): 83–100.

Martin, B.R. (2012). The evolution of science policy and innovation studies. *Research Policy* 41: 1219–1239.

Martín-Peña, M.L., Díaz-Garrido, E., and Sánchez-López, J.M. (2014). Analysis of benefits and difficulties associated with firms' environmental management systems: the case of the Spanish automotive industry. *Journal of Cleaner Production* 70: 220–230.

Martín-Tapia, I., Aragón-Correa, J.A., and Rueda-Manzanares, A. (2010). Environmental strategy and exports in medium, small and micro-enterprises. *Journal of World Business* 45: 266–275.

Masurel, E. (2007). Why SMEs invest in environmental measures: sustainability evidence from small and medium-sized printing firms. *Business Strategy and the Environment* 16 (3): 190–201.

Mathiyazhagan, K., Govindan, K., NoorulHaq, A., and Geng, Y. (2013). An ISM approach for the barrier analysis in implementing green supply chain management. *Journal of Cleaner Production* 47: 283–297.

McKeiver, C. and Gadenne, D. (2005). Environmental management systems in small and medium businesses. *International Small Business Journal* 23: 513–537.

Meredith, J. (1993). Theory building through conceptual methods. *International Journal of Operations & Production Management* 13 (5): 3–11.

Michelsen, O. and Fet, M.A. (2010). Using eco-efficiency in sustainable supply chain management; a case study of furniture production. *Clean Technologies and Environmental Policy* 12: 561–570.

Mitchell, C.L. (2006). Beyond barriers: examining root causes behind commonly cited cleaner production barriers in Vietnam. *Journal of Cleaner Production* 14 (18): 1576–1585.

Muller, P., Mattes, A., Klitou, D. et al. (2018). Annual Report on European SMEs 2017/2018. The 10th anniversary of the Small Business Act. SME Performance Review 2017/2018. https://ec.europa.eu/growth/smes/business-friendly-environment/performance-review_en (accessed 25 April, 2019).

Murillo-Luna, J.L., Garcés-Ayerbe, C., and Rivera-Torres, P. (2011). Barriers to the adoption of proactive environmental strategies. *Journal of Cleaner Production* 19: 1417–1425.

Narayanamurthy, G. and Gurumurthy, A. (2016). Leanness assessment: a literature review. *International Journal of Operations & Production Management* 36 (10): 1115–1160.

Neto, G.C.O., Leite, R.R., Shibao, F.Y., and Lucato, W.C. (2017). Framework to overcome barriers in the implementation of cleaner production in small and medium-sized enterprises: multiple case studies in Brazil. *Journal of Cleaner Production* 142: 50–62.

Ormazabal, M., Prieto-Sandoval, V., Puga-Leal, R., and Jaca, C. (2018). Circular economy in Spanish SMEs: challenges and opportunities. *Journal of Cleaner Production* 185: 157–167.

Osseweijer, F.J.W., van den Hurk, L.B.P., Teunissenb, E.J.H.M., and van Sark, W.G.J.H.M. (2018). A comparative review of building integrated photovoltaics ecosystems in selected European countries. *Renewable and Sustainable Energy Reviews* 90: 1027–1040.

Perron, G.M. (2005). *Barriers to Environmental Performance Improvements in Canadian SMEs*. Canada: Dalhousie University.

Pimenova, P. and Van der Vorst, R. (2004). The role of support programmes and policies in improving SMEs environmental performance in developed and transition economies. *Journal of Cleaner Production* 12 (6): 549–559.

PwC (Pricewaterhouse Coopers) (2015). SME Funding Need for a New Architecture. www. pwc.com/gr/en/publications/assets/sme-funding-need-need-for-a-new-architecture-en.pdf

Revell, A., Stokes, D., and Chen, H. (2010). Small businesses and the environment: turning over a new leaf? *Business Strategy and the Environment* 19 (5): 273–288.

Seiffert, M.E.B. (2008). Environmental impact evaluation using a cooperative model for implementing EMS (ISO 14001) in small and medium-sized enterprises. *Journal of Cleaner Production* 16: 1447–1461.

Seth, D., Rehman, M.A.A., and Shrivastava, R.L. (2018). Green manufacturing drivers and their relationships for small and medium(SME) and large industries. *Journal of Cleaner Production* 198: 1381–1405.

Sfakianaki, E. and Stovin, V.R. (2002). A spatial framework for environmental impact assessment and route optimization. *Proceedings of the Institution of Civil Engineers Transport* 153 (1): 43–52.

Shi, H., Peng, S.Z., Liu, Y., and Zhong, P. (2008). Barriers to the implementation of cleaner production in Chinese SMEs: government, industry and expert stakeholders' perspectives. *Journal of Cleaner Production* 16: 842–852.

Skouloudis, A., Jones, K., Sfakianaki, E. et al. (2013). EMAS statement: benign accountability or wishful thinking? Insights from the Greek EMAS registry. *Journal of Environmental Management* 128: 1043–1049.

Subramanian, N., Abdulrahman, M.D., and Zhou, X. (2014). Integration of logistics and cloud computing service providers: cost and green benefits in the Chinese context. *Transportation Research Part E: Logistics and Transportation Review* 70: 86–98.

Thollander, P., Kimura, O., Wakabayashi, M., and Rohdin, P. (2015). A review of industrial energy and climate policies in Japan and Sweden with emphasis towards SMEs. *Renewable and Sustainable Energy Reviews* 50: 504–512.

Tranfield, D., Denyer, D., and Smart, P. (2003). Towards a methodology for developing evidence informed management knowledge by means of systematic review. *British Journal of Management* 14 (3): 207–222.

Wetzstein, A., Hartmann, E., Bentonjr, W.C., and Hohenstein, N.-O. (2016). A systematic assessment of supplier selection literature – state-of-the-art and future scope. *International Journal of Production Economics* 182: 304–323.

Williamson, D., Lynch-Wood, G., and Ramsay, J. (2006). Drivers of environmental behaviour in manufacturing SMEs and the implications for CSR. *Journal of Business Ethics* 67: 317–330.

Wilson, C.D.H., Williams, I.D., and Kemp, S. (2011). Compliance with producer responsibility legislation: experiences from UK small and medium-sized enterprises. *Business Strategy and the Environment* 20 (5): 310–330.

Wilson, C.D.H., Williams, I.D., and Kemp, S. (2012). An evaluation of the impact and effectiveness of environmental legislation in small and medium-sized enterprises: experiences from the UK. *Business Strategy and the Environment* 21 (3): 141–156.

Worthington, I. and Patton, D. (2005). Strategic intent in the management of the green environment within SMEs. An analysis of the UK screen-printing sector. *Long Range Planning* 38: 197–212.

Xia, B., Olanipekun, A., Chen, Q. et al. (2018). Conceptualising the state of the art of corporate social responsibility (CSR) in the construction industry and its nexus to sustainable development. *Journal of Cleaner Production* 195: 340–353.

Zhang, B., Bi, J., Yuan, Z. et al. (2008). Why do firms engage in environmental management? An empirical study in China. *Journal of Cleaner Production* 16: 1036–1045.

Zorpas, A. (2010). Environmental management systems as sustainable tools in the way of life for the SMEs and VSMEs. *Bioresource Technology* 101: 1544–1577.

Section II

Governing and Protecting Natural Resources

5

Bulk Water Pricing Policies and Strategies for Sustainable Water Management

The Case of Ontario, Canada

Guneet Sandhu[1], Michael O. Wood[1], Horatiu A. Rus[2], and Olaf Weber[1]

[1] School of Environment, Enterprise, and Development, University of Waterloo, Waterloo, Ontario, Canada
[2] Department of Economics and Department of Political Science, University of Waterloo, Waterloo, Ontario, Canada

5.1 Introduction

As echoed in Sustainable Development Goal 6, sustaining sufficient availability and quality of water is crucial for human survival and for sustaining vital natural ecosystems and economic productivity (UN 2015). Thus, any lapses in water resource management can have social, economic, and environmental repercussions, making water a core component of sustainable development (Russo et al. 2014). Water scarcity is a nuanced concept that is not only contingent on the quantity of water available to fulfill human and ecological demands, but also includes the necessary quality of water resources to be sustained for various uses (Brooks 2006).

Due to the impacts of climate change on water availability and quality, coupled with growing anthropogenic demands, no city, province, or country is truly immune to water scarcity. Thus, the objective of "sustainable water management" is to assure that all social, economic, and ecological demands of current and future generations are fulfilled while sustaining the integrity (quality and quantity) of water resources (Russo et al. 2014).

Even the province of Ontario, Canada – a region perceived to be water-abundant – is expected to witness increased episodes of water scarcity owing to stressors like climate change as well as population and economic growth. While measures for sustainable water management are gaining momentum across the globe, Ontario's economy is likely to remain water-intensive with a burgeoning water demand (Disch et al. 2012; Mitchell 2017). These rising concerns over the sustainability of water resources have been the impetus behind the provincial moratorium placed on new groundwater extraction permits for the water bottling sector until January 2020[1] (Ontario Ministry of Environment, Conservation and Parks 2018).

As the focus on provincial water resources is gaining traction, a discussion on comprehensive policy approaches, including economic instruments that can foster sustainable use of water resources and mitigate identified water risks, is timely. This chapter explores different facets of water management, including an overview of policy approaches as

Environmental Policy: An Economic Perspective, First Edition. Edited by Thomas Walker, Northrop Sprung-Much, and Sherif Goubran.
© 2020 John Wiley & Sons Ltd. Published 2020 by John Wiley & Sons Ltd.

well as the rationale for using economic policy instruments. In order to arrive at best practices for pricing water as a resource, some key global and Canadian provincial examples are investigated and synthesized. Finally, the province of Ontario, Canada, is presented as a case study where the best practices are operationalized to arrive at a regionally tailored bulk water pricing framework. The proposed framework is designed not only to address the gaps in current regulatory frameworks, but also to communicate water scarcity and optimize the consumption behavior of self-supplied water using sectors in Ontario.

5.2 Background

The complexity of water as a resource arises due to the dichotomy of it being perceived as both a public good crucial for human survival and an economic good used as a material input in production sectors. The International Conference on Water and Environment (1992) held in Dublin recommended a set of principles highlighting the criticality of water resources. In addition to acknowledging fresh water as a "finite" and "vulnerable" resource, Principle 4 emphasizes the recognition of water as an economic good that provides economic value to its users (Hanemann 2006).

Conversely, arguments have also been made for water being a shared common resource with an intrinsic cultural value. Water bodies have been spiritually significant and a source of leisure in many cultures. Therefore, there is an intangible social value that is associated with water (Hanemann 2006; Lant 2004). Moreover, as stated at the United Nations Rio+20 Summit (2012), drinking water is a basic human right that should be accessible to all (Mitchell 2017). Nonetheless, given the burgeoning stressors on availability and quality, water is a scarce productive resource, and its demand by different users needs to be appropriately managed (Hanemann 2006; OECD 2017). As long as water is considered a free and abundant resource, there is no incentive for a sustainable use of water resources. Without a check on demand, the chances of overconsumption and impending "tragedy of the commons" are reinforced (Hanemann 2006; Hardin 1968).

According to the World Health Organization, the average per capita requirement of water is 50–100 liters (L)/day/person for basic human and residential needs (Howard and Bartram 2003). By multiplying this amount by the relevant population and subtracting it from the total demand, the quantity of water as a human right can be distinguished from the quantity of water demanded by other use-sectors. This "residual" demand can then be efficiently managed using different policy instruments (Lant 2004). Consequently, concerns regarding the access to drinking water as a need can be addressed by prioritizing and providing the "lifeline" volume of 50–100 L/day/person to all (OECD 2017).

Bulk or self-supplied water is the raw or untreated surface water or groundwater that is extracted directly by different water-using sectors, including manufacturing, thermal power generation, agriculture, commercial, and municipal water utilities (Renzetti and Dupont 1999), while drinking or potable water is the treated and supplied municipal utility for which users are charged a tariff. The focus of this chapter is on the raw water resources allocated to self-supplied users by the government.

5.2.1 Policy Instruments for Sustainable Water Management

The availability of freshwater resources is contingent on the spatial and temporal hydrological conditions. However, the allocation of water for social, economic, and environmental uses is managed and regulated by public authorities by means of water policies through an array of instruments (Mitchell 2017). With growing demand and impacts of climate change on water resources, efficient use of water and conservation are the underlying objectives of water management policies (OECD 2013, 2017).

As a publicly governed resource, there are three main types of policy approaches that are employed for water management.

5.2.1.1 Command and Control Approach

This is a popular regulatory approach, which is based on enforced restrictions imposed by public authorities on the use of natural resources and limits on pollution based on human health and environmental impacts. The permits/licensing for water allocation as well as seasonal water use restrictions are included in this approach for water resource management (Cantin et al. 2005; OECD 2013). Monitoring and enforcing compliance of these prescribed regulations form an important aspect of ensuring the efficacy of this top-down approach. Thus, the regulators need to invest significant resources to ensure the policy outcomes are achieved by imposing fines or penalties on users for noncompliance (Cantin et al. 2005; Renzetti 2005).

5.2.1.2 Economic Instruments

A core principle for sustainable water management is recognizing the economic value of water resources. Economic instruments are based on pricing theory, where monetary signals via prices can effectively change consumption behavior and hence drive the demand for a scarce resource (Hanemann 2006; OECD 2017). When a resource is underpriced, there is a risk of overconsumption due to the misconception of excess supply (Vandierendonck and Mitchell 1997). To correct this consumption behavior, water pricing is increasingly being employed to incentivize sustainable water use and recover costs of water management initiatives (Bruneau et al. 2013; Cantin et al. 2005; Renzetti 2007).

Water abstraction charges based on cost recovery principles are one of the commonly used economic instruments for sustainable water management (OECD 2017). Pricing raw water using an extraction charge does not imply the privatization of water resources since allocation, as in the case of Ontario, is governed and regulated under public jurisdiction (Renzetti 2017).

As permitted by certain legal frameworks, pure market-based approaches, like competitive water markets and trading of rights between users, have been used globally and in the Canadian province of Alberta to allocate and price water resources. Nonetheless, the design, implementation, and monitoring of subsequent transactions are inherently complex and cost-intensive (Cantin et al. 2005). Thus, approaches like cap and trade are deemed to be more suitable for pricing water pollution and extraction in arid regions (OECD 2017).

5.2.1.3 Voluntary Stewardship or Compliance

In this self-regulation-based approach, measures in the form of stewardship, eco-labeling, and awareness programs are also gaining momentum for promoting sustainable water use in different sectors. Given the regulatory, economic, and reputational risks associated with water scarcity, many industries are taking voluntary steps to improve their water performance and partake in reporting initiatives for water use efficiency and conservation. An important bottom-up approach, these voluntary compliance initiatives are more prevalent in areas where the threats to water resources are more pronounced (Christ and Burritt 2017). Additionally, the effectiveness of these initiatives can be challenging to measure as well as monitor and require significant financial investments upfront (Cantin et al. 2005).

5.2.2 Efficacy of Economic Instruments in Water Demand Management

Even though water is used as a material input in industries, it is substantially undervalued compared to other inputs like energy, material, and labor. Thus, there is no economic rationale to invest in technologies that are water efficient or produce less waste (OECD 2013; Renzetti 2007).

Bulk water pricing refers to assigning a monetary value to raw water that is extracted directly from either surface water or groundwater sources by different sectors (European Environment Agency 2013). This monetary value signals various risks of using a scarce resource, hence changing the consumption behavior of users. The response to price changes is measured through price elasticity of water demand, which measures the change in water demand when the price changes by a unit. If different sectors exhibit negative price elasticity, pricing becomes a viable demand management strategy (Griffin 2016).

An econometric study by Renzetti and Dupont (1999) establishes the efficacy of volumetric water extraction charges in contrast to flat charges to reduce the demand of self-supplied sectors in Ontario without significant economic implications. Rivers and Groves (2013) also simulate a 25% decrease in water intake by imposing an extraction charge of CAD $13/million liters on all self-supplied sectors in Canada with a negligible gross domestic product (GDP) loss. Although to different extents, all water use sectors (residential, industrial, commercial, and agricultural) are sensitive to changes in charges and consequently reduce demand. Also, if the revenue generated is recycled to reduce taxes, these charges may be welfare-neutral or improving (Rivers and Groves 2013).

Such econometric studies are necessary to alleviate concerns about adverse economic impacts and build the rationale for using economic instruments along with conventional regulatory and voluntary stewardship policy instruments for sustainable water management (Renzetti 2017). The results consistently emphasize that water extraction charges are a key instrument to incentivize water use-efficiency and conservation for self-supplied, high-volume users. The revenue thus generated can financially sustain all water management activities including planning, monitoring, and implementation (Cantin et al. 2005).

5.2.3 Costs Under Consideration for Bulk Water Pricing

In line with the European Union (EU) Water Framework Directive (WFD) and the Organisation for Economic Cooperation and Development's (OECD) full cost recovery principles, the price of water should reflect the full economic and environmental costs arising from water abstraction, use, and discharge (European Environment Agency 2013; OECD 2013). An efficient water price also ensures complete recovery of public resource management costs, demand reduction, and signals spatial as well as temporal water risks to the end-use sectors (Cantin et al. 2005; Renzetti 2007).

5.2.3.1 Economic Costs

These are administrative and operating costs associated with permitting, regulating, and administering various water management programs, monitoring, and evaluation costs of quantity and quality of water sources, drought management programs, environmental assessments, and planning initiatives. These costs also include the capital costs of providing infrastructure to regulate/maintain flows like reservoirs as well as the equipment for monitoring the water quantity and quality in streams and wells (Drafting Group ECO2 2004; European Environment Agency 2013).

5.2.3.2 Resource Costs

These are marginal opportunity costs reflecting the costs incurred due to overextraction of the resource resulting in loss of economic benefits for future water-dependent sectors. The loss of economic value/benefits of allocating the resource to an inefficient user (instead of alternate users) resulting in depletion of the resource can be estimated and accounted by using certain hydroeconomic models (Drafting Group ECO2 2004; European Environment Agency 2013).

5.2.3.3 Environmental Costs

These are costs associated with environmental damage and subsequent loss of ecosystem services due to anthropogenic extraction and pollution of water resources. For instance, if inadequately treated wastewater or contaminants are discharged into water bodies that impair the ecological health or ecosystem services (recreation, fisheries, productive wetlands), the remediation costs or loss of benefits can be used to arrive at the environmental cost (Drafting Group ECO2 2004; European Environment Agency 2013; OECD 2017).

5.2.3.4 Environmental Protection Costs

In many countries, significant investments are made to proactively protect water resources and hence avoid future ecological damages caused by abstraction or pollution. From the context of cost recovery, the expenditures of these preventive measures, to avoid possible environmental damage, are accounted for as part of the larger water resource management initiatives. They are internalized and reflected in the water price (Drafting Group ECO2 2004).

The methodology for designing efficient pricing schemes that cater to objectives of equity, economic efficiency, and environmental sustainability is highly nuanced and variable across the globe. The different categories of costs described above are not mutually exclusive and cannot be simply added (Drafting Group ECO2 2004). Different economic valuation methodologies are adopted to calculate resource and environmental costs including willingness to pay surveys (contingent valuation method), replacement/remediation cost assessments, hydroeconomic modeling, and hedonic pricing (willingness to pay for pristine or high-quality environment) (Dupont and Adamowicz 2017).

Nonetheless, each method has limitations as well as constraints of time and resources required to conduct comprehensive valuation studies. Unlike direct market valuation methods used to price regular economic goods and services, valuation of water resources requires a combination of different methods, studies, and approaches (CCME, 2010).

5.3 Global and Provincial Best Practices for Bulk Water Pricing Policies

To gain a better understanding of how bulk water pricing and subsequently extraction charges can be practically designed, we undertake a global and provincial overview of pricing practices. Article 9 of the European WFD serves as a guiding model for introducing resource and environmental costs as part of full cost pricing for the services of water resources above and beyond municipal water treatment and supply. Tailored at the sub-watershed scale, member states need to recover not only the opportunity costs of extracting water, but also the costs arising from degradation in water quality due to effluents discharged into water bodies (European Environment Agency 2013; OECD 2017).

Countries like Israel and Australia that are grappling with water scarcity have effectively employed economic instruments like pricing to efficiently allocate water, reduce freshwater demand, promote water reuse, and induce technological innovation for water-efficient production practices. Thus, there is a growing global momentum toward employing economic instruments to achieve sustainable water management policy objectives (Becker 2015; OECD 2017). After reviewing key policy documents published by the European Environment Agency, OECD, and individual government agencies, the practices and approaches of some global and provincial examples are synthesized in Figure 5.1. As discussed earlier, the EU WFD is a model framework for introducing economic instruments for water management. Thus, for the purposes of arriving at best practices for designing bulk water extraction charges, some EU member states, as well as OECD countries like Australia and Israel, have been included and synthesized below (OECD 2013, 2017).

In Canada, water governance including management, allocation, and conservation of water resources is primarily the responsibility of individual provinces. In this highly decentralized set-up with minimum federal involvement, each province has established its own legislative framework for water resource management, including the design of water

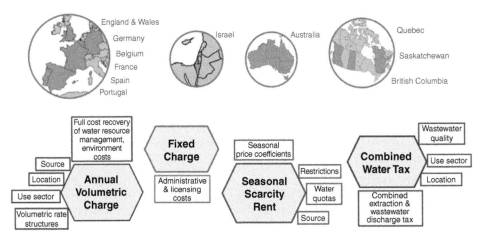

Figure 5.1 Synthesized best practices for designing bulk water extraction charges. *Source:* Maps generated by the authors from Mapchart.net.

extraction charges (Horbulyk 2017). For the purposes of this study, three provincial examples have been included to provide insights on some particular approaches to price bulk water extraction.

The key theme prevalent in most countries is the price differentiation based on the location, type of source (groundwater priced higher), quality of sources, and type of use sector. Season-based charges for drought conditions are also common with subsidies on recycled water as seen in Israel. Interestingly, a combined wastewater quality tax was considered in tandem with water extraction charges in Portugal, yielding a more holistic water resource pricing. However, it has been suggested in the literature that the design of the pricing framework is highly contingent on the existing legislative and regulatory framework of the region under consideration (OECD 2013, 2017).

It has been emphasized throughout the literature that high-volume users are sensitive not only to the magnitude of charges, but also to the rate structure (Cantin et al. 2005; Renzetti and Dupont 1999). Price differentiation based on water source conditions (availability, quality, sector risks) has been used to cater to water conservation and efficiency goals by effectively reducing water demand (OECD 2017; Renzetti 2007).

To address affordability or equity concerns and provisions for accessibility to drinking water, a concessional discount factor or minimum lifeline volume of water can be provided to certain water-using sectors free of charge. Thus, to achieve multiple objectives of sustainable water management, like economic efficiency, equity, and environmental sustainability, a dynamic pricing framework needs to be designed (OECD 2017; Renzetti 2007).

5.4 Case Study: The Province of Ontario, Canada

This section aims to unearth gaps, tease out specific areas of reform, and suggest practical measures that necessitate a sound bulk water pricing framework as a key strategy for sustainable water management in Ontario.

Canada, in comparison with other OECD countries, not only has a water-intensive economy, but also has shown little improvement over the years in water-use efficiency. In contrast, countries like Australia that are similar to Canada's resource-dependent economy but are water scarce have continuously improved their water-use efficiency (Canada's Ecofiscal Commission 2014). The intense and inefficient water use in Canada is attributed in part to the perception of water resources' abundance, leading to underpricing and hence overextraction of water (Debaere 2014).

The province of Ontario, as depicted in Figure 5.2, is surrounded by the Great Lakes and many regional freshwater sources due to which many industries have flourished over the years. However, and contrary to the widely held myth of water abundance in Ontario, when analyzed in detail at finer spatial and temporal resolutions, the industrialized and populous Great Lakes drainage basin has not been immune to water availability issues (Disch et al. 2012; Mitchell 2017).

Certain parts of south-western Ontario have been historically susceptible to droughts and reduced water flow in creeks and rivers due to low precipitation in the summer. With potential increased surface temperatures due to climate change, there will be more evaporation, lower surface water flows, and additional demand for irrigation (Disch et al. 2012;

Figure 5.2 Map of the Province of Ontario, Canada, with major water resources. *Source:* Natural Resources Canada (2002).

Kreutzwiser et al. 2004). Additionally, there have been growing water quality issues due to contamination events from industrial and agricultural activities. Therefore, the threat to the overall sustainability and productivity of freshwater resources is evident (Morris et al. 2008).

Due to various provincial and municipal initiatives including metering, tariff revisions, technology innovation, rebates on water-efficient plumbing, and awareness programs, the residential municipal water use in Ontario did improve from 267 L/day/capita in 2006 to 201 L/day/capita in 2015 (Bruneau et al. 2013; Environment Canada 2010; Statistics Canada 2016). While there has been a significant focus on managing residential drinking water demand, the self-supplied industrial, commercial, and agricultural sectors that are highly water intensive have been outside the policy radar (Mohapatra and Mitchell 2009).

Academic research and analysis on water use in Canada and Ontario have concluded that the water consumption behavior by diverse industrial sectors is inefficient (e.g., NRTEE

2011; Renzetti 2017; Renzetti and Dupont 1999; Rivers and Groves 2013). Consequently, there is a need for improvement in water demand from different sectors, especially with the use of economic policy instruments (Bruneau et al. 2013; Renzetti 2007). For instance, thermal power generation accounts for nearly 85% of the annual water extracted in the Great Lakes drainage basin (Statistics Canada 2014). While the majority of the water extracted by the sector for purposes of cooling or generating steam is returned to the source, the quality and temperature of effluent water are nevertheless altered (Renzetti and Dupont 1999). Thus, thermal power generation is a high-volume water extractive sector that should be included in water demand management policies.

The agriculture sector in Ontario is not a major extractor of water compared to industrial and residential sectors. However, it is a key area for water management policies due to the water quality issues posed by nutrient pollution as well as the highly consumptive use (about 85%) of fresh water, which is inversely related to precipitation patterns (Ontario Ministry of Natural Resources 2014). Since agricultural production is extremely sensitive to water availability, especially during the summer, many of the water conflicts during periods of reduced precipitation are among agricultural and industrial permit holders (Morris et al. 2008).

Thus, water management policies not only need to be calibrated for the temporal and spatial hydrological variability in water resources, but also need to account for the heterogeneous extraction patterns of different water-using sectors.

5.4.1 Current Regulatory Framework for Water Resource Management

While there is considerable variability across provinces pertaining to legislative frameworks for water allocation, the regulated common law riparian system in Ontario enables the use of public policy instruments to manage water taking by different sectors (Brandes and Curran 2017). The Ontario Water Resources Act of 1990 provides the regulatory framework for the allocation and management of water resources within Ontario (Kreutzwiser et al. 2004). As defined in Section 34, any user[2] proposing to extract more than 50 000 L/day of water directly from either surface or groundwater sources needs to obtain a "Permit to Take Water" from the Ontario Ministry of Environment[3] (MOE) (Province of Ontario 2004).

Under Regulation O. Reg. 450/07, a volumetric "water conservation charge" of CAD $3.71/million liters is imposed on select high water consumption industrial sectors[4] that partially recovers water quantity management costs (Ontario Ministry of Environment 2007). In addition to the moratorium, the extraction charge for existing water bottlers has also been hiked to CAD $503.71/million liters as a strategy to incentivize conservation (Ontario Ministry of Environment and Climate Change 2017a).

The Low Water Response Program, implemented by the Ministry of Natural Resources and individual conservation authorities, has been instituted to manage water demand during drought events. During seasonal low-flow conditions, permit holders are required to comply with restrictions imposed on water use. This program first relies on the voluntary reduction of water use after an actual low-flow event, thus becoming a "reactive" strategy instead of proactively bringing long-term improvements in water demand (Disch et al. 2012; Kreutzwiser et al. 2004).

The federal and provincial government under the transboundary water sharing agreement with the United States and the Ontario Clean Water Act (2006) have invested significantly in many water quantity and quality management initiatives, primarily in the Great Lakes basin. However, instead of a financially sustainable water resource management program funded equitably by all water use sectors as beneficiaries, the initiatives in Ontario rely on the pool of general tax revenues (Renzetti 2007). While this practice spreads the costs of water-related initiatives, it fails to signal scarcity and promote conservation for water users.

5.4.2 Gaps and Opportunities for Sustainable Water Management

Even though Regulation O. Reg 387/04 on water taking and transfer aims to ensure sustainable water extraction in high use watersheds, concerns about the sustainability of water resources as well as conflicts among water use sectors have not been alleviated. This brings to force serious deficiencies in current water management policies to *proactively* communicate water risks and manage the demand of different sectors (Disch et al. 2012; Kreutzwiser et al. 2004; Morris et al. 2008).

The current permit fees and volumetric charges imposed on a few industrial sectors recover approximately CAD $200 000 annually. At the very least, when costs attributable only to Permit to Take Water (PTTW) program and water quantity management are considered (CAD $17.5 million/year), the current charges fall substantially short of even partial cost recovery (Ontario Ministry of Environment and Climate Change 2017b). Furthermore, if the costs of all water management initiatives are accounted for, these charges will need significant revision. From an economic standpoint, bulk water is a valuable resource, yet it is provided nearly free of cost to different using sectors affecting both water availability and quality (Bruneau et al. 2013; Environmental Commissioner of Ontario 2015; Renzetti 2007).

Under the Water Resources Act, the hydroelectric power sector is exempt from paying volumetric charges, due to it being an in-stream, nonconsumptive user (Ontario Ministry of Environment 2007). However, hydroelectric power plants pay a separate "water rental charge" (9.5% of annual gross revenue) under the Ontario Electricity Act, 1998, for "using provincial water resources" for the purposes of power generation (Ontario Ministry of Finance 2016). Thus, from an equity perspective, if an in-stream user like hydroelectric power generation is liable to pay a charge for its water use, other water extractive sectors including thermal power generation should also be liable for water extraction charges (Renzetti and Dupont 1999).

While Australia, Israel, and most EU countries are taking measures to sustain the productivity of water resources by transitioning to a water-efficient economy, Ontario has yet to show significant improvements in sustainable water management (Canada's Ecofiscal Commission 2014; Disch et al. 2012). The legislative framework and academic rationale for economic instruments like extraction charges exist, but they are currently not designed to communicate different water risks and costs of water quality and quantity management to beneficiary use sectors in Ontario (Renzetti 2007, 2017).

The lack of a regionally tailored bulk water pricing framework leading to these extraction charges represents a gap in the literature (Cantin et al. 2005; Renzetti and Dupont 1999;

Rivers and Groves 2013). Nonetheless, it is a timely opportunity to address the gaps in the current water management approaches using well-designed economic instruments. Consequently, economic instruments like water extraction charges are warranted to reflect the value of water resources as well as incentivize different sectors to invest in water-efficient and environmentally friendly technologies.

5.5 Conceptual Bulk Water Pricing Framework Proposed for Ontario

As outlined in the Canadian Council of Ministers of the Environment guidance document for water pricing, Ontario currently only partially recovers the costs of administering the PTTW and water quantity management programs. At the same time, the various federal and provincial remediation initiatives, source water protection, nutrient management, monitoring, and evaluation programs are covered through general tax revenues (CCME 2015).

Preferably, full economic, resource, and environmental costs associated with the use of water resources should be considered in the calculation of water charges. However, given the information gaps, the data reliability issues, the paucity of extensive hydroeconomic data at the segregated sub-watershed level as well as the practical limitations of implementing these costs into the existing regulatory frameworks, full cost pricing may be foreseen as an incremental process (Dupont and Adamowicz 2017; OECD 2017). The approach of recovering the actual costs of water management initiatives from all use-sectors is in line with the existing rationale used for designing water charges in many provinces in Canada (CCME 2015).

Thus, for the case of Ontario, accounting for all economic costs spent on the protection, management, and remediation of water resources may be a more practical starting point for arriving at an average provincial base water charge. These costs can be sourced and averaged on an annual basis from the provincial public accounts, the Auditor General of Ontario's reports on remediation costs, and other federal and provincial water program audit reports. The average annual cost can then be combined with Statistics Canada average annual volume of water withdrawn data to calculate the base volumetric extraction charge ($/m^3 of water withdrawn).

The spatial water quantity risks have been duly identified in the Source Protection Assessment Reports prepared by individual conservation authorities at the sub-watershed scale (Conservation Ontario 2016). Thus, if the location of a high-volume water extraction permit is in a moderate or high-risk quaternary watershed, a price multiplier can be applied to the base provincial charge, sending a higher price signal for the area. Similarly, to account for variability in water consumption among different water-using sectors, categories can be assigned to sectors based on high, moderate, and low water consumption as classified in the MOE's proposal of water conservation charges (Ontario Ministry of the Environment 2007).

To account for the sensitivity and high quality of groundwater resources, wherein water extracted from groundwater wells is seldom recharged into the original source but discharged into surface water bodies, a price multiplier can be assigned based on the proposed source (groundwater or surface water) of water extraction (Mohapatra and Mitchell 2009;

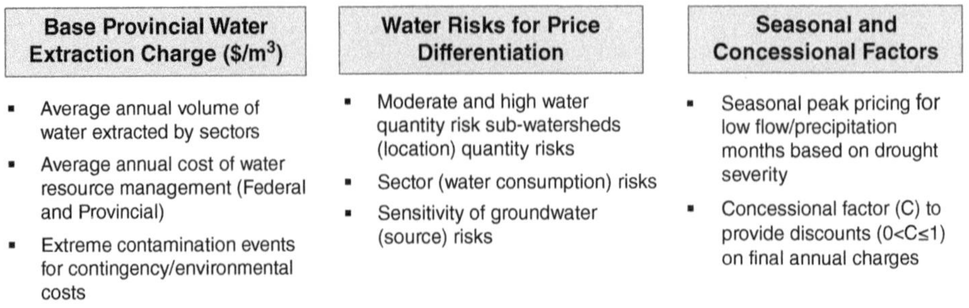

Base Provincial Water Extraction Charge ($/m³)	Water Risks for Price Differentiation	Seasonal and Concessional Factors
▪ Average annual volume of water extracted by sectors ▪ Average annual cost of water resource management (Federal and Provincial) ▪ Extreme contamination events for contingency/environmental costs	▪ Moderate and high water quantity risk sub-watersheds (location) quantity risks ▪ Sector (water consumption) risks ▪ Sensitivity of groundwater (source) risks	▪ Seasonal peak pricing for low flow/precipitation months based on drought severity ▪ Concessional factor (C) to provide discounts (0<C≤1) on final annual charges

Figure 5.3   Proposed bulk water pricing framework for Ontario.

Morris et al. 2008). Therefore, similar to seasonal surcharges used in England and Wales, a multiplier can kick in for low precipitation months. This surcharge is imposed only for the low precipitation months and for the rest of the year, only location, source, and sector-specific risk price multipliers apply. The outline of the proposed framework with key attributes is summarized in Figure 5.3.

Regular Price multiplier (for regular flow months) $M = M_{Watershed} \times M_{Sector} \times M_{Source}$

Seasonal Price multiplier (for low flow months) $M = M_{Watershed} \times M_{Sector} \times M_{Source} \times M_{Seasonal}$

Number of low flow months entered $= N_{Low}$

All Year Charges ($) = {Seasonal Price Multiplier $\times$ (N_{Low} /12) + Regular Price Multiplier $\times$ [(12− N_{Low})/12]}$\times$ Base Provincial Volumetric Charge $\times$ Annual Volume of Water Extracted

---

For increased policy flexibility, a fractional concessional factor can also be integrated to provide relief to water-efficient stewards or on grounds of affordability. Thus, based on the specific inputs on location, sector, source of water, and season, different multipliers or coefficients can help in generating price differentiation to optimize consumption behavior (Morris et al. 2008; OECD 2017).

5.6   Discussion and Conclusion

There has been paucity in the literature that addresses the methodological process of arriving at water extraction charges that cater to various objectives for sustainable water management, especially in the case of Canada. The key characteristics of water charges like social equity, economic efficiency, and environmental sustainability have been discussed theoretically. However, a unified framework for the evaluation of these charges based on established pricing principles and spatial and temporal considerations has not yet been put forth for Ontario (Cantin et al. 2005; Renzetti and Dupont 1999).

From the discussion presented in this chapter, it is evident that water resource management is a multidimensional construct and that the full costs of water quantity and quality programs need to be communicated to all beneficiary water extractive sectors, instead of being sourced from general tax revenues (Renzetti 2007). While regulatory measures like moratoriums and voluntary restrictions are important steps in water management, proactive measures are needed for all water-using sectors that are currently "locked" in locations with potential water risks (Disch et al. 2012; Water Canada 2016). Also, by underpricing water extraction, not only are the users oblivious of the economic value added by all water resource management initiatives, but also low-volume users can subsidize high-volume users (Renzetti 2007).

Given the uncertain hydrological conditions as well as the population and economic growth in the Great Lakes basin, the current low static volumetric charge of CAD $3.71/million liters will not suffice if dual goals of economic cost recovery and environmental sustainability are to be attained (Bruneau et al. 2013; Renzetti 2017). Addressing this policy as well as the methodological academic gap, a transparent pricing mechanism is warranted, where all water extractive sectors and beneficiaries of water resources are charged equitably on a volumetric basis.

As demonstrated by previous econometric studies for Canada, volumetric extraction charges can be effective policy instruments to manage water demand in diverse sectors without major economic implications (Rivers and Groves 2013). Drawing on best practices as well as the regulatory context for Ontario, a pricing framework is objectively proposed on the basis of public cost recovery and hydrological information available. While the base provincial charge can cater to the cost recovery objective of pricing, the price multipliers based on the quaternary watershed quantity risks, the use of groundwater sources, the water consumption-based sectors, and the seasonal severity of flow provide price signaling to serve the environmental sustainability objectives (Morris et al. 2008; OECD 2017).

The proposed framework builds on the existing water extraction charges, hence providing compatibility with existing regulatory frameworks, ease of administration, and implementation (Cantin et al. 2005; Renzetti and Dupont 1999). Besides, the surplus revenue generated through the multipliers can provide enough contingency for ongoing resource management initiatives as well as funding future sustainability initiatives, including providing detailed hydrological assessments, upgrading municipal infrastructure, funding conservation authorities, and other provincial water-related initiatives that tend to be underfunded (Mohapatra and Mitchell 2009).

Given the ongoing provincial review of water policies and renewed interest in economic instruments that manage water demand, an opportunity for conversation and policy reform is evident. The arguments for accessibility to water as a human right, affordability concerns as well as economic implications are important considerations that can be addressed using a complementary mix of regulatory, economic, and voluntary instruments rather than maintaining the status quo (OECD 2017; Renzetti 2007).

This chapter's main contribution is to provide insights into managing a complex, common, and economic resource like water, using pertinent economic policy instruments. Based on best practices and contextual considerations, a conceptual pricing framework is proposed that can be further operationalized to arrive at bulk water extraction charges to fund future water management initiatives and trigger Ontario's transition into becoming a more water-efficient economy.

Acknowledgment

The authors would like to thank the University of Waterloo Water Institute Seed Grant Program for supporting this work.

Notes

1 On December 21, 2018, via Regulation O. Reg. 529/18, the Ontario Ministry of Environment, Conservation and Parks extended the original moratorium until January 1st, 2020.

2 Except water used for domestic, livestock, poultry, emergency services like firefighting, wetland conservation, or temporary diversions for construction purposes.

3 As of July 2018, the Ontario Ministry of Environment and Climate Change (MOECC) has been renamed the Ontario Ministry of Environment, Conservation and Parks (MECP).

4 Water bottling, beverage manufacturing, fruit/vegetable canning, certain chemical manufacturing where majority of the water extracted is incorporated in the final product and not returned as wastewater to watershed.

References

Becker, N. (2015). Water pricing in Israel: various waters, various neighbors. In: *Water Pricing Experiences and Innovations* (eds. A. Dinar, V. Pochat and J. Albiac-Murillo), 181–199. Cham: Springer International Publishing.

Brandes, O.M. and Curran, D. (2017). Changing currents: a case study in the evolution of water law in Western Canada. In: *Water Policy and Governance in Canada* (eds. S. Renzetti and D.P. Dupont), 45–67. Cham: Springer International Publishing.

Brooks, D.B. (2006). An operational definition of water demand management. *International Journal of Water Resources Development* 22 (4): 521–528.

Bruneau, J., Dupont, D., and Renzetti, S. (2013). Economic instruments, innovation, and efficient water use. *Canadian Public Policy/Analyse de Politiques* 39: S11–S22.

Canada's Ecofiscal Commission. (2014). Smart, Practical, Possible: Canadian Options for Greater Economic and Environmental Prosperity. https://ecofiscal.ca/reports/smart-practical-possible-canadian-options-for-greater-economic-and-environmental-prosperity

Canadian Council of Ministers of the Environment. (2015). Water Pricing Options. Report Number PN 1536. www.ccme.ca/files/Resources/water/water_valuation/Principles%20for%20Water%20Pricing%201.1_e%20PN%201536.pdf

Cantin, B., Shrubsole, D., and Aït-Ouyahia, M. (2005). Using economic instruments for water demand management: introduction. *Canadian Water Resources Journal* 30 (1): 1–10.

Christ, K.L. and Burritt, R.L. (2017). Water management accounting: a framework for corporate practice. *Journal of Cleaner Production* 152: 379–386.

Conservation Ontario. (2016). Conservation Ontario Submission on the 'Proposed Permit to Take Water Moratorium'. http://conservationontario.ca/resources/?tx_fefiles_files%5Bfile%5D=305andtx_fefiles_files%5Baction%5D=showandtx_fefiles_files%5Bcontroller%5D=FileandcHash=6e77af21b2d5e3a0717c5c7bd065e1d6

Debaere, P. (2014). The global economics of water: is water a source of comparative advantage? *American Economic Journal: Applied Economics* 6 (2): 32–48.

Disch, J., Kay, P., and Mortsch, L. (2012). A resiliency assessment of Ontario's low-water response mechanism: implications for addressing management of low-water under potential future climate change. *Canadian Water Resources Journal* 37 (2): 105–123.

Drafting Group ECO2. (2004). Assessment of Environmental and Resource Costs in the Water Framework Directive. www.waterframeworkdirective.wdd.moa.gov.cy/docs/OtherCISDocuments/Economics/ECOResouceCosts.pdf.

Dupont, D.P. and Adamowicz, W.L. (2017). Water valuation. In: *Water Policy and Governance in Canada* (eds. S. Renzetti and D.P. Dupont), 181–199. Cham: Springer International Publishing.

Environment Canada. (2010). 2010 Municipal Water Use Report, 2006 Statistics. http://publications.gc.ca/collections/collection_2010/ec/En11-2-2006-eng.pdf.

Environmental Commissioner of Ontario. (2015). Annual Report 2014/2015. Small Things Matter. http://docs.assets.eco.on.ca/reports/environmental-protection/2014-2015/2014_2015-AR.pdf.

European Environment Agency. (2013). Assessment of cost recovery through water pricing. https://doi.org/10.2800/93669

Griffin, R.C. (2016). *Water Resource Economics: The Analysis of Scarcity, Policies, and Projects.* Cambridge, MA: MIT Press.

Hanemann, W.M. (2006). *Water Crisis: Myth or Reality* (ed. P.P. Rogers), 61–91. Taylor and Francis Group.

Hardin, G. (1968). The tragedy of the commons. *Science* 162 (3859): 1243–1248.

Horbulyk, T. (2017). Water policy in Canada. In: *Water Policy and Governance in Canada* (eds. S. Renzetti and D.P. Dupont), 29–43. Cham: Springer International Publishing.

Howard, G. and Bartram, J. (2003). *Domestic Water Quantity, Service Level and Health.* Geneva: World Health Organization.

Kreutzwiser, R., de Loë, R.C., Durley, J., and Priddle, C. (2004). Water allocation and the permit to take water program in Ontario: challenges and opportunities. *Canadian Water Resources Journal* 29 (2): 135–146.

Lant, C. (2004). Water resources sustainability: an ecological economics perspective. *Water Resources Update* 127: 20–30.

Mitchell, B. (2017). The hydrological and policy contexts for water in Canada. In: *Water Policy and Governance in Canada* (eds. S. Renzetti and D.P. Dupont), 13–28. Cham: Springer International Publishing.

Mohapatra, S.P. and Mitchell, A. (2009). Groundwater demand management in the Great Lakes Basin – directions for new policies. *Water Resources Management* 23 (3): 457–475.

Morris, T.J., Mohapatra, S.P., and Mitchell, A. (2008). Conflicts, costs and environmental degradation – impacts of antiquated groundwater allocation policies in the Great Lakes Basin. *Water Policy* 10 (5): 459–479.

National Round Table on the Environment and Economy (NRTEE) (2011). *Charting a Course: Sustainable Water Use by Canada's Natural Resource Sectors.* Ottawa: NRTEE.

Natural Resources Canada. (2002). Political Map of Ontario. Reference Maps – Provincial and Territorial. http://ftp.geogratis.gc.ca/pub/nrcan_rncan/raster/atlas_6_ed/reference/bilingual/ont_new.pdf

OECD (2013). *Water Security for Better Lives*, OECD Studies on Water. Paris: OECD Publishing.

OECD (2017). *Water Charges in Brazil: The Ways Forward*, OECD Studies on Water. Paris: OECD Publishing.

Ontario Ministry of Environment. (2007). Water Conservation Charges Proposal. Report Number PIBS 6134e. www.ontla.on.ca/library/repository/mon/16000/272421.pdf

Ontario Ministry of Environment and Climate Change. (2017a). Regulation Establishing a New Water Bottling Charge. EBR Registry Number: 012–9574 released on January 18, 2017. www.ebr.gov.on.ca/ERS-WEB-External/displaynoticecontent.do?noticeId=MTMxNTQwand statusId=MjAxNzE3

Ontario Ministry of Environment and Climate Change. (2017b). Report Notice on the Regulatory Water Charges Review. EBR Registry Number: 013–2020 released on December 30, 2017. www.ebr.gov.on.ca/ERS-WEB-External/displaynoticecontent.do?noticeId=MTM0MTU1and statusId=MjA0MDkzandlanguage=en

Ontario Ministry of Environment, Conservation and Parks. (2018). Extending the moratorium on water bottling permits. EBR Number: 013–3974 released on October 30, 2018. https:// ero.ontario.ca/notice/013-3974

Ontario Ministry of Finance. (2016). Ontario Public Accounts Volume 1, 2016–17, pp. 1-3:1-11. www.ontla.on.ca/library/repository/ser/15767/2016-2017//V.1.pdf

Ontario Ministry of Natural Resources. (2014). Technical Backgrounder on Consumptive Use Coefficients: Reporting under the Great Lakes – St. Lawrence River Basin Sustainable Water Resources Agreement and Basin Mapping for Ontario. http://waterbudget.ca/consumptiveuse

Province of Ontario. (2004). Ontario Regulation: O.Reg. 387/04: Water Taking and Transfer. www.ontario.ca/laws/regulation/040387

Renzetti, S. (2005). Economic instruments and Canadian industrial water use. *Canadian Water Resources Journal* 30 (1): 21–30.

Renzetti, S. (2007). Are the prices right? Balancing efficiency, equity, and sustainability in water pricing. In: *Eau Canada: The Future of Canada's Water* (ed. K. Bakker), 263–279. Vancouver: University of British Columbia Press.

Renzetti, S. (2017). Water pricing in Canada. In: *Water Policy and Governance in Canada* (eds. S. Renzetti and D.P. Dupont), 201–212. Cham: Springer International Publishing.

Renzetti, S. and Dupont, D. (1999). An assessment of the impact of charging for provincial water use permits. *Canadian Public Policy* 25 (3): 361–378.

Rivers, N. and Groves, S. (2013). The welfare impact of self-supplied water pricing in Canada: a computable general equilibrium assessment. *Environmental and Resource Economics* 55 (3): 419–445.

Russo, T., Alfredo, K., and Fisher, J. (2014). Sustainable water management in Urban, agricultural, and natural systems. *Water* 6 (12): 3934–3956.

Statistics Canada. (2014). CANSIM Table 153-0079. Water use parameters in mineral extraction and thermal-electric power generation industries, by region. www150.statcan.gc.ca/t1/tbl1/ en/tv.action?pid=3810006701

Statistics Canada. (2016). CANSIM Table 153-0127. Potable water use by sector and average daily use for Canada, provinces and territories. www150.statcan.gc.ca/t1/tbl1/en/tv. action?pid=3810027101

United Nations. (2015). Transforming Our World: The 2030 Agenda for Sustainable Development. A/RES/70/1. https://sustainabledevelopment.un.org/content/documents/ 21252030%20Agenda%20for%20Sustainable%20Development%20web.pdf

Vandierendonck, M. and Mitchell, B. (1997). Water use information for sustainable water management: building blocks and the Ontario situation. *Canadian Water Resources Journal* 22 (4): 395–415.

Water Canada. (2016). Water Bottling is a Sideshow: Bigger Issues Need to be Addressed. www. watercanada.net/feature/water-bottling-is-a-sideshow-bigger-issues-need-to-be-addressed

6

The Role of Water Pricing Policies in Steering Urban Development

The Case of the Bogotá Region

Camilo Romero

Technical University of Berlin, Berlin, Germany

6.1 Introduction

In May 2012, the then-mayor of Bogotá, Gustavo Petro, announced that the water utility company Empresa de Acueducto y Alcantarillado de Bogotá (EAAB) would stop selling bulk water (hereafter block water) to the rural zones of nine municipalities in the Bogotá area. This decision prompted a political debate involving national and municipal regulatory agencies as well as municipal governments, which opposed the mayor's decision. Since the year 2000, the 20 municipalities that surround the Bogotá region have tripled their population, while reducing population density by 54% (IDOM 2018). Petro considered that restricting the circulation of block water was an appropriate measure to stop the urban sprawl and conurbation process, which were taking place across Bogotá and the urban municipalities of Chía, Cota, Cajicá, Funza, and Mosquera.

Researchers in Colombia have attributed urban sprawl and conurbation in the Bogotá region (seen in Figure 6.1) to a variety of factors. From an economic perspective, the urban sprawl of surrounding municipalities is the result of the "economic dynamism of the Capital city" (Rendón 2009). Researchers have also analyzed the role of transportation networks to explain how Bogotá merges with nearby urban areas (Moreno 2009). Other authors have focused on the role of land-use plans and land-use changes in favoring the appearance of urban land in the surrounding municipalities (Contreras 2017; Rincón 2014). As Contreras shows, the formulated land-use plans for Bogotá's neighboring municipalities were the result of poor planning, underpinned by the private ownership of urban land, caused by market pressures and the necessity to overcome housing deficits (2017). These land-use plans have provided numerous profitable capital investment opportunities (Colmenares 2007).

Other researchers have considered there to be a link between block water and the urbanization of Bogotá's surrounding municipalities; however, this connection is not yet fully understood and thus can only be considered speculative at this point in time (Caicedo 2016).

Environmental Policy: An Economic Perspective, First Edition. Edited by Thomas Walker, Northrop Sprung-Much, and Sherif Goubran.
© 2020 John Wiley & Sons Ltd. Published 2020 by John Wiley & Sons Ltd.

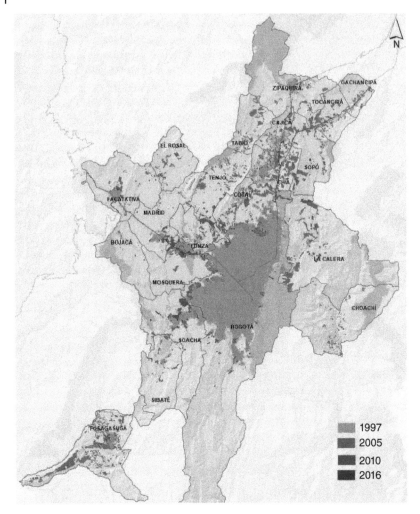

Figure 6.1 Expansion of the urban footprint in Bogota and 20 surrounding municipalities. *Source:* IDOM (2018).

This chapter therefore considers how water flows can help steer urban development, and what links exist between water flows and the appearance of particular urban patterns.

We will address these questions using the case of the Bogotá region and the controversy surrounding the use of block water, its pricing, and its trading policies. We will begin by presenting urban political ecology (UPE) as a theoretical framework for urbanization. We will then introduce the debate around block water in the Bogotá region, followed by a brief historical account of the water regulation framework in Colombia, and the multiple itera-tive political and economic processes that have shaped it. This framework has been the determinant factor for the production of two types of water in the Bogotá region: block water and public service water. We will discuss the pricing mechanisms, their underlying rationale, and the pricing policy regulations for the two water types. Finally, we will examine the relationship between water pricing, trading policies, and urban sprawl.

6.2   The Urbanization and Materiality of Water

From the perspective of UPE, urbanization is understood as a "dynamic socio-ecological transformation process that fuses the social and natural in the production of distinct and specific built environments" (Heynen et al. 2006). Urbanization can thus be conceptualized as a metabolism – a continuously circulating process of socioecological transformation. Hence, the city "can be understood through the analysis of the circulation of socially and physically metabolized 'nature'" (Kaïka 2005, cited in López 2015).

In this light, the metabolism is seen not as an anatomical analogy, but as a simultaneous and interweaving biophysical and social process. Water as a natural element is brought into the city, becomes standardized in cubic meters, is transformed into potable water, then waste, and is ultimately disposed of (López 2015). In addition to the biophysical transformation, water is also socially metabolized, becoming commodified and acquiring value. The commodification of water implies its insertion into the dynamics of capital accumulation and circulation. This double circulation of water and capital is sustained and organized by infrastructure networks, which are the "mediators through which the constant transformation of Nature into City takes place" (López 2015). Through this metabolic transformation, water is no longer seen as just a resource and a neutral object, but is referred to as "socio-nature" (Swyngedouw 2004). Urbanization is therefore seen as a metabolic process that produces new forms of urban nature and should be understood as a process of socioenvironmental change, rather than merely a sociotechnical change.

The circulation of metabolized water in any city is primarily embedded in broader processes of political and economic interest. When managing and governing natural resources, formal (institutional) and informal decision-making processes are marked by imbalanced power configurations. At the same time, these decisions are embedded within regulatory legal frameworks with various power configurations. Within UPE approaches, the concept of waterscape has been used to illustrate how water, power, and capital fuse together. It is then possible to analyze the circulation of metabolized water as a process which has both physical and social dimensions (López 2015).

By acknowledging the simultaneous dimensions of metabolized water, it is also easy to notice the heterogeneous materiality of water. The concept of materiality with regard to water, or nature, is used to frame water as a socio-natural configuration with simultaneous biophysical, spatial, social, and political dimensions (López 2015; López and Bruns 2018). Water is seen as an actor whose varied materiality shapes the relationship between social actors, and poses challenges and opportunities in different political-economic contexts (Sultana 2009). As its materiality affects a number of different dimensions, we can conclude that water flows affect a number of power relations, social practices, and behaviors – and in turn, are affected by unequal structures within these fields.

6.3   Methods and Scope

Following the theoretical framework of the UPE, water pricing policy can be understood as a important part of water materiality's social, political, and economic dimensions. Water pricing policies and contracts are formulated according to the conditions and technical bases established by a legal and regulatory framework. Through this framework, water acquires different social meanings, resulting in different political discourses and decisions. In turn, these meanings and discourses shape the institutional practices, discourses on urban development, and urban development policies of Bogotá's surrounding municipalities.

Using press notes from debates about block water in Colombia, we will use a qualitative approach to understand the social perceptions and implications of water pricing policies. We will also review various scientific articles on the subject to illustrate how water pricing policy, as a dimension of materiality, has affected discursive representation and social meanings.

6.4 The Debate About Block Water

The city of Bogotá, capital of Colombia, has a population of approximately 7 million inhabitants. It extends out onto a flat area known as the Savanna of Bogotá, where most of its surrounding municipalities can be found. Together, Bogotá and its surrounding municipalities account for one-fifth of Colombia's total population, with approximately 10 million inhabitants. While Bogotá has its own governing body, the surrounding municipalities belong to the department of Cundinamarca, and each has its own independent municipal authority. However, the region has no legal status, but instead is interlinked with the capital city and its surrounding municipalities by the circulation of resources, people, and goods.

One of these links is the trading of water, which takes place across Bogotá and its municipalities. This area's waterscape is made up of the metabolic circulation of different types of water, distinguished due to legal and technical factors. One of these types of water is block water, which is sold through the EAAB to nine neighboring municipalities; public service water, on the other hand, is sold by the EAAB to the inhabitants of Bogotá. The trade of block water dates back to the 1970s, when the EAAB received approval to sell the water to the urban areas of five municipalities (Chía, Cajicá, Mosquera, La Calera and Soacha). Eventually, the sale of block water spread to new housing estates and industrial complexes, located outside the urban areas.

The pronouncement made by Bogotá's then-mayor Gustavo Petro in May 2012 to suspend the sale of block water was supported by various arguments. Petro argued that the sale of block water was the driving force behind the urban sprawl of Bogotá's surrounding municipalities. Hence, restricting the circulation of block water would curb urban sprawl and the consequential destruction of rural and natural lands ("Enfrentados por el agua," 2012). He also argued that the sale of block water largely benefitted construction stakeholders, which profited from the expanding urbanization. Petro added that this decision was not only based on a short-term vision to change the urbanization pattern of the Bogotá region, but on a long-term vision to take care of the water and establish governance mechanisms for this resource ("Bogotá no venderá agua," 2012).

The initial pronouncement was reversed a couple of days later, and instead a decision was made to increase the price of block water. Although the price increase never came into effect, this controversy significantly affected the viability of construction ventures in the Bogotá area municipalities (Barreto 2014). In September 2012, the then-Minister of Housing, German Vergas, and the governor of Cundinamarca, Álvaro Cruz, stated that the mayor's decision would cause profound consequences in the development of the Bogotá region. Cruz added that Petro was "exceeding his powers when trying to order the territory of other municipalities" ("Enfrentados por el agua," 2012). Furthermore, Juan Manuel González, a lawyer who represented more than 40 construction companies affected by the decision, said that the "EAAB became the spearhead of Petro, who [sought] to impose his idea of urban development on the municipalities of the Savanna" ("La Guerra por el Agua," 2012). The Superintendent of Public Services (SPD) also initiated an investigation against the EAAB to assess whether the company had violated the law. On September 14th, 2012, the Superintendent of Commerce (SIC) imposed a fine on the EAAB for abuse of a dominant position in the market.

6.5   The History of Water Supply Regulation

The extensive legal framework that regulates the water supply in Colombia has evolved over the last 110 years. By examining the legal milestones for water supply regulation through a historical perspective, we can better understand the underlying political and economic processes that have shaped water supply regulation in Colombia. The evolution of the water regulations in Colombia fits into "well-known trends" of historical development for the water supply in high-income countries (Guerrero et al. 2016). This evolution can be divided into three periods: the privatization period after the mid-nineteenth century, the municipalization or nationalization at the turn of the twentieth century, and the decentralization and neo-liberalization of water utility services after the 1970s.

During the mid to late nineteenth century, the water supply in Colombia was entirely organized by private companies and characterized by limited amounts of water treatment and coverage. But as cities began to grow, this weak water system worsened, leading the Colombian national government to take control of the water supply. The changeover from private water utilities to government-run utilities was achieved through municipalization, a process that began in 1910 with the first water utility in Bogotá. With the establishment of Law 4 in 1923, the municipalization of water utilities was incorporated into the legal framework at a national level. This law gave more autonomy to local governments by allowing municipal councils to create water service administration boards. These boards were entitled to appoint a general manager, who would have full autonomy to organize and manage water utilities (Guerrero et al. 2016). However, local governments lacked the funds to buy private water utilities, which consequently had to be municipalized with the aid of the national government. Despite this, local governments still lacked the necessary resources for the provision and expansion of water infrastructure networks. Therefore, municipal councils were allowed to raise funds for the expansion of the water service, establishing fees, taxes, and extra fees for nonpayment. This funding strategy was also used by the municipal council of Bogotá, with the establishment of Law 72 in 1922. However, their funds remained insufficient, and expansions had to be financed via loans from international and domestic banks, which were only finally paid in 1968 (Guerrero et al. 2016). In Bogotá, the 1929 municipal Accord 25 gave banks, together with the mayor and city council, the power to appoint the management board of the water utility company. The same accord introduced strategies for full cost recovery, cross-subsidization, and volumetric metering to curb high consumption.

After the 1936 constitutional reform, water tariffs were subject to the approval of the national government, whose aim was to prevent utility companies from charging excessive fees in contravention of "society's collective well-being" (Guerrero et al. 2016). In 1953, the Health Code was issued, which established procedures for water treatment and purification. That same year, the Ministry of Public Health started regulating water quality and approving the development of the water infrastructure. Water utilities also had to keep a record of levels of chlorination, bacteriological analysis, and consumption data. In 1954, a group of businesspersons led by the National Business Association successfully lobbied the National Congress for the approval of Legislative Act 5 (Guerrero et al. 2016). This act helped create a new legal concept for the management of public services, which defined water utilities as public and autonomous entities with an independent legal status. In 1955, the 1816 Decree ensured technical and apolitical management for water utilities. In the aftermath of these reforms, the multi-utility companies Empresas Públicas de Medellín (EPM) and the EAAB were founded.

In 1959, Law 155 established the basis for competition law in Colombia, appointing the Ministry of Development and the Superintendent of Economic Regulation (SRE) to enforce its mandates. For instance, the SRE ensured adequate funding for water supply extension by regulating prices to ensure cost recovery (Guerrero et al. 2016). This changed in 1968, when the SIC was created and, as a result of Legislative Act 1, the National Tariffs Board (JNT) was formed. The SIC was then charged with enforcing competition and antitrust law and monitoring pricing policy, and the JNT assumed the SRE's responsibilities in regulating water tariffs and ensuring cost recovery. In the following years, a series of legal regulations further changed the landscape for water supply. The Legislative Act 1 of 1968 established a national cross-subsidy system based on property value. In 1974, Law 2811 introduced environmental fees as part of water tariffs. Finally, in 1974, the 2811 Decree defined water as a public domain natural resource, strengthening users' rights to have access to water; this decree also established the concession as the agreement granted to water utility companies to supply water to customers.

After the 1970s, water utility services transitioned from municipalization and nationalization to decentralization and neo-liberalization. The constitutional reform of 1991 made public services subject to free-market competition, guaranteeing both private and public operators the same market access conditions. In 1993, Law 60 authorized municipalities to privatize the water supply, and in 1994, Law 142 required cities wishing to retain public ownership to justify their choice (Guerrero et al. 2016). One of the main principles of Law 142 was to ensure economic cost recovery, improve efficiency, and increase competition; it thus created a national legal framework for the operation of domestic public services – namely water, sewage, garbage, electricity, gas – under the principle of market competition. It also created two new regulatory agencies: the Commission for the Regulation of Water and Basic Sanitation (CRA), which replaced the JNT. These new agencies became responsible for defining cost recovery methodologies, water pricing, ensuring efficiency, evaluating performance, and approving tariffs. The Superintendent of Household Public Services (SPD) was also created and became responsible for monitoring the public services sector and protecting consumers.

In 1995, a new pricing policy was put in place for water distribution. This policy resulted in users having to pay for the costs of capturing, transporting, and delivering water – cost-reflective prices – under the regulation of the SIC, the CRA, and the SPD. That same year, under Law 142 and District Agreement Number 6, the EAAB adopted commercialization as a new institutional model, transforming it from a municipal company to a state-owned industrial and commercial company. As a result of this transformation, EAAB remained a public company but was regulated by private law.

Though the legal framework for water supply in Colombia changed considerably during this time, the regulatory framework was not significantly modified until June 2012, when the debate about block water started. Indeed, Law 142 did not envisage clear regulations for the procurement and price setting of block water. In fact, since the 1980s, block water trading had not been regulated through the formal regulatory structure and institutional arrangements for public services, but through private law contracts between the EAAB and Bogotá's municipalities. There were thus no guidelines for the transportation, treatment procedure, and disposal site of block water; it was disposed of without treatment in rivers, lagoons, and wetlands (Barreto 2014).

However, in 2012, the sale of block water – as well as being regulated by private law – became subject to regulation via the SPD and CRA. The latter issued Decree 608 to define the price methodology for the sale of block water. Furthermore, through the 2012 Resolution 53 992, the SIC perceived the sale of block water as a submarket within the market of "public water utilities" (Barreto 2014). This resolution meant that, for the EAAB, two relevant markets existed: the market for the provision of public service water, which was delivered directly to users, and the submarket for the provision of block water, which was delivered to an intermediary who then transformed it into public service water, before finally delivering it to users.

6.6 The Politicization of Water Supply Regulation

Throughout the evolution of water supply regulation in the Bogotá region, various processes have shaped water's governance. As we have seen, these processes – namely privatization, commercialization, centralization and decentralization – have ultimately metabolized the water flows in the Bogotá waterscape into two types of water: block water and public service water.

Processes such as commercialization were present in Colombia from the municipalization period at the start of the twentieth century and not only from the 1970s neo-liberalization, as is often indicated (Guerrero et al. 2016). The commercialization of water services was characterized by the introduction of autonomy principles and cost recovery strategies. Instituted by the creation of water service management boards, autonomy principles sought to assure the independence of water utilities from local governments and avoid the "politicization of the water supply" (Guerrero et al. 2016). Full cost recovery measures, on the other hand, were embodied by the introduction of volumetric metering; water prices started to reflect the full costs of the water infrastructure, and consumers started to pay for the quantity they used. Charging cost-reflective prices was assumed to be an effective measure in reducing wasteful consumption behaviors. Thus, water was metabolized from an abundant and public good into a scarce and economic good. The ideological representation of water as a scarce resource has been signaled as the foundation for market mechanisms in the water supply sector (Swyngedouw 2004). Furthermore, as a result of the financial dependency of water utility companies on banks caused by loans for infrastructure expansion, water pricing policies were set in place to ensure these loans were reimbursed. The commodification of water during this time then encouraged water utility companies to increase profits and shift their focus away from the public good.

Private sector participation in water governance was contingent upon this redefinition of water from a public resource to a marketable commodity, and the creation of administrative boards – which later lead to the decentralization of the water supply sector. Indeed, water management boards never obtained autonomy from bankers, as bankers started to appoint their own representatives to the administrative boards (Guerrero et al. 2016). Their introduction intensified the power relations between banks and water utility companies, orientating Bogotá's water utility operations to simultaneously provide a public service while meeting the needs of financiers. Contrary to the spirit of the 1923 Law 4, which sought to avoid the politicization of water supply, this process of administrative decentralization inherently politicized the water supply system. Therefore, the creation of water management boards has been interpreted as the effort of private sector actors to redefine the geographical scope and participants of the decision-making process to be more in line with their interests (Budds and Hinojosa 2012). This also reveals how the water sector in Colombia was not only governed by the state, but was already subject to particular, market-focused forms of governance, in which nonstate actors played a crucial role.

However, water supply governance during the municipalization period was centered not only on market notions, but also on the promotion of social goals. After the National Health Code was established in 1953, the Colombian state played a central role in centralizing the regulation process for the water sector by securing social rights and public health protection, especially as environmental protection and public health fees influenced water prices. For instance, the Ministry of Public Health started to control water quality and approve water infrastructure expansion.

The bond between water provision and public health during this period was further emphasized with the development of the ideal sanitized city and the implementation of a municipal hydraulic water management paradigm. The idea of a sanitized city emerged in response to concerns over increasing health problems and deteriorating environmental conditions. The municipal hydraulic paradigm was the approach introduced by most countries to manage water supply systems at the beginning of the twentieth century (López 2015).

The reforms of 1954 and 1955, which led to the creation of the modern water utility companies like the EPM and the EAAB, intensified the administrative decentralization of the water sector. Business associations actively lobbied for these reforms under the arguments of autonomy, "depoliticization," and the technical management of water utilities. The administrative decentralization of the water sector can thus be understood as a process carried out and influenced by the political agendas and interests of powerful groups and business associations.

Through national institutionalization, the constitutional reform of 1991 also played an essential role in the process of neo-liberalization in Colombia. This reform legitimized the profit-seeking intentions of private initiatives providing public services. In 1994, Law 142 institutionalized the neo-liberal principles of commercialization, administrative decentralization, and regulatory centralization for private and public water operators on a national scale. The creation of new national regulatory agencies, like the CRA, repositioned the state on a national scale, primarily through being a regulator rather than an operator of the market. This repositioning is contrary to the often-repeated mantra that neo-liberalization weakens state power by delivering public services and contributes to the deregulation of markets; neo-liberalization in water supply should instead be understood as a re-regulation process, initiated and led by the state (López 2015). As a result, water utilities that remained publicly owned, such as the EAAB and the EPM, were required to behave like private water operators, adopting commercial principles (such as cost recovery, competition, and efficiency), commercial methods (such as cost–benefit assessments and performance contracts), and commercial objectives (such as profit maximization) (López 2015).

The SIC's interpretation of the sale of block water as a submarket within the public services market in 2012 implied the re-regulation of this economic activity. The sale of block water, previously regulated exclusively by private law, was then simultaneously regulated by both private law and the formal regulatory structure for public services in Colombia. This meant that governmental institutions such as the CRA, SIC, and SPD were involved in its regulation. Unfortunately, the re-regulation of the sale of block water stemmed from a narrow understanding of regulation, as the SIC exclusively conceptualized block water as a marketable commodity, ignoring its other dimensions as a public good, a fundamental right, a collective right, a resource granted in concession, or an organizer of the territory (Barreto 2016). Therefore, the social and environmental dimensions of water were not considered as aspects that required regulation.

As we have seen, water supply regulation has been underpinned by the simultaneous and iterative processes of commercialization, administrative decentralization, and regulatory centralization. Throughout the regulatory framework's evolution, the influence of nonstate actors such as the banks and business associations is striking – evidence of the unbalanced power relations through which water in Colombia has been governed. As a result of their influence, water governance has been arranged around market notions. The social goals through which water was governed during the municipalization period were overcome by the economic and commercial goals after the beginning of neo-liberalization (Guerrero et al. 2016). This shift in objectives also implies that the organization of water infrastructure has prioritized the political and economic interests of nonstate actors, therefore creating social and spatial inequalities.

6.7 Water Pricing in the Bogotá Region

The metabolic process of water in the Bogotá region has been embodied by the production of two types of water – block water and public service water – separated into two markets with distinct legal, technical, and procurement processes and, most notably, distinct pricing policies. These are set according to the cost recovery methodology established by the CRA, wherein water is delivered to customers at a cost-reflective price. The methodology is based on reference costs, which are founded on a regular fixed rate and a pay-per-use rate. The regular fixed rate includes the average cost of administration; the pay-per-use rate includes average operation (transportation), investment, and environmental costs.

Public service water prices also vary due to a cross-subsidization scheme. This scheme, which was introduced during the municipalization period to reduce unequal access to public services, only operates in residential areas. The subsidized amount of water varies according to the social strata of the residential area, ranging from Strata 1, which is the lowest, to Strata 6, which is the highest. Users from the wealthiest strata, located on land classified as Strata 5 and 6, subsidize the water usage for users from low socioeconomic strata (Strata 1, 2, and 3). Hence, users from Strata 5 and 6 are charged more than the average cost. Strata 4 remains neutral, and consumers do not pay more than the marginal costs of water provision.

As for block water, before 2012, prices were freely agreed between the EAAB and the water operators of municipalities which received block water, abiding only by the market conditions (Barreto 2014). The construal of block water sales as a submarket within the market of public water utilities motivated the CRA to issue Decree 608 in 2012, which stipulated its contractual activity and price-setting methodologies. Thus, the sale of block water came to be regulated by interconnection contracts (block water contracts) and not by water supply contracts, as was the case with public service water. The CRA also instituted the key components that should be considered when entering into such a contract, which included the water prices and connection points to the network, among others. The price of block water was therefore regulated according to the CRA's cost recovery methodology which, in the case of block water, included the average operation (transportation) and investment (network infrastructure expansion) costs. It also added an interconnection cost, defined as a toll payment to access the EAAB's water network. Environmental costs, however, were not taken into consideration. Barreto mentions that conceptualizing block water as a market, as the SIC did, constitutes a significant oversight, as it denies the multiple materialities of water beyond its economic dimension (2014).

6.8 The Rationale of Water Pricing Mechanisms through Environmental Fees

The environmental fees included in the cost-reflective prices of water are in place to make up for any environmental impact caused by water users and companies. These costs include two types of fees: a retribution fee for water pollution and a water usage fee that reflects the environmental impact on water sources. Environmental fees, also referred to as green taxes, were established with the intention of offering a "double dividend" (Rudas 2008). This means that, firstly, environmental fees can help to improve the environment by creating and supporting projects that conserve water sources and by stimulating a more rational behavior in users. Secondly, these fees can generate resources to be used by the environmental authorities.

Water prices are ultimately differentiated by a spatial parameter like land use, a decisive dimension when it comes to environmental regulation. When setting a water pollution fee, it is assumed that industrial zones dispose of more contaminated water than residential zones do. Hence, environmental fees vary according to land use, and are typically higher in industrial zones. As a result, public service water prices for residential users in Strata 6 are higher than in Strata 1, especially for the users who reside on industrial land (Rudas 2008).

Environmental fees – concerning either the amount of water used to encourage a conscious consumption or the amount disposed to reduce the pollution of rivers – are driven by the neo-classical economic paradigm of perfect rationality, called *homo economicus*. This paradigm assumes that market subjects are endowed with a universal principle of rationality and always act to maximize their utility. Thus, the behavior of individuals is assumed to align with specific goals or desirable behavior. Policymakers assume that the environmental fees send price signals to consumers and drive them to use natural resources more sparingly, compared to consumers who can access natural resources free or at a lower price, who are going to being more wasteful. In Bogotá, environmental fees for water contamination and water usage are higher for residential users from the wealthiest strata. According to the *homo economicus* paradigm, one would expect these fees to curb water consumption. However, the real behavior of these users challenges the notion of the "double dividend" and the effectivity of the environmental fees; despite the elevated prices of the public service water, the wealthiest strata consume more water than lower-income groups ("Salvar el agua" 2014).

6.9   Water Pricing Mechanisms and Urban Sprawl

Before 2012, pricing policies for block water were not subject to the formal regulation mechanisms for the provision of public services, namely from the SIC and the SPD. Block water trading policies were freely set between the contracting parties, the EAAB, and the water operators in the surrounding municipalities, regulated only by private law. In fact, in 2012, Comptrollership found that block water had been sold at equal prices in all nine municipalities. The EAAB's pricing policy for block water did not consider the varying transportation costs to the different municipalities, nor did it consider the final users to which the water was being delivered. These particularities in setting block water prices were caused by the weak regulation of private law and acted against the economic principles of efficiency and cost recovery.

The agreements between the EAAB and the municipalities established a lax control on prices and loose stipulations on the traded flow rates of block water. Based on their contracts, the municipalities supplied with block water formulated their own regulations to encourage urban development. The agreement between the EAAB and the municipality of Chia illustrates this point, as the EAAB agreed to supply the required block water flow rates to Chia without restriction. In other words, Chia could ask for block water upon demand from its inhabitants, activities, and industries, and the EAAB would provide it. Comptrollership concluded that based on this premise, the municipality of Chía promoted its urban development by delivering construction licenses which guaranteed access to water provisions (2012). The municipality of Chia adopted tacit water provision for new housing estates as an institutional practice, even if the estates were constructed outside the urban area or other perimeters. Combined with lax control on other laws and building regulations, this practice provided a big incentive for construction stakeholders to move their activities to Chia and other municipalities in Bogotá's surroundings with similar provisions. As a result of this movement, Bogotá's savanna landscape became characterized by gated communities and scattered housing estates.

In 2012, then-mayor Gustavo Petro interpreted the heightened construction activity and the extension of Bogotá's surrounding municipalities' urban areas as city encroachment onto rural land, blaming the sale of block water as the driving force for the urbanization of the Bogotá region. Public authorities viewed block water prices that favored construction activity as the catalyst for urban sprawl, and perceived the block water trade as only serving the interests of construction stakeholders (Barreto 2014). In addition, Petro argued that the increased urbanization in Bogotá's surrounding municipalities and the resulting heightened water consumption jeopardized Bogotá's water supply. Therefore, water was represented as a scarce resource that was being threatened by the heightened demand caused by new housing estates. This exemplifies the several social meanings and discursive representations of block water's materiality; weak regulations for block water prices gave rise to particular social meanings of water, which in turn shaped urban regulations and discourses on urban development.

The representations and meanings assigned to block water constituted the first step in creating a policy narrative, framed by discourses on managing industrial and urban growth (Horn 2019). This policy narrative, therefore, represented the urban sprawl of Bogotá's municipalities as a public problem, and hoped to likewise justify the need for public action. Block water prices created incentives for particular building regulations, which in turn encouraged the construction activity and urban sprawl. At the same time, water prices were also identified as a tool to limit scattered urban development. Nowadays, urban growth management consists of containment policies or growth boundaries, limiting sprawl and encouraging higher-density urban development. These policies are classified into three main groups: greenbelt policies, urban growth boundaries, and urban service boundaries (Pendall et al. 2002).

As the policy narrative gained traction, the EAAB initially decided to suspend the sale of block water to stop urban expansion. This approach, which can be regarded as a "command-and-control regulation," would create an urban service boundary beyond which block water was not provided. In contrast, the decision to increase block water prices aimed to introduce into the water market an economic tool for addressing the environmentally harmful impacts of urban sprawl. Heightened water prices meant changing the basis of the municipal and building regulations which had enabled low-density and scattered urban development in Bogotá's municipalities. Thus, municipal authorities and construction stakeholders face a new contractual reality – namely, the impact of increased block water prices on extant policies.

However, the EAAB's unilateral decision to increase block water prices is not to be confused with an environmental fee meant to account for the impact of urban sprawl, though the underlying rationales are certainly similar. The increase in price was guided by the idea that environmental protection is best achieved by incorporating environmental externalities, applying price mechanisms to nature services, and trading these services within the market. This policy orientation has been referred to as market environmentalism (López 2015). Thus, the creation of policies and pricing mechanisms for protecting rural land in the Bogotá region was underpinned by the representation of water as a scarce resource which, when provided at low to no cost, could trigger negative environmental effects.

6.10 Conclusions

The decisions made by the EAAB in 2012, encouraged by the then-mayor Gustavo Petro, to cut off and then to increase block water prices were an attempt to bring the behaviors of different political-economic agents in line with new and emerging economic paradigms. This new paradigm has been typified as the shift to a low-carbon economy to mitigate the harmful effects of human activity. In other words, Bogotá's mayor sought to impose a new mode of social and economic regulation to reduce the harmful environmental effects of urban sprawl.

In formulating policies and analyzing their constitution, water and nature can no longer be framed as static objects subjected to human manipulation. As shown in this case study, water pricing and trading policies appear as the basis for formulating urban development policies, despite being frequently subsumed and neglected by urban development. This oversight has also affected water utility companies, which have been subordinated to private and political interests. As shown in the case of the EAAB and block water trading, the EAAB was for many years basing its operations on the urban development plans of Bogotá's municipalities and, later, on the development policies of Bogotá's mayor. Acknowledging the impact of water pricing and trading in urban development highlights the potential role that water utilities can have in both the development and regulation of urban growth. Therefore, it is necessary to strengthen the processes of administrative decentralization of the devolution type.

Water utilities must also act to ensure water is recognized as a right for all, and thus should regain the autonomy to represent the interests of citizens and construct a notion of public good. However, this must be encouraged by the regulatory framework and the notions used to conceptualize the trade of water. In addition, the debate about block water raises questions regarding the limits of a water utility company with regard to geographically expanding its economic activity. These limits are currently established by the private law that regulates the operation of water companies in Colombia.

Moreover, an extensive network of actors, including independent government agencies, water utilities, and state regulatory agencies, play an essential role in regulating trading and prices, ultimately establishing the conditions for urban policy making. These policies are underpinned by an understanding of water as a commodity, as well as the regulation of water utility companies under antitrust laws. Furthermore, the regulatory framework that sustains water pricing policies reveals how juridical systems appear to regulate not only individual social behavior but also political life. The efficacy of an environmental pricing mechanism on urban sprawl can be called into question if we consider the effects of environmental water fees in the Bogotá region. These environmental fees aim to regulate individual behaviors but have not prompted any behavioral change among Bogotá's wealthiest socioeconomic strata. We can thus conclude that the current pricing mechanism maintains social inequalities with regard to access to water.

Through this use of a water pricing mechanism, the debate on block water in Bogotá provides a benchmark attempt at controlling urban sprawl by introducing environmental pricing on land. This environmental pricing seeks to regulate institutional practices, as well as the entrepreneurial activity of construction stakeholders. This case study highlights how environmental pricing mechanisms can be useful when regulating entrepreneurial activities that are harmful to the environment, as entrepreneurial stakeholders – such as construction companies – will more willingly respond to economic incentives and surrender to the hegemony of calculating economic rationality. Therefore, water pricing policies and environmental fees for managing urban growth are useful tools for incentivizing proper resource management, such as land and water, by entrepreneurial stakeholders.

References

Barreto, A. (2016). La Estructura Reguladora del Servicio Público Domiciliario de Aguas, Más Alla del Enfoque Mercantil. In: *El Estado Regulador en Colombia* (eds. H. García and E. Lamprea), 75–97. Bogotá, D.C.: Universidad de los Andes.

Bogotá no venderá agua en bloque a municipios aledaños [Bogotá will not sell block water to surrounding municipalities] (May, 2012). El Espectador: www.elespectador.com/noticias/bogota/articulo-348339-bogota-no-vendera-agua-bloque-municipios-aledanos

Budds, J. and Hinojosa, L. (2012). Restructuring and rescaling water governance in mining contexts: the co-production of waterscapes in Peru. *Water Alternatives* 5 (1): 119–137.

Caicedo, S. (2016). Agua y Territorio en la Bogotá Humana [Water and Territory in the la Bogotá Humana]. Bogotá, D.C.: Editorial Bonaventuriana.

Colmenares, R. (2007). El Agua y Bogotá: un panorama de insostenibilidad [Water and Bogotá: an unsustainable panorama]. Bogotá, D.C.: Foro Nacional Ambiental: Documento de políticas públicas.

Comptrollership of Bogotá. (2012). Evaluación de los efectos generados por las decisiones tomadas por el gobierno distrital frente a la suspensión de la venta en agua en bloque [Evaluation of the effects generated by the decisions taken by the district government on suspending the sale of water in block] www.contraloriabogota.gov.co/sites/default/files/Contenido/Informes/Auditoria/Direcci%C3%B3n%20Sector%20H%C3%A1bitat%20y%20Ambiente/PAD_2012/CicloII/EAAB%20-%20AGUA%20EN%20BLOQUE.pdf

Contreras, Y. (2017). Estado de la vivienda y del espacio público en el municipio de Chía [State of housing and public space in the municipality of Chía]. Colección Ciudades, Estados y Política. Editorial Universidad Nacional de Colombia, Bogotá D.C.

Enfrentados por el Agua [Confronted for the water] (2012). El Espectador. www.elespectador.com/noticias/bogota/enfrentados-el-agua-articulo-352734

Guerrero, T., Furlong, K., and Arias, J. (2016). Complicating neoliberalization and decentralization: the non-linear experience of Colombian water supply, 1909–2012. *International Journal of Water Resources Development* 32 (2): 172–188. https://doi.org/10.1080/07900627.2015.1026434.

Heynen, N., Kaïka, M., and Swyngedouw, E. (eds.) (2006). *In the Nature of Cities: Urban Political Ecology and the Politics of Urban Metabolism*. London: Routledge.

Horn, A. (2019). The history of urban growth management in South Africa: tracking the origin and current status of urban edge policies in three metropolitan municipalities. *Planning Perspectives* 34: 959–977. https://doi.org/10.1080/02665433.2018.1503089.

IDOM (2018), for BDPS (Bogota District Planning Secretary). Estudio de crecimiento y evolución de la huella urbana para Bogotá [Study of growth and evolution of the urban footprint for Bogotá]. www.sdp.gov.co/sites/default/files/4-DOCUMENTO-TECNICO-DE-SOPORTE/Estudio%20de%20Crecimiento%20de%20la%20huella%20urbana%20de%20Bogota%20y%20La%20Region.pdf

La Guerra por el Agua [The war for Water] (2012). Revista Semana. www.semana.com/nacion/articulo/la-guerra-agua/265610-3

La Sabana de Bogota [The Bogota Savanna] (2018). El Tiempo. www.eltiempo.com/colombia/la-sabana-de-bogota-foco-de-grandes-proyectos-urbanisticos-de-amarilo-187832

López, M. (2015). Contested Urban Waterscape: Water, Power and Urban Fragmentation in Medellin, Colombia (Doctoral dissertation). https://refubium.fu-berlin.de/bitstream/handle/fub188/10320/Lopez_Marcela_diss.pdf?sequence=1&isAllowed=y

López, M. and Bruns, A. (eds.) (2018). *Urban Water Management: A Critical Handbook*. Trier: Trier University.

Moreno, A. (2009). El proceso de conurbación Bogotá- Soacha a través del estudio de la movilidad: análisis del comportamiento del transporte público en el corredor de la autopista sur como eje de integración regional [The Bogotá-Soacha conurbation process: study of mobility. Analysis of public transport behavior in the south highway corridor as a regional integration axis]. Master's thesis, Pontificia Universidad Javeriana. https://repository.javeriana.edu.co/handle/10554/218

Pendall, R., J. Martin, and W. Fulton. (2002). Holding the line: urban containment in the United States. Center on Urban and Metropolitan Policy, Brookings Institution, Washington, D.C.

Rendón, J. (2009). Industrialización y dinámicas espaciales en Bogotá: Las urgencias de la gestión territorial [Industrialization and space dynamics in Bogotá: The urgencies of territorial management]. *Semestre Económico* 12 (24): 93–112.

Rincón, A. (2014). Análisis de la expansión Urbana del municipio de Facatativá desde las políticas de ordenamiento territorial en el periodo 2002-2011 [analysis of the urban expansion of the municipality of Facatativá from the land use policies in the period 2002-2011]. *Perspectiva Geográfica* 17: 123–146. doi:10.19053/01233769.2265.

Rudas, G. (2008). Instrumentos Económicos en la Política del Agua en Colombia: Tasas por el uso del agua y tasas retributivas por vertimientos contaminante [Economic Instruments in Water Policy in Colombia: Fees for water use and remuneration rates for polluting dumping]. www.cepal.org/ilpes/noticias/paginas/6/40506/8_Rudas_2008_Instrumentos_Economicos_Politica_Agua.pdf

Salvar el agua [Save the water] (2014). El Tiempo. www.eltiempo.com/Multimedia/especiales/salvar_agua_bogota

Sultana, F. (2009). Community and participation in water resources management: gendering and naturing development debates from Bangladesh. *Transactions of the Institute of British Geographers* 34 (3): 346–363.

Swyngedouw, E. (2004). *Social Power, and the Urbanization of Water: Flows of Power*. Oxford: Oxford University Press.

7

Effective Environmental Protection and Regulatory Quality

A National Case Study of China

Sharanya Basu Roy

University of Derby, Derby, UK

7.1   Introduction

Since the initiation of market-oriented reforms in 1978, the growth rate of the Chinese economy has been unprecedented. Economic growth increased production but unfortunately it also increased waste creation, air pollution, and noise pollution. Since the 1990s, China's environmental health has deteriorated steadily which has jeopardized public health and the safety of millions of people (Sun et al. 2013).

Water is one of the natural resources which has faced acute degradation in China. For instance, more than three-quarters of China's surface waters in urban areas are considered unsuitable for drinking and fishing (Turner 2007). In 2009, the country's Ministry of Environment Protection (MEP) declared that approximately 42.7% of all rivers are unsuitable for drinking and fishing (MEP 2009).

China's water pollution control is basin and region based, from both a policy and governance perspective (Shen 2009). Therefore, for the purpose of water resources management, rivers in China are categorized into nine major river basins. As water-related issues and problems are increasingly recognized in China, the river basin approach is gradually being acknowledged as the most appropriate approach for water resources management.

7.2   Understanding China's River Basin Management Policy and Governance Gaps

7.2.1   Policy Gaps

China's history of water legislation is fairly recent and the laws governing the river basins are even newer. The fundamental law dealing with water resources management in China is known as the country's first Water Law, which lays out the river basin management structure's legal framework. The earlier version of the Water Law (1988) did not define river basin management clearly, but it was later amended in 2002, where it established

Environmental Policy: An Economic Perspective, First Edition. Edited by Thomas Walker, Northrop Sprung-Much, and Sherif Goubran.
© 2020 John Wiley & Sons Ltd. Published 2020 by John Wiley & Sons Ltd.

the basics of the river basin management system[1] including the planning, allocation, saving, and protection of water resources (Shen 2009). Furthermore, Article 12 established that river basin management should be based on the basin and regional boundaries. This law delegates the primary responsibility for supervision and administration of water resources to the Ministry of Water Resources (MWR) under the State Council; it also holds the provincial water resources protection agencies accountable for the monitoring of water resources (Art. (26) Water Law of the People's Republic of China (2002)). Unfortunately, numerous rules and regulations devised under the previous Water Law have not been revised or updated in the recent Water Law, resulting in confusion over administrative responsibilities (Wouters et al. 2004).

The Water Law is often criticized for determining "economic growth" as the principal reason to establish water resources management (He 2016). It is also criticized due to its failure in reducing water pollution, improving water shortage, and providing high-quality water governance (Zhou et al. 2008). One of the main reasons attributed to the inability of the Water Law to improve the water quality and governance in China is the "dual management" model of water pollution control and water utilization.[2] This inability is further exacerbated by the country's obsolete legislative philosophy and technique, whereby only general principles and frameworks are provided for water resources management. China's Water Law lacks instruments and procedures for implementation, and clear definitions of critically important terms; for instance, it has failed to define "water right" clearly. Consequently, some decipher "water rights" as the right to own and use water while others refer to it as an authorized acquisition of the property right to use ground or surface water (Zhou et al. 2008). The flawed separation of responsibilities and the ambiguous language employed among various levels of government (River Basin Commissions (RBCs)) as well as the water administrative authorities have further resulted in numerous overlaps and vacuums (Wouters et al. 2004).

When comparing laws and regulations at the national and local levels, progress in terms of legislation at the river basin level is lagging. The Tarim Basin Water Resources Management Regulation[3] is the only comprehensive legislation on river basin management among the seven river basins. Consequently, the RBCs' role in river basin management is limited (People's Republic of China Changjiang Water Resources Commission 2004). The idea for a "river basin law" has been widely advocated by legal scholars to improve administrative efficiency and cooperation at the local level (Boxer 2001). However, even after many years of effort, no formal legislative progress has been made because of the complex nature of establishing a comprehensive river basin law.

At the sub-basin level, on the other hand, after a man-made ecological disaster in the Taihu Lake and criticism that followed that disaster, the Taihu Lake Basin Management Regulation (TBRM 2011) was enacted as a river basin law (He 2016). It took 10 years of negotiation and research on river basin regulation before enacting the law in 2011 (He 2016).

The TBRM states that the objective of regulation is to protect water resources, prevent water pollution, and improve the ecological environment of the Taihu Lake Basin (Art.(1) Regulation on the Administration of the Taihu Lake Basin (2011)). This is appreciably different from the Water Law, where water resources are considered only as a tool for promoting economic development (Art.(1) Water Law of the People's Republic of China (2002)). Though this new regulation fails to provide fully integrated river basin management and a consolidated administrative management regime, it advocates a more collaborative mechanism in terms of administration of the Taihu Lake Basin (Art.(4) Regulation on the Administration of the Taihu Lake Basin (2011)).

Ultimately, the Taihu Lake Basin Management regulation law represents a crucial shift from a fragmented water management toward an integrated river basin management system (He 2016).

7.2.2 Governance Gaps

As mentioned before, China has adopted a dual environmental management system. This system separates the water-related and nonwater-related responsibilities into two divisions, where the MWR handles the water-related responsibilities and the MEP takes care of other environmental management responsibilities. This division is done in an attempt to improve coordination and supervision.

China's river basin approach is a very complicated multilevel administrative system as it includes various levels of local governments and water authorities (at the county, provincial, and prefecture levels). Other authorities such as the MEP and the Ministry of Agriculture (MOA) are responsible for their respective sectoral interests. The present institutional setting for water management can be described as "vertically fragmented and subject to primarily sectoral management" (He 2016).

Responsibilities are distributed among different levels of government and water authorities, ranging from the central to regional and local levels. For river basins and lakes of national significance, the central government plays a dominant role in the centralized administrative system. At the national level, the MWR is responsible for preparing water plans and issuing water abstraction permits. The 2002 Water Law declares the water resources of China to be state owned (Art. (3)), therefore as per the laws and regulations of the MWR, RBCs are under state administration and are responsible for water resources management and supervision (Art. (12) Water Law of the People's Republic of China (2002)). The RBCs, in turn, are in charge of implementing these water plans and laws at the basin level. They are also provided with certain powers with respect to establishing key plan elements for sub-basins and transprovincial tributaries along with the provincial water authorities.[4]

At the local level, where local governments are responsible for the development of water resources, the RBCs are required to collaborate with local governments. Similarly, provincial water authorities are responsible for water resources development at the provincial level under the MWR's professional and technical guidance. However, due to lack of administrative hierarchy between RBCs and provincial water authorities (and ambiguous language in the 2002 Water Law defining the RBCs' responsibilities), it is challenging for RBCs to get involved in water management at the local level. Therefore, this entire process is far too dependent on voluntary collaboration, which hinders the RBCs' interests in managing water resources for the entire basin. In addition, as water authorities are part of the local government, they are highly influenced by the corresponding local government decisions as the local government provides financial resources and, more often than not, economic growth takes precedence over pollution issues when allocating funds at the local level (He 2016).

The powers provided to the RBCs are inadequate. Firstly, they are not provided with the authority to control pollution at source. For instance, though the Hai River Basin

Management Commission is responsible for interprovincial flood control, the responsibility for operative control lies with the county, prefecture, and provincial governments in most cases. Secondly, the RBCs do not have any authority over economic or administrative issues. This is because various levels of local authorities are responsible for water resources management along with the RBCs (He 2016) and the latter lies at the lowest tier of the hierarchical administrative structure.

Besides the issue of RBCs' lack of powers, there exists inadequate coordination in water management among the environmental protection agencies, provincial water resources bureaus, and other agencies involved (Dawei and Jingsheng 2001). This situation is further complicated as laws, policies, and regulations adopted by the central government face difficulties in being implemented at the local level. Furthermore, factors such as regional conflicts relating to costs of storing and transferring water, free-riding on the benefits, bureaucratic inadequacies, corruption, and discrepancies in terms of capacity all threaten to derail the water quality maintenance initiatives. Interjurisdictional rivalries also influence water-related authorities' decisions where the needs of individual jurisdictions are given preference over the interests of the basin as a whole (Moore 2013). These factors and the lack of clear governance have undermined China's efforts in implementing its water management plans (Moore 2013).

The institutional system at the horizontal level is referred to as the "Nine Dragons Governing the Water" (Yan et al. 2006). The nine dragons are the MWR, MEP, State Oceanic Administration, Ministry of Housing and Urban and Rural Construction, Ministry of Finance, MOA, Ministry of Land and Resource, Ministry of Transportation, and State Forestry Administration and National Development and Reform Commission (NDRC). As these nine authorities are from different sectors and have varied interests in water resources management, it results in fragmented water management (He 2016). The lack of clarity regarding the boundaries between the institutional jurisdictions results in internal conflicts and overlapping responsibilities.

Given the fragmented structure of river pollution policies and governance in China, this chapter attempts to analyze China's river basin management, using an econometric approach.

7.3 Methodology and Variable Description

This study attempts to understand the effectiveness of river pollution abatement policies, governance, and rule of law (RL) on the country's river water quality. It also attempts to identify the factors influencing the RQ and/or policy of river pollution abatement in China. For this purpose, two models were formulated. The results have been estimated using time series analysis over the period 1990–2016.

Model 1 determines the impact of each river pollution control measure on pollution abatement. At time "t," China's river pollution is expressed as a function of RQ, RL, government effectiveness (GE), free media (FM), industrialization (Indst), and population (Pop). The model is expressed as follows:

$$waterpoll_t = \beta_0 + \beta_1 RQ_t + \beta_2 RL_t + \beta_3 GE_t + \beta_4 FM_t + \beta_5 Indst_t + \beta_6 Pop_t + u_t$$

where u is the error term.

RQ has been defined as "implementation of statutory law" for the purpose of this study. As it is extremely difficult to value RQ, a regulatory quality index has been formulated for the purpose of this study. The index is based on the shadow price approach used in Brunel and Levinson's (2013) study of measuring environmental stringency.[5] RL measures the quality of environmental policing. Corruption (Corr) has been considered as a proxy for RL. Corruption perception index measures were employed for measuring corruption. This index, developed by Transparency International, assigns scores to countries depending on the level of public sector corruption. The higher the score, the less corrupt a country is. GE gauges the credibility of the government's commitment to environmental policies. Investment in antipollution projects, specifically "investment in treatment of industrial wastewater (invstindwater)," is taken as a proxy for GE in China. The data source for the same is China's National Bureau of Statistics.

The variable FM intends to capture the role of media (if any) in pollution abatement. To capture FM, the "voice and accountability (VAP)," compiled by the World Bank, has been used as a proxy for FM. VAP measures perceptions of the extent to which a country's residents have freedom of expression and a FM.

Data on population (Pop) is obtained from the United Nations database.

Proportion of industry as a percentage of GDP (Propindst) has been employed to measure industrialization. Data on Propindst has been obtained from the World Bank statistical yearbook.

Model 2 is formulated to comprehend the impact of the river pollution control regime's RQ on institutional/public decision making. In this model, the RQ is taken as the dependent variable (DV). The proxy for RQ is similar to that in Model 1. The independent variables (IVs) include RL, industrialization (Indst), Population (Pop), FM and natural resources rent (NRR). Similar to Model 1, the proxies for RL, Indst, Pop, and FM are corruption (Corr), proportion of industry as a percentage of GDP (Propindst), population growth (Popgrowth), and VAP respectively. The model is expressed as follows:

$$RQ_t = \beta_0 + \beta_1 RL_t + \beta_2 Indst_t + \beta_3 Pop_t + \beta_4 FM_t + \beta_4 NRR_t + u_t$$

where u is the error term.

Data sources for this model are the same as in the previous model.

One new variable, NRR, has been considered in this model. Total NRR is an economic concept. The World Bank defines it as a concept which involves accounting for the contribution of natural resources to economic output. There are very few studies exploring the relationship between NRR and water pollution RQ either for China or for any other country. Therefore, this variable has been considered in the model to obtain an idea about the existing relationship between the two. The data for this variable have been provided by the World Bank.

7.4 Results and Discussions

We start by estimating Model 1 using the Prais–Winstein regression method (Table 7.1). The Durbin–Watson test indicated the presence of first-order autocorrelation and, therefore, we do not use methods such as Ordinary Least Squares. Furthermore, as there was a

Table 7.1 Time-series model estimation results for China (Model 1)

Independent variables (IVs)	Prais–Winstein (first model)
Emi_{t-1}	0.110
$lnCorr_t$	−2.39**
$lninvstindwater_t$	0.29
VAP_t	−0.009
$Popgrowth_t$	0.35**
$Propindst_t$	0.50***

Note: Dependent variable is $lnwaterpoll_t$. According to the variance inflation factor (VIF) estimate, our model does not suffer from multicollinearity problems as the value of the mean VIF is less than 10. ***, **, * represent statistical significance at 1%, 5%, and 10% levels of significance respectively.

possibility of reverse causality between water pollution and RQ, the Granger causality test was employed. This test shows that there is reverse causality between water pollution and RQ. Therefore, lagged values for RQs proxy, emissions per dollar of value added (Emi), are considered.

7.4.1 Emi

The results show that the effect of emissions per dollar of value added (Emi) on water pollution is insignificant, indicating that the regulations in place for river pollution in China have failed to contribute to water pollution abatement. This is unsurprising given that China's environmental legislation suffers from many legislative gaps along with apparent noncoordination between laws and regulations. Laws often cannot be implemented since specific regulations required by corresponding laws are not adopted in a timely manner.

Furthermore, the problem of repetition exists in environmental laws and regulations. For instance, before the Environmental Protection Law was revised in 2014, 31 of the 47 articles were repeated in other environmental pollution control laws which leads to inconsistencies between different laws and regulations with respect to basin procedures and fundamental principles (Mu 2014). Environmental laws also suffer from inconsistencies between different versions of the same law, which can be demonstrated quite starkly when comparing between China's Water Law (2008) and its earlier version in terms of quantity control of pollution discharge, formulating water plans, drinking water protection, water quality monitoring, and information disclosure (Wang 2012). Ambiguity in laws results in intentional exploitation by institutions and other actors (Economy 2010). The issue of ambiguity is also evident in China's water legislation which is routinely criticized for its ambiguous legal provisions. For instance, the 2002 Water Law states that the MWR is responsible for guiding the RBCs in undertaking administrative water management responsibilities for the corresponding river basins. However, as a result of ambiguous language, the law does not clearly state the powers and responsibilities that should be allocated to the RBCs and, instead, their role is now limited to only offering scientific guidance about river basins.

China faces equally challenging issues in drafting environmental laws. The consultative process of law drafting involves lobbying for substantial changes among various levels of administrative personnel, which results in the law being watered down to a level where it is of no actual use any more (Economy 2010). On the other hand, China's judiciary, which also determines the country's environmental RQ, has always operated "under the rule of men [rather] than the rule of law" (Economy 2010). Similar to the bureaucracies for water pollution management, China's judiciary is also highly decentralized. The Supreme People's Court, though the highest judicial authority, only has a supervisory role and does not have any real power over the budgets or personnel of the local courts. Such control lies exclusively with the local governments. As a result, the courts are not independent (Chang 2000).

Additionally, as China's water management is based on river basins, this leads to numerous interjurisdictional issues. For instance, the Water Law provides local governments with sufficient administrative powers to influence the local court's decisions. The local governments, moreover, compete among themselves for water allocation rights and for "local protectionism" of their respective industries to promote economic growth without considering the issues of the basin as a whole.

7.4.2 InCorr

Regarding China's corruption-river pollution nexus, the results suggest that if we decrease the corruption perception index by 1%, water pollution increases by 2.39%. The implication is that corruption positively affects water pollution in China. During the last decade, several corruption cases with regard to water pollution management have been reported. China's National Auditing Office declared that approximately US$59 million which was allocated for water pollution control had been mishandled and misappropriated between 2001 and 2007. Furthermore, US$650 million was unaccounted for, which could have been stolen, misdirected, or never used. In addition, industrial river pollution fines totaling US$300 million were never paid by the polluting companies. It is noted that factories in China quite frequently prefer not to invest in the treatment of wastewater as the penalties are much lower than the cost of operating effective treatment plants (Economy 2010).

Few past studies on China have attempted to understand the effect of corruption on pollution levels. Stoerk (2015) finds that, besides institutional corruption, the institutions responsible for efficient environmental management are also guilty of statistical corruption. Between 2008 and 2013, the institutions responsible for efficient environmental management altered air quality data for Beijing, in an attempt to portray to its citizens and the outside world that China had minimal air pollution.

7.4.3 Invstindwater

Government effectiveness does not show any significant relationship with water pollution. This implies that institutional investments under the river basin management structure have failed to reduce water pollution effectively. We can assume, therefore, that these institutions are inefficient. As accepted by the Chinese government, lack of investment along with poor supervision have contributed to increasing water pollution (Xie and Xie 2016). Management of environmental protection in China is assigned to multiple agencies and different actors, as evidenced in the water management structure where there exists not only a vertical management system but a horizontal system as well. Therefore, instead of cooperating with each other, the different government bodies usually end up competing for limited available resources and influence (Michalak 2005). Furthermore, the Ministry of Finance plays a crucial role in determining the level of water pollution fines for noncompliant factories whereas the Environmental Protection and Resources Conservation Committee (EPRCC) of the National People's Congress (NPC) is responsible for managing and drafting water pollution laws and regulations. As numerous ministries and agencies are involved, the entire water management system suffers from lack of coordination.

7.4.4   VAP

VAP, the proxy for a FM, shows no significant relationship with water pollution in China. This implies that VAP has failed to bring any change in China's water pollution abatement. The media, particularly the newspapers, are considered as an influential means for promoting environmental education. The mass media does not only encourage public concern for environmental issues, it also raises public awareness (Parlour and Schatzow 1978). However, we do not observe any such trend in China. Firstly, most of China's media houses are state controlled; the independent ones are also controlled indirectly through financial incentives, legal restrictions, and surveillance. Apart from a crackdown on local media, China has also strengthened its control over the country's narrative internationally by buying overseas radio stations, sending state media reporters abroad and establishing news start-ups which look and feel like independent news organizations but, in reality, are controlled by the Chinese Communist Party (CCP) (Griffiths 2016).

China's CCP arguably maintains one of the world's most restrictive media systems. The central government censors all forms of media so as to maintain its monopoly over power and information. In fact, China was ranked at 176 out of 180 countries in 2016 by French watchdog "Reporters Without Borders" in terms of freedom of the press. Weekly censorship guidelines are distributed among prominent editors and media providers regarding the Communist Party's propaganda. Further, the Communist Party's Central Propaganda Department also issues media outlets with guidelines and directives which restrict the coverage of politically sensitive issues. For censorship of the media, the government also resorts to lawsuits, fines, arrests, demotions, dismissals, and forced television confessions. Given the huge price that journalists have to pay in China for voicing their opinions about sensitive environmental issues, they usually refrain from publishing news regarding the country's environmental pollution. This is probably the reason why "voice and accountability" in China fails to affect river water pollution by raising environmental awareness among people.

7.4.5   Popgrowth and Propindst

As expected, the results of this study also show that population has a positive relationship with water pollution. China's industrialization also seems to affect water pollution levels positively.

Table 7.2   Time-series model estimation results for China (Model 2)

Independent variables (IVs)	Prais–Winstein (first model)
$Corr_t$	−0.38**
NR_tR	−0.64***
VAP_t	−0.008
$Popgrowth_t$	0.89**
$Propindst_t$	0.659***

Note: Dependent variable is Emi_t. According to VIF estimate, our model does not suffer from multicollinearity problems as the value of the mean VIF is less than 10.
***, **, * represent statistical significance at 1%, 5%, and 10% levels of significance respectively.

Given China's rapid economic growth and the undesirable effects of industrialization on the country's water quality, it is essential to understand why the country's river basin management regulations are unable to contribute significantly toward water pollution abatement, and what determines its (in)effectiveness. The results of Model 2 attempt to provide an explanation.

Model 2 has been estimated using the Prais–Winstein regression to control for autocorrelation, as found by Durbin–Watson test (Table 7.2). Further, as there was a possibility of reverse causality between water pollution and RQ, the Granger causality test was run. The results reveal no evidence of reverse causality.

7.4.6   Corr

The results indicate that if corruption perception increases by one unit, the emissions per dollar of value-added decrease by 0.38 units. In other words, if a country's level of corruption drops by 1 unit (the higher the corruption perception score, the lower the level of corruption a country suffers), then the RQ of water pollution improves by 0.38 units (as the lower the emissions per dollar of value added, the higher the regulatory stringency and better the RQ).

China's river basin management, despite having a sufficient legal basis, suffers because of the fragmented structure of water laws. In addition, despite China enacting many environmental law statutes, there are no Chinese environmental law statutory interpretation cases (Nagle 1996). The underlying cause of this is not only the lack of an independent judiciary, but also because there is corruption at every single level (Nagle 1996). For instance, a corruption scandal at the intermediate court of Hubei Province in 2002 revealed that two vice-presidents of the court, seven mid-rank judges, one court clerk, and three deputy divisional directors received bribes from 44 lawyers on various occasions (Sina 2004).

The underlying reasons for the existence of corruption in China's courts could include the following.

- The CCP rules the country and therefore is alone in determining the content and nature of Chinese law. The Party influences not only the legislative process but also its enforcement. This issue raises particularly critical questions in environmental law as Party intervention is required to ensure local enforcement, but such intervention deprives the law of its independent force (Ross and Silk 1987). The Party's power over the legal system also extends to statutory law. This is because, in China, statutes are used as a means by the CCP to decide what is acceptable and what is not. Though the CCP lacks the constitutional power to enact legislation, it is capable of initiating the process for drafting statutes, including environmental statutes (Ross and Silk 1987). This continuous influence of the CCP explains why there are few disputes concerning interpretation of environmental statutory laws (Nagle 1996).

- China's constitution, like those in most other countries, provides for an independent judiciary. However, this independence is constrained by many factors. Firstly, Chinese judges are very poorly trained and held in low regard in society, unlike in other countries (Economy 2010). In fact, the NPC, which selects the justices serving on the Supreme Court, can even remove appointed justices at any time during their two consecutive terms without giving any valid justification (Leung 1987). This implies two things; firstly, the NPC more often than not would choose someone who has similar political ideologies to the CCP. Secondly, because the justices might lose their position if they go against the Party, they usually do not have any incentives to do so (Nagle 1996).

Besides these issues, corruption acts as an obstacle to private enforcement of Chinese environmental statutes. Private citizens in China, along with organizations and businesses, face numerous difficulties while bringing environmental cases to courts. Chinese courts can be better described as dispute resolution mechanisms as Chinese law places judges closer to law-interpreting bureaucrats, rather than enforcers of the RL. In fact, cases involving Party members are not even heard by the courts; they are only dealt with as internal corrective matters (Lubman 2002).

China's case filing division is also infamous for breaking up collective lawsuits into individual cases (Lubman 2002). In the 1990s, the Chinese courts started establishing case filing divisions with the aim of making the act of bribing judges more difficult. But some litigants confess that this reform has only led to an increase in the number of people who need to be bribed so as to get their cases into court (Li 2012). The underlying reasons for corruption in the judicial institutions include inadequate salaries for judges (Zou 2000), insufficient funding for courts (Xin 2009; Wang 2013), inadequate legal training of judges, lack of judicial independence and local protectionism (Zou 2000). Thus, corruption reduces the stringency of China's water pollution RQ, thereby making the RQ ineffective in water pollution abatement, as our Model 1 result indicates.

7.4.7 NRR

The relationship between NRR as a percentage of GDP and emissions per dollar of value added is positive and significant. According to political economy literature, NRRs interact with state institutions and governance in numerous ways; for example, extraction of natural resources induces deterioration in governance. Even if resource dependence does not degrade governance, it could affect the quality of institutions and governance through the lax natural resource management policies adopted and their implementation throughout the value chain. Therefore, the institutional quality and the government's capacity with regard to drafting and implementing effective natural resources regulations would affect outcomes, not only in the natural resources sector such as fossil fuels from where rents can be earned, but also in other resources sectors such as water, air, and land. This is because as natural resources revenues generate huge profits, the government might be tempted to make policy and public decisions which would have long-term consequences in terms of environmental degradation for other resources (Barma et al. 2012).

In China's case, this theory applies. In 2013, the UN Environment Program declared China the world's biggest consumer of primary materials, including minerals and metal ores. This resulted in a huge investment in infrastructure development and manufacturing. The mineral and metal resources which supported this boom came mostly from domestic reserves (West et al. 2013). In fact, China's coal reserves are hugely responsible for the industrialization and manufacturing sector growth as electricity generated from these coal reserves propelled China to its current position (Greenovation Hub 2014). Besides the opportunities, mining is also associated with degradation of the environment as it has a huge environmental impact on water resources, mainly because large amounts of water are required in the various stages of mining and mineral processing. For instance, mining, washing, and processing of coal are all very water-intensive, while chemicals used in mining pose the risk of run-off and poisoning of rivers and lakes nearby (Greenovation Hub 2014).

Numerous environmental laws and regulations have been adopted to internalize the environmental cost imposed by mining in China. Despite this, China's water quality has been affected significantly due to mining activities and because of the weak environmental controls caused as a result of inefficient regulatory and institutional structure (Zhu and Cherni 2009).

Responsibility for handling the environmental consequences of coal mining rests with the MEP and MWR (which are also responsible for water pollution and air pollution) (Zhu and Cherni 2009). As the ministries handling river basin management and controlling the environmental consequences of mining are the same, a rise in NRR also therefore results in a decline in RQ of river basin management. This is because the failure in controlling the environmental consequences of coal mining has also led to a rise in river water pollution. Therefore, the failure of one automatically guarantees the failure of the other in terms of regulatory enforcement.

7.4.8 VAP

The results further indicate that an FM does not seem to affect the RQ of water pollution significantly. Given that China has media censorship, the media are unable to hold the government accountable for deteriorating environmental quality and, as a result, are unable to affect the stringency of water regulations in any way.

7.4.9 Popgrowth and Propindst

Population growth as well as the proportion of industry as a percentage of GDP have a significant and positive relationship with emissions per dollar of value added. These results further reinforce our Model 1 results which revealed that an increase in industry and population growth as a percentage of GDP lead to an increase in water pollution levels. It is predicted that this is because the regulations and institutions in place were ineffective in water pollution abatement due to corruption and lack of government capacity regarding the legal framework for river basin management. Model 1 showed that RQ and GE were ineffective in affecting water pollution. Further, Model 2 results reaffirm Model 1's results that RQ is ineffective in water pollution abatement. Model 2 results indicate that RQ's ineffectiveness is because of increasing population growth and an increasing industry share as a percentage of GDP. Given that the existing enforcement of environmental regulations is already weak in China, with growing population and industries, it is obvious that the RQ will further weaken if China does not attempt to amend its RQ and GE and its corruption problems.

7.5 Conclusion

China's unprecedented growth since the late 1970s has resulted in severe environmental degradation, including water pollution. This study attempts to understand the underlying causes of China's water pollution in its river basins, and also attempts to evaluate how China's RQ, GE, RL, FM, population, and industrialization interact with its water pollution abatement measures. For this purpose, two econometric models were formulated along with devising a RQ index for water pollution.

The first model attempted to comprehend the existing relationship between river basin policy and governance with river basin pollution. The results revealed that only corruption seems to affect water pollution measures. The second model was formulated to understand the role of river basin policy and governance in the decision making of institutions and the public. The results indicate that corruption heavily influences China's RQ of the RBCs. Moreover, because of the existence of other natural resource endowments such as gas, oil, and minerals, the RQ of the RBCs has declined. Additionally, this study found that FM does not seem to affect RQ at all in China.

7.6 Policy Implications

Given the fragmented, unclear governance and ambiguous water resource management in China, the following changes need to be made in the policy and institutional framework.

Regulations need to be formulated quickly after the introduction of laws, in order to reduce the waiting period and increase the efficiency of the corresponding laws.

China's water legislation also needs to deal with the issue of ambiguity. In the existing Water Law, the MWR is responsible for designating administrative water management responsibility to the RBCs. However, because of ambiguous language and a lack of clear specifications regarding the powers and responsibilities of the RBCs, their role is limited to organizing and formulating river basin plans, therefore, the RBCs also need to be provided with substantive powers.

Another major issue is the lack of an independent judiciary which could prevent the Party's intervention in decision making. Therefore, there is an urgent need to establish an independent judiciary in China. In terms of improving governmental effectiveness, there is a need to increase China's spending on environmental management. Given that the country is growing at a rate of 80–120% every five years, allocating only 1.3% of its GDP for environmental management does not seem enough. Also, China's water management system is divided among multiple agencies, not only vertically but also horizontally. Although the TBRM demonstrates a significant development in terms of river basin legislation by progressing toward more integrated management, there is still a long way to go. This fragmented system has resulted in a lack of cooperation among various agencies as well as competition between them for the limited resources available. The dual leadership system of the MEP and MWR for water resources management has also affected the effectiveness of the local environment protection bureaus.

Ultimately, there is a need to establish a comprehensive river basin law in order to manage the fragmentation and to establish clear governance between the MWR, RBCs and other involved entities.

Lastly, high corruption levels also affect the management of environmental resources significantly. There is a need to increase the transparency of not only the institutional structure for river basin management but also the law drafting process for environmental statutes and the work of the judiciary. Given that China is not a democratic country, bringing about these changes could be difficult, but NGOs could be the way forward for increasing transparency in environmental management. Chinese NGOs might not only increase public participation but could also develop a positive government attitude toward dealing with environmental issues through raising environmental awareness among Chinese citizens.

As China's environmental NGO sector has seen rapid growth in the last 10 years, with student groups increasingly participating in these grass-roots bodies, relaxing government policies for NGOs could be the way forward for China's fight against environmental degradation.

Notes

1 For instance, Article 14 clearly lays down on the basis of river basins and regions, unified plans should be laid out. It also specifies the different categories of plans under both; Article 15 elaborates on the basin, regional comprehensive and special planning systems; Article 17 mentions that the River Basin Organizations (RBOs) are responsible for devising the cross-province comprehensive plans.
2 Explained in detail in the next section.
3 The Tarim river basin is a part of the inland river basins.
4 Article (17) Water Law of the People's Republic of China (2002) [English translation: www.mwr.gov.cn/english/01.pdf].
5 For more details about the index, please get in touch with the author.

References

Article (1), (3), (5), (12), (14), (15), (17) and (23) Water Law of the People's Republic of China (2002) [English translation: www.mwr.gov.cn/english/01.pdf]

Article (1) and (11) Water Law of the People's Republic of China (1988)

Article (1) and (4) Regulation on the Administration of the Taihu Lake Basin (2011)

Barma, N.H., Kaiser, K., Minh Le, T. et al. (2012). *Rents to Riches? The Political Economy of Natural Resource-Led Development.* Washington, DC: World Bank.

Boxer, B. (2001). Contradictions and challenges in China's water policy development. *Journal of Water International* 26 (3): 335.

Brunel, C. and A. Levinson. (2013). OECD Trade and Environment Working Papers, No. 2013/05. Measuring Environmental Regulatory Stringency. www.oecd-ilibrary.org/docserver/5k41t69f6f6d-en.pdf?expires=1583928132&id=id&accname=guest&checksum=2E8084A27AFF0BCD56FC1E50D633323D

Chang, P.L. (2000). Deciding disputes – factors that guide Chinese courts in the adjudication of rival responsibility conduct disputes. In: *Environmental Regulation in China* (eds. X. Ma and L. Ortolano). Lanham, MD: Rowman and Littlefield.

Changjiang Water Resources Commission. (2004). The Introduction of CWRC. www.mwr.gov.cn/english

Dawei, H. and Jingsheng, C. (2001). Issues, perspectives and need for integrated watershed management in China. *Environmental Conservation* 28 (4): 368.

Economy, E.C. (2010). *The River Runs Black: The Environmental Challenge to China's Future,* 2e. Ithaca, NY: Cornell University Press.

Greenovation Hub. (2014). China's mining industry at home and overseas. www.ghub.org/cfc_en/wp-content/uploads/sites/2/2014/11/China-Mining-at-Home-and-Overseas_Main-report2_EN.pdf

Griffiths, J. (2016). From Xi to Shining Xi: China's Propaganda Machine Goes into Overdrive. https://medium.com/cmd-v-cmd-v/from-xi-to-shining-xi-chinas-propaganda-goes-into-overdrive-17b74aa46ddd

He, X. (2016). Developing sustainable water legal framework in China: prepare for adaptation. In: *Legal Methods of Mainstreaming Climate Change Adaptation in Chinese Water Management,* 21–68. New York: Springer.

Leung, F.F.-L. (1987). Some observations on socialist legality of the People's Republic of China. *California Western International Law Journal* 17: 102.

Li, L. (2012). The "production" of corruption in China's courts: judicial politics and decision making in a one-party state. *Law and Social Enquiry* 37 (4): 848.

Lubman, S.B. (2002). *Bird in a Cage: Legal Reform in China After Mao.* California, USA: Stanford University Press.

Michalak, K. (2005). Environmental governance in China. China Governance Project, OECD: Paris.

Ministry of Environmental Protection (2009). State of the Environment Report. Beijing: Ministry of Environmental Protection.

Moore, S. (2013). Issue brief: water resources issues, policy and politics in China. www.brookings.edu/research/issue-brief-water-resource-issues-policy-and-politics-in-china.

Mu, Z.L. (2014). Research on the Interests of Environmental Legislation. PhD thesis. Renmin University of China, Beijing.

Nagle, J.C. (1996). The Missing Chinese Environmental Law Statutory Interpretation Cases. Notre Dam Law School NDLS Scholarship, 517

Parlour, J.W. and Schatzow, S. (1978). The mass media and public concern for environmental problems in Canada, 1960–1972. *International Journal of Environmental Studies* 13 (1): 9.

People's Republic of China Changjiang Water Resources Commission. (2004). Water Legislation Plan in China's River Basin, Order No. 293.

Ross, L. and Silk, M.A. (1987). *Environmental law and policy in the People's Republic of China.* Westport, USA: Praegar.

Shen, D. (2009). River basin water resources management in China: a legal and institutional assessment. *Water International* 34 (4): 488.

Sina (2004). Legal system and news: 13 criminal judges investigating the Wuhan Intermediate People's Court. http://news.sina.com.cn/c/2004-03-12/13423016437.shtml

Stoerk, T. (2015). Statistical corruption in Beijing's air quality data has likely ended in 2012. Working Paper No. 194. Grantham Research Institute on Climate Change and the Environment, London.

Sun, D., Zhang, J., Hu, Y. et al. (2013). Spatial analysis of China's eco-environmental quality: 1990–2010. *Journal of Geographical Sciences* 23 (4): 695.

Taihu Lake Basin Bureau (2011). The Background of Taihu Lake Basin Management Regulation. Beijing: Ministry of Water Resources.

Turner, J.L. (2007). In deep water: ecological destruction of China's water resources. In: *Water and Energy Futures in an Urbanized Asia: Sustaining the Tiger* (eds. E.R. Peterson and R. Posner), 26–35. Washington, DC: Centre for Strategic and International Studies.

Wang, J. (2012). Coordination of legal system between water resources protection and water pollution prevention and control. *Journal of Central South University* 18 (6): 89.

Wang, Y. (2013). Court funding and judicial corruption in China. *China Journal* 69: 43.

West J., Schandl, H., Heyenga, S., et al. (2013). Resource Efficiency: Economics and outlook for China. United Nations Environment Program: Washington, DC.

World Bank. (2009). Addressing China's Water Scarcity: Recommendations for Selected Water Resource Management Issues. http://documents.worldbank.org/curated/en/996681468214808203/Addressing-Chinas-water-scarcity-recommendations-for-selected-water-resource-management-issues

Wouters, P., Hu, D., Zhang, J. et al. (2004). The new development of water law in China. *University of Denver Water Law Review* 7 (2): 243.

Xie, W. and Xie, J. (2016). Analysis on the current status and regulatory measures of water pollution in the Xiang River basin of China. *Nature, Environment and Pollution Technology* 15 (4): 1435.

Xin, H. (2009). Court finance and court responses to judicial reforms: a tale of two Chinese courts. *Law and Policy* 31 (4): 463.

Yan, F., Daming, H., and Kinne, B. (2006). Water resources administration institution in China. *Water Policy* 8 (4): 291.

Zhu, S. and Cherni, J.A. (2009). Coal mining in China: policy and environment under market reform. *International Journal of Energy Sector Management* 3 (1): 9.

Zhou, J., Peng, G. and Zhen, C. (2008). Trading water in thirsty China. www.chinadialogue.net/article/show/single/en/2144-Trading-water-inthirsty-China

Zou, K. (2000). Judicial reform versus judicial corruption: recent developments in China. *Criminal Law Forum* 11: 321.

8

The EU Legal and Regulatory Framework for Measuring Damage Risks to the Biodiversity of the Marine Environment

Ivelin M. Zvezdov

AIR Worldwide, VERISK Analytics Corporation, Boston, MA, USA

8.1 Introduction

During the next two decades, the marine environment will arguably be the location where a significant share of new major industrial developments will take place. At least within Europe, demand for renewable energy will mainly be satisfied from marine sources; there will be an increasing exploration for hydrocarbons both to extract remaining reserves from current oil and gas fields and to move into deeper waters, an increasing demand for marine aggregates, and continuing demand for fish protein, which will be delivered from aquaculture facilities that may move into deepwater regions. Although this is predominantly a European perspective, similar trends are present worldwide.

In both North America and Europe, there is significant new legislation in place or currently being developed to regulate and control these different types of offshore developments. Regulation is aimed at achieving three main objectives: (i) to sustain the health of the marine system in the face of industrial encroachment, (ii) to maintain biodiversity within the marine environment, and (iii) to maximize the productivity of the marine system in economic terms. It is recognized that conflicts will arise between these objectives and, as a result, systems for planning activities must be put in place to optimize across all of these objectives.

Within Europe, the European Commission, through legislation like the Water Framework Directive, the Habitats Directive, the Marine Strategy Framework Directive (MSFD) and the Environmental Liability Directive (ELD), sets the legal precedents. In the UK, the Food and Environmental Protection Act 1985 (FEPA) is the primary vehicle for the regulation of offshore development, but licenses issued under this legislation take account of the UK national and regional legislation that enacts the European environmental directives. Where power generation is concerned, the Electricity Act (1989) is also a relevant legal instrument. In cases where European law is breached (infracted), the European Commission can place hefty fines (£millions/day) upon member states until the problem is rectified.

Environmental Policy: An Economic Perspective, First Edition. Edited by Thomas Walker, Northrop Sprung-Much, and Sherif Goubran.
© 2020 John Wiley & Sons Ltd. Published 2020 by John Wiley & Sons Ltd.

Because the penalties for infraction are substantial, there are strong incentives placed upon member states to be highly precautionary in the regulatory regime put in place. For commercial enterprises wishing to use the marine system, this may result in two forms of risk: (i) regulatory risk and (ii) infraction risk.

- *Regulatory risk* is associated with regulators taking the view that development cannot be licensed because, in their view, there is a possibility that environmental damage will occur, with a risk of infraction.
- *Infraction risk* is the risk that once a commercial enterprise has developed its infrastructure and begun an operation, a form of previously unknown and unpredicted environmental damage occurs and that this results in an infraction being brought by the European Commission on the member state. In such a case, the member state will almost certainly withdraw the operating license for the facility and may require the removal of any structures put in place. In a worst-case scenario, the threat of infraction could result in the withdrawal of the operating license.

Within Europe, different member states take differing approaches to the application of the spirit of the European directives. In the UK, the approach is one of upfront regulation; in other countries, such as Spain, the approach involves punitive downstream sanctions on developers if an infraction occurs. In one case, there is risk mitigation through bureaucratic oversight and in the other, there is mitigation through punitive deterrence, although of course, most legislative systems rely on graduation from one extreme to the other. This graduation is often misinterpreted as differing levels of compliance across cultural transitions within Europe, but the results and risks are similar. Although they are shared differently, they are potentially higher in the case of the approach involving downstream actions without upfront mitigation.

8.2   Regulatory Mechanics

Very broadly, there are four different pieces of environmental legislation in Europe that underpin the regulatory risks associated with offshore marine development.

8.2.1   Water Framework Directive

This directive is constructed to protect aquatic environments and, in particular, to regulate pollution in the near-shore region.

8.2.2   Habitats Directive

This directive regulates the effects on natural habitats by designating areas called Special Areas of Conservation (SACs). Although SACs are often quite small areas of critically important habitat for designated species, the risks from this legislation derive from the possibility that anything can affect that "special conservation status" of the featured species in that small area and lead to an infraction. For large mobile species like marine mammals and seabirds, virtually anything that is done anywhere could affect the special conservation status of these species. Moreover, SACs for these species have been placed in such a way to ensure the total coverage of Europe's regional seas and coastlines. A further feature of this legislation is that those exploiting the environment must show their commercial and industrial activities, which will not affect the advance of any implementations and developments. So, they must prove a negative. This is, of course, impossible but it adds to the overall risks attached to development in the offshore region.

8.2.3 Marine Strategy Framework Directive

This directive was adopted in 2008, and it regulates all issues related to the protection of natural resources used and involved in marine commercial, industrial, and social activities. The directive provides legislative coverage for four European marine regions – the North-East Atlantic Ocean, the Baltic Sea, the Mediterranean Sea, and the Black Sea. The main policy contribution of the MSFD is the definition of a concept called Good Environmental Status (GES: Commission Decision (EU) 2017/848), with five distinct process components (see Figure 8.1) and two critical subcomponents or submetrics.

First, the establishment and measurement of initial assessments, or initial conditions, was accomplished in the 2012–2014 time period, Second, from 2015 through 2018 came the development of a monitoring program and a measure plan to achieve the environmental "goodness" status by 2020. Lastly, the cycles of monitoring and measures are repeated from 2018 to 2021. These overall monitoring and measuring strategies are driven by 11 quantitative metrics, which cover baseline species status, nonindigenous species introduction, commercial exploitation of species, seafloor resources, the concentration of contaminants, and others (MSFD, Annex 1).

In practice, these 11 quantitative indicators drive the processes of the monitoring program, defined in 2014 (see Figure 8.1), of the resilience and functioning of the whole GES initiative, as well as structural changes of GES 2020 due to human, commercial, industrial, and recreational activities. MSFD 2008 also enhances the agility of EU regulation

Figure 8.1 Timelines, components, and dependencies in GES 2020.

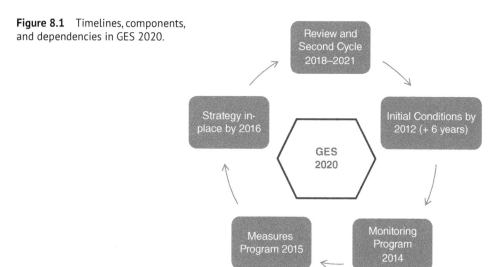

implementations by serving as a bridge with other legislation pieces and directives and by requiring the continuous assessment and monitoring of the functioning of marine ecosystems, such that environmental changes are linked to human activities and pressures. The desired outcome is a process which allows causes and consequences to baseline parameters of the marine environment to be identified, controlled, and managed with a multitude of legislative instruments to ensure that each community member state meets the goals of GES 2020, which will trigger acceptance and commencement of the second cycle of the initiative to be complete by 2021 (see Figure 8.1). These goals represent a balance of the social and economic benefits of marine activities and the fitness of indicators and targets of the health of marine biodiversity and ecological systems. The proposed model measures and reports changes in these biodiversity goodness of health targets and indicators.

8.2.4   Environmental Liability Directive

This directive has been in force since April 2007. This liability legislation covers only certain industrial activities, such as the transportation of dangerous products, the handling of hazardous substances, waste management operations, and plant protection products. The fundamental contribution of the directive is the legal establishment of the "polluter pays" principle, that is, the public sector should not provide resource funds for remediation of environmental damage by private commercial and industrial operators. The directive requires both prevention of immediate threat and post damage restoration by liable operators, who bear the entire cost. A satisfactory outcome after environmental damage has occurred is the recovery to baseline of predamage environmental conditions, including biodiversity assets. For this proposal, it is significant that the directive does not apply in the case of *force majeure*, and notably in cases of maritime oil disasters, and cumulative forest and natural habitat decline. Damage and loss of protected species and natural habitats caused by officially approved infrastructure projects, such as motorways, bridges, ports, offshore industrial and energy facilities, are not considered as environmental damage, provided that the operator and the state authority have taken them into account during the process of their official authorization. However, this rule highlights another subset of regulatory risk involving the need to ensure that any developments are accompanied by appropriate and thorough upfront risk assessment and evaluation with due diligence.

It is also at the discretion of the national legislative procedures to exempt companies from the cost of penalties and remediation of the damage that has occurred during normal and approved operations. For this exemption to take place, the company must prove both that the damage has not occurred through a fault of its own and that there was no previous technical and scientific knowledge that could have predicted such negative outcomes.

At present, the directive does not require any mandatory private insurance coverage. Yet, all the EU member states have translated this directive into national law at various paces since 2007. In 2010 the European Commission provided a report (COM.2010 #581) to the European Council and European Parliament on the general adequacy and availability of environmental insurance products. The Commission report recommended that it is up to member states to introduce a mandatory financial security system, including mandatory requirements for environmental insurance coverage. However, the EU member states were requested to encourage the development of financial security practices, instruments, and markets. By and large, national governments limit their "encouragement" of building up insurance markets to periodical consultations with trade organizations and large insurance firms. The insurance firms have been the principal driving force in the development of environmental liability markets, even in conditions where national governments legislate mandatory financial security systems.

The implementation continues to be refined and to evolve, which can be seen in the recent development of the Multi-Annual Work Programme (2017–2020) – "Making the Environmental Liability Directive more fit for purpose." The Commission encourages and requests (i) improvement in tools and measures, which describe the technical evidence base of cases subject to the ELD; (ii) the actual implementation of cases to be vigorously supported by further synchronization of understanding, terms, and concepts among the Commission and member states' civil servants, national, and local elective bodies and practitioners by enhanced capacity building and training; (iii) the availability of financial security and remediation to insolvency needs to be ensured for actual remediation to take place.

8.3 Market Mechanics

Commercial enterprises operating in the marine system are becoming more receptive and willing to carry insurance against both regulatory risk and infraction risk. They may wish to have an improved definition and more granular and explicit quantitative metrics of near- and long-term financial risks associated with compliance with environmental protection legislation.

Both regulatory and infraction risks are associated with the risk of damaging the environment, but both have, in effect, the same type of financial consequences associated with acute environmental events, including chronic environmental exposures. These risks may be revealed through new knowledge of the effects, technological developments in measurement, or the gradual ramp-up of statistical and computational power to detect patterns. It is becoming possible to model such risks and to develop appropriate theoretical underpinning for assessing and quantifying their impacts with increasingly reliable metrics. Such metrics immediately raise the quality of advice about the financial viability of proposed developments in the marine system and additionally provide the basis for a market that can insure commercial enterprises against this form of low-grade, high-frequency risk as well as more rare, low-frequency catastrophic risk.

For operators of hazardous industrial and energy facilities and processes, the issue of cost due to environmental damage becomes central to their basic view on the viability and profitability of the business proposition. The costs associated with such industrial and energy operations without environmental insurance coverage have been discussed above. At the same time, such coverage, when purchased, carries the cost of its premiums, operational expenses along with the risks of its adequacy, the default of the insurance provider and legal operations, all of which are well understood by commercial operators.

Quantifying the cost/risk schemes and trade-offs of the above framework becomes a necessary task for operators both currently engaged in such situations and planning to expand or enter new opportunities in the physical marine offshore and coastal systems.

While a lot of legal, operational, and policy level research has been done in this field, less has been accomplished in developing statistical and numerical risk methodologies to quantify the probability of damage and cost/risk trade-offs and likely scenarios in terms of financial and insurance risks to operators. Practice shows that estimating quantitative risk and using probability-based scenarios are difficult tasks for central and local regulators. For example, the environmental liability cost of the Chemie Pack BV accident in Holland in 2011 was 65 million euros, with an evident gap to a mandatory model, that is, the maximum environmental security instrument under the mandatory Spanish model is 20 million euros.

8.4 Classification of Environmentally Hazardous Activities and Damage

For a quantitative risk methodology and metrics to be consistent and accepted on the market, it is necessary to define environmental activity-to-damage liability relations (see Figure 8.2) in specific terms. One possible approach is to follow the definition of the EU Commission White Paper on Environmental Liability (2000), where it is defined as damage resulting from industrial accidents and gradual pollution and deterioration by hazardous substances, waste, or operations from strictly identifiable sources.

In cases where the damage itself is identifiable and limited to biodiversity and contamination of sites, it is defined as a strict liability (see Figure 8.2) where a precise economic quantitative valuation of damage is essential, mainly when it is irreversible, although this remains a challenging proposition. When restoration of damage is still possible, the estimation of costs is a necessary tool for both the commercial operator and the public state body. In both cases, if partial restoration is the only option, alternative solutions and

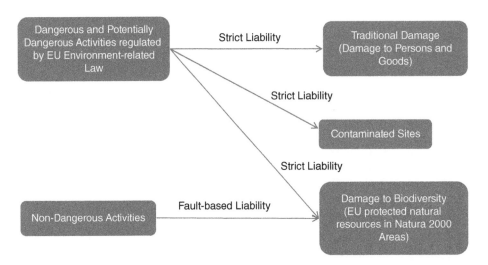

Figure 8.2 Activity-to-damage liability relations: EU Commission White Paper on Environmental Liability (02.09.2000).

replacements may need to be evaluated. The outcomes will very much depend on the soundness of the methodology used to assess risks and measure impact. The community directives provide a spatial component in the classification of environmentally hazardous activities by requiring states to measure the ecological impacts of commercial operators beyond their borders on preestablished transborder marine areas. Such measures are also multidisciplinary and range from the physical and technical to the socioeconomic and socioecological impacts on populations most affected by the deterioration of environmental services (Art.13.3, DG Environment 2015, MSFD CIS).

The process of creating a market risk community with a classification knowledge, systems, tools, and funds for estimating and insuring biodiversity and environmental liability is both continuous and laborious. Similar processes at the national and international level have developed over many years with relevant examples from marine cargo and hull and space satellite telecommunications market segments. Society, with its academic, regulatory, and market institutions, is capable of developing risk classification and knowledge systems for complex multinational and large geospatial hazard risk problems. The European Commission recommends that national governments should encourage and facilitate the development of markets for mandatory environmental liability coverage. Simultaneously, a gap may be already emerging in readiness, theoretical understanding and metrics, and systems in place between the insurance industry and best practice recommendations from EU institutions.

8.5 Developing Standard Insurance Instruments for Environmental Biodiversity Liability

Various market-based financial instruments, including bank guarantees, municipal and sovereign secured funds and bonds, are used to provide coverage for environmental liability. Insurance products have emerged as the most effective and popular coverage instrument. National insurance pools have also emerged in France, Italy, and Spain. In principle, market insurance coverage for the ELD is effectively achieved with third party liability and environmental impairment liability policies. Still, the challenges in developing standard insurance instruments for environmental biodiversity liability are significant. An obvious definition of scope and scale of coverage, and the triggering mechanisms of its indemnification are critical for acceptance by all market players – insurers, commercial policyholders, and regulators. The very nature of large-scale environmental catastrophe brings the risk of bankruptcy upon the offending party. This creates the need for a solution that remediates to the possibility of corporate bankruptcy, such as (i) mandatory insurance, (ii) the intervention of the state by providing full financial coverage, and (iii) the private–public sector fund pools, which are insulated from litigation and are thus available to provide uninterrupted financial security.

The definition of unpredictable and random insurable events and scenarios that span over longer time periods, covered by general liability policies, is still in process from both an economic and legal perspective. This definition and explicit socialization are performed within the framework of the ELD 2008 (see Figure 8.3). The ELD 2008 and four additional components as defined in Figure 8.3 are deemed indispensable to the effective functioning

Figure 8.3 Requirements for operational insurance market for marine biodiversity liability.

of an insurance market for marine biodiversity liability. This first step of putting in place legal definitions has been in progress since 2008 at many levels for national and EU bodies. Once a stable and acceptable legal framework is functioning, then inevitably follows the task of production and dissemination of modeled statistical risk metrics. Such metrics require that the probability and severity of marine biodiversity damage and loss must be quantifiable, which is tremendously challenging since the scientific expertise in this field is just beginning to emerge. A market risk community of commercial operators, researchers, and insurers and financiers, which is to appreciate required dependencies among these five components (see Figure 8.3), needs to emerge to identify, share, diversify, and ultimately turn this risk into a transferable commodity. Only then will private insurance and reinsurance firms be able to estimate the risk and charge a premium for its transfer and accumulation.

The scarcity of available historical data from past events, the unique nature of the processes, which may not have historical precedence, and the current generation of mathematical models of marine biological diversity damage make the production and distribution of risk metrics a serious challenge. To begin with, the very definitions of hazard risk events, responsible operators, protected habitats, and risk exposures influence the magnitude of potential losses. The combination of high potential losses with little opportunity for diversification, remediation, or recovery presents another challenge for all involved in the process. Some case studies are available concerning pollution events, but not many have been compiled concerning acute and permanent environmental damage. Cost estimates associated with risk mitigation may be constructed around experience gained by the oil and gas industry. Case studies exist for scenarios involving discharged byproduct water, separated from the oil and gas extraction in the North Sea.

Modeled probability of industrial accidents from known existing processes, installations, and materials is more straightforward to estimate due to accumulated knowledge by insurers and actuaries, and the availability of historical statistical data. Comparative and transferable modeling and estimation of probabilistic frequency and severity of biodiversity hazard events connected to new industrial processes, new installations, and retrieval of new materials in newly developed geospatial areas are necessary practices for the academic and insurance industry modeler. Continuously changing legislative requirements and definitions of exposed and protected habitats, thresholds, and boundaries of damage make building coherent, biodiversity damage models a challenging proposition. Nevertheless, regulators, academic institutions, and insurance firms are taking action to accumulate data about the severity, scale, and estimated economic costs of environmental impact. The same effort is also occurring in gathering data on the impact of already known environmental hazards to human health, soil, water resources, and biodiversity as a measurable benchmark with which we can estimate new and unknown damage occurring from prospective industrial activities. Opportunity and hope lie in the methodologies of inference on processes with minimal information and stochastic nature that mathematical and statistical sciences have developed.

Measuring the likelihood of damage and economic and social losses following hazard events becomes a critical factor in determining if environmental and biodiversity assets are insurable. Inevitably, for some market practitioners, the notion of damage to biodiversity will be considered vague and hence not suitable for insurance coverage. However, measurable monetary metrics of the cost of remediation, opportunity and economic costs, and the costs associated with having operating licenses withdrawn may prove more practical to be translated to liability and insurance policy coverage. All such insurance markets' growth depends on the development of coherent methodologies for accurate monetary and legal estimation of such costs.

From the perspective of an insurance industry firm, the fundamental insurability of environmental and biodiversity liability must be legislated and justified by market dynamics. Along with the multiple challenges presented above, there are still positive developments in both the legislative and market climates that support the process of developing such insurance products and instruments. Faure and Grimeaud (2010) observe at least three positive developments in legislative environments that would encourage market developments. The authors argue that, first, the ELD (2007) does not impose a general strict liability rule for any environmental harm but has a balanced approach to limit strict liability to specific types of activities and geographical areas that are considered hazardous. Second, the directive establishes that the new environmental and biodiversity regimes should not be retroactive. Third, several defenses are available to operators, of which the most notable is *force majeure*. The same authors advocate that additional defense should be included in the regime in order to improve its predictability and attractiveness to insurers.

8.6   Financial Insurance and Security and Cost of Infraction

A range of costs arises for operators practicing in the regulatory and legislative framework of the environmental liability and marine strategy directives of 2007 and 2008. Decontamination and restoration of environmental habitat and species after incurred damage constitute a significant "remediation cost." In the case of water, marine habitat, and species, the directive classifies remediation costs into (i) primary: returning natural resources toward baseline; (ii) complementary: compensation for any deficit from baseline, and (iii) compensatory: compensation for interim loss until primary remediation has taken place. The assessment of the baseline condition of the bioenvironmental asset becomes of primary importance; here, all available scientific and technical data, and measurements are used. Subjecting the baseline condition to a time-discounting function is also necessary, as this provides the estimated time progression condition of the assets as if a hazardous impact has not occurred. The employment of best measures, information, and scientific practices in this process is critical. The EU Habitats Directive and Water Framework Directive are valuable information sources for this estimation.

Significant environmental catastrophe events such as the Bouches-du-Rhone oil spill of August 2009, the Hungarian toxic sludge spill of October 2010, and the Deepwater Horizon in April of the same year, invariably and expectedly had impacts upon European civil service thinking. The Commission encourages and continues to consider a wider compulsory environmental liability framework for all member states. The current ELD does not require any financial security or insurance purchase against the deterioration of bioenvironmental diversity and assets from established and measured existing conditions. It does, however, encourage such developments by national and local governing bodies and market insurers. Local financial security requirements have already been introduced in some countries. Such requirements can be satisfied with the purchase of insurance, bonds, bank guarantees, and other instruments. The traditional European insurance and reinsurance markets and instruments cover basic liabilities resulting from property damage and physical harm, which result from pollution and environmental damage by third parties. With the increased importance of the ELD, the risk and liability exposure of many commercial and industrial operators will not be covered by the existing general liability or environmental liability insurance policies. First of all, operators and local governments are in the process of reviewing and quantifying liability concepts such as "significant damage" to environmental assets and biodiversity and "imminent threat." In parallel, the leading insurance firms have begun to develop local and global insurance products for public law claims on biodiversity deterioration and damage.

Further refinement and customization of insurance coverage policy details and instruments allow firms with low capital funds to obtain protection against the cost of significant environmental damage and to provide protection against insolvency. With flexible and customizable insurance coverage available on the market, large energy, transportation, and other strategic corporations will not have the incentives to transfer high-risk hazardous operations to smaller, marginally capitalized firms to avoid the responsibility and cost of environmental liability. However, the growth of environmental liability and biodiversity damage insurance markets is in its very early stages of development. There is still a very low level of sophistication in markets and products, especially in the area of remediation and restoration of environmental damage. Notably lacking are techniques to evaluate the cost of restoration, ecological damage, and biodiversity damage, as well as the cost of pure economic loss. Working papers from the EU Commission and large insurers support the notion that in more than half of the EU member states, there is no mature liability market at all and in only two-thirds of the countries is there liability coverage for property damage and bodily injury as a result of industrial environmental accidents. At the level of global insurance markets, the availability of such standard instruments is even lower.

Practice in some countries shows that solutions are sustainable and available. The case of Spain is a valid example. The Spanish environmental ministry has extended the coverage of the ELD 2007 directive, which is limited to protected areas and species, equal to 14% of the territory of Spain, to full coverage of the territory. In the area of financial security, Spain proposes a mandatory system which would create the necessary incentives and pressure for operators to adopt more cautious practices and preventive rules. The financial security scheme established that if the economic operator carries the potential to cause more than €2 million of damages, it must purchase an insurance policy or provide financial guarantees such as bank or independent fund guarantees. For damages between €300 000€ and €2 million, operators must provide financial guarantees or join the EU Eco-Management and Audit Scheme. Similar legislation from the Czech environmental ministry establishes a mandatory financial security system for operators in the form of insurance or re-insurance policies or bank guarantees. France has strengthened its "environmental liability insurance framework," known as CARE, and established an insurance pool, known as Assurpol. The French experience shows that having more insurance firms involved in the environmental liability market, accompanied by strong policies of information sharing and public awareness of environmental and biodiversity hazards, has had a very positive effect on lower and competitive premiums, and increased insurance coverage penetration.

Nonetheless, these approaches are not replicated across all member states. There is often an implicit need for insurance coverage when it is not traditional for national or local legislation to specify how a company should protect itself from financial risks. Uneven and unequal interpretation and implementation may discourage market participants. Still, understanding the mechanics of the EU environmental liability legislative process and its implementation in different member states will help insurance policy providers to work toward a well-functioning market.

8.7 Quantitative Modeling Approach

Designing a quantitative risk model to estimate the damage to environmental assets and biodiversity needs to meet several challenges and requirements. The first of these is to transfer concepts established within traditional qualitative risk assessment, typically involving some form of conjugation of the likelihood of exposure and the severity of the response, to a quantitative framework. In order to develop a quantitative model of that nature, the risks to the reduction in biodiversity need to be deemed as having a sufficiently probabilistic character. This conclusion involves multiple measuring, modeling, and statistical algorithmic stages. A typical marine environment risk model, described and recommended by Smith et al. (2016), involves numerical components which model the present state of variables of the marine habitat; pressures and drivers of degradation change upon these measurable variables; and a measurable output defining impacts and quantitative data changes to the initial state of marine and biodiversity metrics. In particular, the driver for baseline change and impact metrics in such models should be anchored to the definitions of the MFSD (2008).The general components of such an environmental risk model are described schematically in Figure 8.4.

The model in Figure 8.4 captures and consumes historical data of existing commercial, industrial, and social activities within the marine domain and simulates measurable hazard

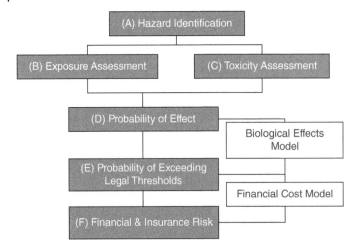

Figure 8.4 Measuring, modeling, and statistical algorithmic stages in a biodiversity damage quantitative risk model.

drivers of impact upon the baseline variables of biodiversity status. Long-term and good-quality geospatial data sets in this context are invaluable in the determination of baseline status of environmental and biodiversity metrics and the definition of their degradation through the workings of drivers of human economic and social activity.

8.7.1 Hazard Identification Module

In this module, industrial activities would be classified to present some level of hazard risks, but the majority is likely to be benign. Classification and general assessment of the hazards in the context of marine renewable energy are the normal product of environmental impact assessments (EIAs). The outputs from EIAs are the starting point of definition of existing conditions, and baseline scenarios of biodiversity status in a selected geospatial zone. Measuring upon this baseline, the academic researcher and industry practitioner can identify and classify hazard risk scenarios and short- and long-term hazard impact events. For example, in case studies on the impact of offshore renewable energy on marine mammals, the hazards will vary depending upon the industrial devices, facilities, and geophysical characteristics of the installation. In general, these will fall into three categories: (i) habitat exclusion through physical exclusion; (ii) habitat exclusion through habitat modification because of acoustic or physical disturbance, change in the disposition of food, or loss of habitat critical to a life-history function; (iii) direct physical damage to populations and single species. Given the baseline of the EIAs and the classification of hazard risks, the hazard identification module selects the most pressing set of risks presented by current commercial and industrial activities by market operators.

8.7.2 Exposure Assessment

This function measures the number of individuals exposed or the proportion of the population exposed to the hazard. This measure depends on the disposition of the hazard in relation to the distribution of sensitive species or habitats. Similarly, based on reviews of existing data and case studies, enough information is available about the distribution of marine mammals around the European continental shelf to allow the construction of exposure distributions. It may be necessary to develop movement models of different species to allow simulation of exposure. Such movement and migration models develop an exposure assessment that describes the probability of exposure for different selected sensitive species to each hazard type, focusing on those deemed most important under stage A. The exposure assessment includes varying sensitivity between sedentary and mobile species and uses the predicted disposition of industrial, commercial, and energy facilities. The response or vulnerability functions of each species are thus highly sensitive to their chronological and geospatial location within the model domain. This dependence of the species response functions and, in more general terms, of the baseline environmental and biodiversity metric creates a premise for a complex multivariable, multidimensional model, with explicit temporal and geospatial domains of metrics and simulation and modeling functions (see Rose et al. 2010). In this context, it becomes critical to define the temporal and geospatial granularity and scale of the model domain, which at present can be reliably supported by advances in hardware resources management and computational algorithms.

These complex types of models by their very nature create result sets which could be nontrivial for review and interpretation, and in turn this premise will create demand for adequate results and data validation, visualization, and interpretation tools.

8.7.3   Toxicity Assessment

This module and algorithmic library serve to identify the dose–response characteristics of species, which is generally not well characterized. Data that already exist would be used to develop probability distributions across all sensitive species for the three general categories of hazard listed under stage A. Again, in the context of offshore energy facilities developments and hazard risks for marine mammals, work has already been done to model some hazard factors. The relationship between the body size of an animal, the blade diameter and speed, and the probability of an animal striking a tidal turbine blade can be modeled and quantified. The effect of direct damage upon the viability of individual animals may be modeled under different scenarios to examine the sensitivity of the risk assessments to different assumptions about toxicity until measured data become available. Similar approaches can be taken for other hazards, such as noise from pile driving. In principle, at this stage of the process, a range of dose–response functions is developed for selected sensitive species exposure to different hazards that should include the effects of habitat exclusion and physical damage. Where possible, these functions should be based upon any experimental data available but where these are not available, scenarios should be used to explore the sensitivity to the toxicity of each hazard.

8.7.4   Probability of Effect

A fully probabilistic module measures the joint probability of animals being exposed to all hazards, some of which may occur simultaneously. The toxicity of these hazards provides an overall risk probability for the sensitive species or habitats involved. This, in effect, provides a fully quantitative risk quotient, including uncertainty around the quotient. For example, this measure defines the number of animals of a particular population likely to be exposed to hazardous impacts. In situations where there is insufficient information to feed a reasonable risk profile derived from first principles, academics and actuarial practitioners typically use current data sets to develop a generalized risk distribution for environmental hazards that could then be applied in context. An example could be to use stranding data for marine mammals around the UK and European Atlantic coastlines, or fisheries data to examine the possibility of developing underlying generalized functions for extreme events. This use of data is particularly important in the context of offshore renewables because there will always be an underlying probability of effects occurring, which are unrelated to the effects of the activities being modeled. It is essential to understand the capacity as we must distinguish the effects and signals of the development of renewable energy from the background of natural stochasticity noise of variability in environmental biodiversity.

In summary, at this stage of the modeling process, the probability distribution of the effects of a hazard upon selected sensitive species or habitats is constructed by using available time-series of species-based data. Generalized information about the background distribution of natural hazards is derived with an estimated likelihood of detecting effects on sensitive species. The probability of effect is measurable by category and will vary by hazard driver and by the geospatial resolution of the model. The measurability of the impacts of the change drivers in most general terms has two aspects. First, it captures the sensitivity of the disturbance and degradation upon the whole geospatial habitat, and second, it captures in a more detailed manner the impact on species of interest (Smith et al. 2016). Furthermore, the complexity of modeling is increased by interdependence and correlations between hazard drivers and the time variability element of such impacts (Smith et al. 2014).

8.7.5 Probability of Exceeding Legal Thresholds

Social, economic, and legal measures are a much-needed estimate in the definition of insurable loss under the framework of the ELD 2007. Unlike for many forms of chemical pollution where discharge concentrations have been set as thresholds that should not be exceeded, thresholds for the effects on sensitive species have generally not been agreed to, despite recent legislation developments. Since all commercial and industrial procedures are likely to have some level of impact, it is critical to distinguish effects, which are likely to be biologically significant. Although some European Commission directives call for there to be demonstrably no effect before a development can proceed, this is impractical, and such an extreme position will likely shift to one where effects must not exceed some predetermined level of biological significance. The outputs from a methodology measuring the likelihoods of effects (D) are translated into a biological significance through specific biological models. These often include generalized demographic models for sensitive species. Past assessments of biological significance are available for some species, as for harbor porpoises concerning fisheries bycatch. These estimates, frequently provided by academic institutions, are used by national and local governments, which lead to the development of sensible legal thresholds, by embedding them in fundamental biological theory.

Differences between regulatory risk and infraction risk are apparent, and they do constitute different definitions of likelihoods of exceeding a legal threshold. Without a clear case law or precise guidance and in the presence of high uncertainty, regulators may take a precautionary approach to set the acceptable thresholds, and local sociopolitical drivers may also drive this threshold to some extent. This regulatory risk is likely to differ from infraction risk. Regulatory risk may also show volatility through time depending upon the political process. However, it should not be assumed that regulators will be more precautionary with legal thresholds that will lead to infraction because, under some circumstances, regulators may be forced to apply "national and local interest" criteria to their judgments, and these may force decisions about licensing that exceed legal thresholds.

In summary, a methodology for estimating likelihoods of exceeding legal thresholds provides a generalized approach to define the probability of a hazard or a set of hazards resulting in an infraction. The likelihoods of exceeding thresholds in terms of precautionary levels address regulatory risk, while defining thresholds in terms of legal terminology cover infraction risk.

8.7.6 Financial and Insurance Risk

Monetary losses, impact, and policy outcomes are always the superior methodology products of research and modeling strategies for defining and measuring marine biodiversity risks. Having defined the probability of exceeding legal thresholds, actual, or perceived, in section 8.7.5, the next methodology task is to express this in terms of financial and insurance risk. In the industry, as well as consulting and academic case studies, it is achieved through an economic model of the offshore renewables industry that calculates the trade-offs involved in different scenarios. The functions relating to biodiversity hazard risk translated to regulatory and infraction risk and mapped to economic cost are critical to making value judgments about the financial viability of projects. These methodologies define the insurance and financial risks being carried by commercial and industrial operators and in turn inform their demand for the proper kinds of insurance schemes to provide the most appropriate coverage for offshore oil, renewables, and other industries operating in coastline marine environments. Standard insurance and financial modeling methodologies and actuarial knowledge of available and proposed insurance and reinsurance instruments and policies are used in this practice to estimate losses and costs to the economic operators and the insurance sector. As EU environmental regulations and national government implementations are continuously developing, ELD is one of the fastest growing legislative fields. Financial cost models need to reliably define the insurance risks associated with compliance with current and proposed EU environmental regulations.

In the overall development and growth of these methodologies, innovation is required and is continuously emerging in many areas. These advancements are summarized as follows: (i) creating data sets of risk parameters for bioenvironmental assets; (ii) deriving mathematical-statistical methodologies to describe the severity of environmental and biodiversity hazard events; (iii) designing relation functions to translate the impact of hazard events on the baseline of the risk parameters of bioenvironmental assets into physical and monetary costs to the public domain; and (iv) creating mathematical insurance models to estimate the cost and loss distributions to economic operators and insurance companies.

However, traditional mathematical-statistical methods have limitations because of the low availability or simple lack of technical historical and statistical data, and lack of experience in economic operators, which otherwise provide valuable information for modeling risk distributions under the assumption that these are stationary. The use of insurance claims experience and data is limited for the same reason and due to current changes in regulatory practices, which are compensated by alternative methods of mathematical and statistical inference for problems of low initial information and data availability.

8.8 New Developments and Challenges, Further Work

Developments in legislative, research, and modeling processes that enhance the growth of marine biodiversity liability insurance markets are too recent to generate some definitive pronouncements from the data on their effectiveness. Nevertheless, progress is evident in the increased exchange in information such as in discussion of methodologies and proposed risk metrics, raw historical data, policy papers, and proposals among key stakeholders, which include operators, government experts and agencies, academic and research institutions, and financial and insurance firms. The very publication of research articles, White Papers on policy, and empirical studies serves to raise awareness of needed actions toward the development of this market segment. As with any new market, the definition of initial conditions – "baseline condition" – needs to be critically and equivalently interpreted among national and local agencies and governments, and by local and multinational operators. The same critical acceptance and agreement hold true for more complex concepts and risk metrics, including modeled and probabilistic metrics on "significant environmental damage" and "marine biodiversity damage."

At the same time, local and national governments have a less well-discussed role as facilitators, and one critical aspect of this is their role in the collection of data and physical metrics of ELD registered cases and the building of databases available to academic and research institutions. On a positive note, between 2007 and 2014, member states reported 1245 registered cases, albeit unevenly across nation-states, of environmental damage of both marine, water, and land origin, which invoked the legal application of the ELD. At the EU level, a centralized ELD case data repository system with public and academic access is much needed and will invariably further academic research, including the development of quantitative risk models and policy proposals. These are all components that encourage growth in insurance markets and coverage penetration for commercial and industrial operations.

On the legislative side of the market, the all-important choice to make is whether to move toward a mandatory system of environmental and marine coverage liability. The European Commission has taken the position that at present, there is not enough evidence that a mandatory system of financial security is necessary and could be made successfully operational. After all, experience has proved that the enforcement of legislation across countries is a challenging and demanding process prone to disappointment rather than developing and adopting policy proposals. The reevaluation of options will continue, particularly for the option of local and national governments managing mandatory financial security systems instead of doing so at the supranational level. It is evident that there is a need for convergence on definitions and quantitative metrics, such as "thresholds of significant environmental damage."

A more pressing need than the evolution toward mandatory environmental financial security is the closure of gaps in the present system of marine environmental coverage. The damage to biodiversity and requirements for remediation of the marine environment due to oil spills and energy industry industrial accidents are not yet adequately addressed. Current EU legislation defines environmental damage as damage to soils, water, natural habitats, and protected species, which leaves unprotected species and the general marine coastal environment outside of many requirements and regulations.

Of the three critical costs for marine environmental damage – infraction, administrative, and remediation – the latter can be reduced by financial and insurance instruments. The purchase of environmental liability insurance by European companies, including coverage for the marine environment, has increased by 13.6% since 2006. Still, demand is low, and insurance market growth and penetration are slow. The registration and reporting of insufficient numbers of ELD cases in many countries are a possible impediment to insurance market growth. Eleven member-states have not reported any ELD cases, since its legislative implementation in 2007. This nonreporting by member-states is undoubtedly linked to the harmonization of the implementation, and actual or perceived lack of conceptual clarity. The directive among national legislators is, in fact, the second impediment to insurance market growth. Harmonization and synchronicity play an intricate role not only among national and regional legislation within the EU, but also across the conventions of the International Maritime Organization (IMO). For marine biodiversity hazard risk factors, such as oil pollution from tankers, there is more than one level of conventions and mechanisms of compensation and remediation, which introduce a need for additional efforts in harmonization.

References

European Commission (2000). *White Paper on Environmental Liability*. Luxembourg: Office for Official Publications of the European Communities.

European Commission (2010). Report on the Environmental Liability with regard to the prevention and remedying of environmental damage. COM (2010) 581. https://eur-lex.europa.eu/LexUriServ/LexUriServ.do?uri=COM:2010:0581:FIN:EN:PDF

Faure, M. and Grimeaud, D. (2010). *Financial Assurance Issues of Environmental Liability*. Maastricht: Maastricht University Press.

Marine Strategy Framework Direction (MSFD) (2008). https://ec.europa.eu/environment/marine/eu-coast-and-marine-policy/marine-strategy-framework-directive/index_en.htm

Rose, K.A., Allen, J.I., Artiol, Y. et al. (2010). End-to-end models for the analysis of marine ecosystems: challenges, issues, and next steps. *Marine and Coastal Fisheries Dynamics, Management, and Ecosystem Science* 2: 115–130. https://doi.org/10.1577/c09-059.1.

Smith, C. J., Papadopoulou, K.-N., Barnard, S., et al. (2014). Conceptual Models for the Effects of Marine Pressures on Biodiversity. DEVOTES Deliverable 1.1. www.devotes-project.eu/wp-content/uploads/2014/06/DEVOTES-D1-1-ConceptualModels.pdf

Smith, C.J., Papadopoulou, K.-N., Barnard, S. et al. (2016). Managing the marine environment, conceptual models and assessment considerations for the European marine strategy framework directive. *Frontiers in Marine Science* 3 (144) http://irep.ntu.ac.uk/id/eprint/28356/1/PubSub5885_Little.pdf.

9

Redefining Nature and Wilderness through Private Wildlife Ranching

An Economic Perspective of Environmental Policy in South(ern) Africa

Tariro Kamuti

Department of Public Law, University of Cape Town, Cape Town, South Africa

9.1 Introduction

This chapter is a critique of wildlife policy in southern Africa, which is partly driven by the private wildlife ranching practice (also known as game farming) that has become widespread in the region. Wildlife policy, as part of environmental policy, has a bearing on the practice of game farming and has strongly influenced the southern African agricultural and environmental landscapes. To some extent, this policy also shows how society relates to nature. Societies have different and sometimes unique views of how to optimally relate to nature – a critical and inescapable relationship due to the inherent impact of one on the other, more so during this Anthropocene period (Borrini-Feyerabend and Hill 2015; Büscher and Fletcher 2019). This relationship is bound by governance arrangements based on the appropriation and use of power at different levels of organization and circumstance which impact nature and human livelihood, as shaped by policy and practice (Borrini-Feyerabend and Hill 2015).

The focus of this chapter is the governance of wildlife issues as part of the wilderness, and therefore the governance of nature at a national level. The chapter looks at the governance of the private wildlife sector in southern Africa,[1] which is anchored on private ownership of land and the relations of the private wildlife sector with the state. This chapter argues that private wildlife ranching has caused a cumulative shift in the attitudes that participants and other stakeholders in the sector hold toward wildlife, wilderness, and nature in general. This shift can be broadly attributed to the appropriation of common resources and their subsequent incorporation into private holdings through the institution of property rights across the southern African region.

The chapter draws from the author's study and research experience of the governance of the wildlife ranching sector in South Africa over the past decade. Apart from use of data collected through interaction with industry players, government officials, other scholars, and interested stakeholders, empirical experiences through interviews, field observations in the countryside, attending game auctions, and presentations at work-

Environmental Policy: An Economic Perspective, First Edition. Edited by Thomas Walker, Northrop Sprung-Much, and Sherif Goubran.

shops, seminars, local and international conferences, the study also draws from analyses of documentary evidence that relates to the wildlife ranching sector. As southern African countries interact on areas of common interest at various levels of trade and exchange of ideas, so do the cross-cutting and transboundary characteristics of wildlife conservation and consumption spread. Many of these interactions show nuances of great historical bearing on the development of wildlife ranching in southern Africa. As a result, this analysis has now started to extend to other southern African countries, which will be referred to in this chapter.

9.2 From Precolonial to Colonial Systems of Natural Resource Management

The growth in wildlife ranching is part of a wave of postproductivism which has transformed the countryside by engaging in activities that are outside the realm of agriculture in general (Kamuti 2017). This transformation has been driven by political and socioeconomic changes over a long period and has its foundation in the shift of scientific and cultural orientations toward wildlife (Carruthers 2008). In precolonial times, southern African people lived in harmony with their natural environment and practiced their own way of hunting. This changed after their lands were taken away from them through conquest (Guy 1980). The advent of settler hunters caused a well-orchestrated slaughter of wildlife to satisfy the Victorian way of life. By the mid-1870s there was a massive decline and even disappearance of various herds of animals (Duminy and Guest 1989; Kamuti 2017). Backed by different driving factors, this decline has continued, such that species like the black rhino and African elephant (population decreased by 30% for the period 2007–2014) have become endangered (Hill 1991, p. 29; Bale 2019). Some of this slaughter was sanctioned as a tsetse fly control measure to allow for the expansion of conventional agriculture, particularly livestock production (Steele 1979; Carruthers 2008). Furthermore, the forcible incorporation of the African population into the Western capitalist system through expropriation and fencing off of land became a critical part of the enclosure of wildlife into this market economy system, to the exclusion of the commons.

Societies in southern Africa share a common heritage when it comes to the management of natural resources from the precolonial to the colonial eras. This period was characterized by the widespread and violent subjugation of indigenous people, and the exploitation of their resources for net gains in the primary productive sectors (Gewald et al. 2019). This also extended to the enclaves created to cater to nature, which were made into private nature game reserves and public protected areas. Thus, in present-day southern Africa, there is a compelling impression of how the state played a critical role in permitting the appropriation of resources and shaping the trajectory for the governance of nature conservation (Gewald et al. 2019, p. 3). New systems of land and natural resource governance were instituted, based on the search for profit and implemented through segregationist laws that benefited the conquerors. They substituted the value systems of the local people, which were based on their traditions and heritage (Kamuti 2017, p. 48). The current characteristics of nature conservation in the region, where a select few continue to benefit at the expense of the majority, reflect this governance system (Gewald et al. 2019).

In the same vein, various practices have, under the banner of nature conservation, become a lucrative way for an elite to achieve "accumulation by conservation," which mirrors primitive accumulation under capitalism (Büscher and Fletcher 2014, p. 273). Since the 1960s, through "wilding the farm or farming the wild," this accumulation has increased by embracing scientific principles (Carruthers 2008, p. 160). Although it first started as an experiment by two American Fulbright scholars at a cattle ranch in Zimbabwe, where they proved that wildlife was a potentially viable enterprise that could coexist with livestock, this shift significantly shaped the industry as it is known today (Carruthers 2008). This model was eventually adopted in the region, contributing to the increase in wildlife population under private property ownership. Today, this model anchors a well-developed wildlife-based value chain, which constitutes the so-called wildlife economy. Further developments to the model include changes which point toward increased market orientation, such as selective breeding (for instance, for horn length and color variation), and the formation of specialist clusters in the wildlife sector to cater for certain flagship species. Wildlife ranchers argue that they have revolutionized conservation by not only saving wildlife but increasing it to sustainable levels. Through active participation in wildlife ranching and government lobbying in various capacities and fora, the private sector has played a pivotal role in shaping the governance of wildlife conservation. While Zimbabwe was credited for pioneering wildlife ranching under private property in southern Africa (Carruthers 2008), South Africa ushered in the conservancy movement that later spread to other countries in the subregion, especially back to Zimbabwe and Namibia (Kamuti 2014; Wels 2015).

The entrenchment of capital and property rights on land, and therefore on wildlife, have had a long-lasting effect on various facets of natural resources management. This, in turn, impinges on the livelihood and socioeconomic well-being of the majority population within the region. Hence, issues in these postcolonial societies concerning, for example, land, socioeconomic transformation, and human–wildlife conflict, bear the hallmark of this checkered history and remain topical. The consequences of this history persist in various levels of governance, from local to national, and in different spatial scales. The changes in how people view wildlife stem from the different values that people attach to wildlife, depending on a wide range of factors such as ownership, access, benefits, governance, abundance, and sociocultural issues. These changes, witnessed across temporal-spatial scales, amount to redefinitions of wildlife, wilderness, and nature. Along with the enabling hand of the state, this is where we can locate private wildlife ranching and its influence in shaping environmental policy.

9.3 The Rise of Private Wildlife Ranching in Southern Africa

Since the 1960s, a growing trend in southern Africa has seen the state facilitate private wildlife ownership through legislative measures (Kamuti 2019). This trend is what Muir-Leresche and Nelson referred to as "the Southern African experiment," namely emphasizing Botswana, Namibia, South Africa, and Zimbabwe (2000, p. 1). For example, in Zimbabwe (previously called Rhodesia), the Rhodesian Parks and Wildlife Act of 1975 paved the way for the devolution of wildlife to property owners (Brink et al. 2011). Through this landmark legislation, the state formally acknowledged landowners as the proprietors of game on their land. This legislation also extended to communal areas, which technically ceded ownership of wildlife to farmers across rural areas (Hill 1991). However, through the skewed ownership of land, the major beneficiaries of this development were large-scale commercial farmers (Hill 1991). In these examples, we can see the changes in the land tenure system under various statutes of land reform in postcolonial Zimbabwe. This is especially visible in the drastic and controversial "fast track" land acquisition program which, since the 2000s, has greatly shifted the character of wildlife conservation to the current state. Now, we will look at private wildlife conservation and the arrival of the new elite which coexists with new communal land owners.

In South Africa, private landowners started embracing wildlife ranching from the 1960s and this has been a growing trend since then (Smith and Wilson 2002; Kamuti 2014). During the transition to democracy, and since the end of apartheid, the government has made considerable efforts to restructure the South African state and economy in order to reduce the imbalances of the past and pave the way to an inclusive and cohesive society that coexists with the natural environment. However, there have been misgivings that the adopted neoliberal macroeconomic policies are inadequate to stem the tide of unemployment, poverty, and inequality. These policies have, to some extent, been coupled with threats to biodiversity, as the state reduces its role in nature conservation in a manner that promotes the commodification of natural resources (Castree 2003; Higgins-Desbiolles 2008; McGranahan 2008).

The neoliberal macroeconomic policy context in South Africa is also influenced by the inequal distribution of land ownership – a minority of the population own the largest proportion of land – and the dualistic structure and deregulated nature of the agricultural sector (Cousins et al. 2008; Bond et al. 2009; Aliber and Cousins 2013). Structural and racial inequality in land ownership has been a hallmark of the colonial legacy in South Africa and Namibia, while in Zimbabwe it was present prior to the "fast track" land reform or displacement of white landowners (Kariuki 2009). The role of the state in continuing the structural and racial inequality in land ownership is also evident in the context of South Africa on the back of neocolonial connotations of globalization, which increasingly entrenches the hegemony of the West (Fraga 2006; Torgerson 2007). In 1991, the passing of the Game Theft Act No. 105 gave farmers the right to own game, dependent on the provision of appropriate fencing (Child 2009; Snijders 2012). Prior to this act, all game – even that found on private land – was considered as state property. This was a milestone for the wildlife ranching industry in South Africa. While the South African state, through this act, played an enabling role in growing the wildlife ranching industry (Cousins et al. 2010), it is now confronted by some of the industry's negative practices, which do not augur well for nature conservation (Kamuti 2014).

9.4 The Changing Nature of the Conservation Narrative in Southern Africa

The policy and practice around wildlife ranching is categorized as "governance by private individuals and organisations" (Borrini-Feyerabend and Hill 2015, p. 180). The boom in wildlife ranching in southern African has been achieved through its practice as an economic activity which, despite its limited and controversial contribution to nature conservation, nonetheless offers huge returns on capital. The transition from conventional agricultural enterprises to wildlife ranching was driven by a host of factors, primarily for leisure; however, due to the financial outlay required to sustain it, the practice became increasingly profit oriented (Cloete et al. 2007; Kamuti 2014). In this sense, individual participants in the wildlife ranching sector give the impression that they are contributing to conservation, when in reality, given these financial motives, it is difficult to measure the extent to which they indeed contribute to conservation.

It is wildlife ranching's contribution to nature conservation which, I argue, brings about the redefinition of the meaning and role of wildlife, wilderness, and nature, as it conforms to a capitalist mode of production. Nature is defined in monetary terms, with its various aspects treated as commodities and services for exchange. For example, the concept of "if it pays, it stays" has gained widespread acceptance and has become an underlying basis for the practice of wildlife ranching across the region. Its core meaning is that a species will sustain itself through generating income, which will in turn be reinvested for its conservation (Kamuti 2019). In wildlife ranching, animal species that receive special treatment – for example, through breeding to produce desirable and market-competitive characteristics – have become portfolios for investment.

Wildlife policy in South Africa is situated in the country's broad macroeconomic framework and subsumed in the various regulations pertaining to agricultural, environmental, tourism, rural development, and land reform. However, it must be said that South Africa's macroeconomic disposition is by and large neoliberal, where the market is a fundamental determinant. Other institutional platforms aimed at accomplishing social goals by using wildlife resources have had to comply with and gain from the dictates of the market through the "financialization" of conservation (Sullivan 2013; Marsh 2004). This orientation has been entrenched since the country's transition to democracy in the early 1990s and its subsequent adoption of neoliberal economic programs, which pander to the dictates of capital and property rights, as enshrined in the Constitution. Moreover, this bears the hallmark of the World Bank and other multilateral agencies, as they played a key role in economic policy and land reform during the negotiations that ushered in a new era for South Africa (Bernstein 2013). The World Bank Research Committee and the World Bank's Africa Region also sponsored a research project, entitled "Nature Tourism and Conservation," undertaken between 1999 and 2002 (Aylward 2003).

The dilemma of wildlife as a fugitive resource causes challenges to those who are involved in wildlife ranching, as well as to the various other stakeholders affected by the sector. For example, habitat fragmentation and loss to other land uses, together with negative human influences, all have implications on species' survival and diversity (Farvar et al. 2018). Subsequent compartmentalization of wildlife through high fences – a huge capital investment – has become part of an increasingly controversial and militarized conservation (Duffy et al. 2019).

Wildlife ranching, as an economic activity inherently anchored to land, has spawned a number of issues related to competition over access to resources. For example, wildlife ranching has been caught up in the land debates in southern Africa, as land under private ownership that is teeming with wildlife has become subject to a demand for redistribution, restitution, or governance changes. Another example is that people from surrounding areas – or even far afield – find themselves in conflict with private landowners due to trespassing, illegal hunting, or plundering (Kamuti 2018b). The intricacy of the wildlife-land nexus across southern Africa highlights the unresolved question of unequal resource distribution.

In South Africa, for instance, a study in the Gongolo area of the KwaZulu-Natal province has shown that some land that had been designated for transfer under restitution as part of the country's land reform program was instead earmarked for conversion to ranches; part of it has already been converted to wildlife ranching, thus causing disputes (Kamuti 2018a). As a result, there has been some tension among various stakeholders, with private landowners accusing local communities (as prospective landowners) of scuttling capital investment by introducing or continuing with wildlife ranching (Brooks et al. 2012; Kamuti 2018a). The landowners even had lofty ideas of reworking landscapes to support a new kind of wilderness that resembles "pristine" levels, which would then be packaged for international tourists (Brooks et al. 2011). Wildlife ranching is purported to bring benefits to the economy, which in turn contribute to the welfare of people in general. On the other hand, the local communities destined to benefit from land restitution long to own land, so they see the introduction or continuation of wildlife ranching as a hindrance, barring their chances of empowerment through the land ownership and access that would let them determine their own livelihood strategies. Meanwhile, the government appears to be indifferent to the goals of these previously disadvantaged people, too bureaucratic and constrained to implement progressive policies and transform the economy to address unemployment, poverty, and inequality. A notable symptom of this ambivalence is the slow rate of land reform as part of the broad agrarian question subsumed within broader socioeconomic policies.

The persisting imbalance of land ownership, under which a huge proportion of prime agricultural land has remained in the hands of the minority population and private capital, has been a topical issue since the colonial era. This imbalance increased further in the late 1990s; in Zimbabwe, it triggered a radical shift in macroeconomic policy, focused primarily on changes in land ownership through the fast track land reform (FTLR) program. This program led to the reconfiguration of all institutions linked to the wildlife sector, characterized by the expropriation of land without compensation, a practice that intensified at the turn of the new millennium. As a result, the FTLR program had huge implications for the broader economy, property rights, and ultimately the country's significance to wildlife ranching in the southern African region. This was a significant blow to the institution of property rights. The reform was seen as a reversal of the cumulative gains to wildlife management that had been pioneered in Zimbabwe and had acted as a role model to both the region and the world (Muir-Leresche and Nelson 2000).

As for South Africa, the specter of land reform, through the expropriation of land without compensation, has heightened speculation that a similar economic disaster will occur there as well. At a national level, the position of the wildlife ranching sector in South Africa is unclear. Although game farming is recognized as a legitimate enterprise on paper, the way the state interacts with the wildlife sector signifies otherwise. There is a clash of two major imperatives: the need to maintain biodiversity integrity through nature conservation and the economic drive to earn a living from natural resources through trading wildlife, wildlife-based products, and developing ecotourism. This clash happens despite the fact that people are inherently linked to nature and that there is a deep connection between biodiversity and development (Morrison 2014; Drutschinin et al. 2015). The dilemma, then, lies in which of the two conservation should prioritize (Minteer and Miller 2011). However, the push for "game farming" to resemble conventional farming causes many challenges for environmental regulation authorities. The numerous issues occasioned by wildlife ranching include, for instance, intensive (captive) wildlife breeding for color variants, trophy breeding, increased need for predator control, illegal dog hunting, "canned hunting," and the introduction of species not naturally occurring in a particular area (extralimital species). These issues bring new dimensions to nature conservation, redefining our understanding of nature and the wilderness.

The hunting sector, a powerful anchor of the wildlife ranching sector, also has concerning and ethically doubtful hunting practices, such as "canned hunting"[2] and "put and take" hunting (Cadman 2009; Lindsey et al. 2009; Cousins et al. 2010). In South Africa, the issue of canned hunting has caused a great deal of concern in the wildlife industry and in the public sphere – concerns premised on the welfare of the animals that are bred and kept in captivity and around the ethics of hunting itself. The Department of Environmental Affairs and Tourism expressed worries regarding violations of fair chase principles and the humane treatment of animals (2005). The persecution of predators is also rampant, causing further concern for the authorities (Cousins et al. 2008, 2010). These concerns demonstrate a greater attention to the business imperative of game ranches than to conservation, to the detriment of biodiversity (Cousins et al. 2008, 2010; Lindsey et al. 2009). Consequently, government responds with strict regulations, like those dealing with the Threatened and Protected Species Regulations (TOPS). Indeed, Jacques Malan, a former president of the organization Wildlife Ranching South Africa (WRSA), said at their congress that "TOPS was introduced because of canned hunting" (speech at the Inaugural WRSA Congress, April 2013). Generally speaking, in Africa, the weakness of government, the decline in rural incomes, increased availability of hunting technologies, and growth in the human population have all contributed to declining numbers of bush meat species (Crookes and Milner-Gulland 2006). Meanwhile, the human–wildlife conflict in Africa is on an upward trend. There is then an obvious need for appropriate policy decisions to ameliorate such conflict and boost conservation efforts (Browne-Nuñez and Jonker 2008).

9.5 Toward Redefining Nature and Wilderness Through Private Wildlife Ranching

Büscher and Fletcher, noting the historical trajectory of conservation, describe it as a shift from "mainstream conservation," through "new conservation," "neoprotectionism," and eventually to what they term "convivial conservation" (2019, p. 3). In South Africa, there is increasing investment in private game farming, despite the regulatory instability in the sector. These investors largely see no need to go beyond market considerations as a way of valuing nature (Mace 2014; Nel 2014; Sukhdev et al. 2014). However, scholars have called all those involved in conservation "to analyze the justice of ecosystem governance in addition to their effectiveness and efficiency" (Sikor et al. 2014, p. 1).

As argued earlier, wildlife ownership is so closely linked to land ownership that "the political determination of property regimes is critical to conservation, especially with regard to wild fauna" (Naughton-Treves and Sanderson 1995, p. 1265; Snijders 2012). In South Africa, the stronghold of private landowners on land has been enhanced to extend their ownership to the wildlife thereon, causing the tension noted in the land reform program. Land reform's inherent connection to the governance of the private wildlife sector causes further complications, as issues relating to property rights over land remain unsolved, inherently affecting the character of any ensuing nature conservation (Naughton-Treves and Sanderson 1995). Key players in the wildlife ranching sector use the prevailing governance arrangements (anchored in private property rights) to strategically position themselves to maximize benefits, even though some of their activities cause tension. The imbalanced implementation of existing laws and the anticipated changes in the regulatory regime through rhetoric, for instance related to land reform, create an environment of uncertainty for game farmers, who are now major players in the wildlife sector. These constant changes contribute to an unstable environment, and wildlife ranchers argue that this hinders them from playing an active role in both conservation and the much-needed transformation of the economy.

The broader context of "radical uncertainty" caused by, for example, the specter of loss of land has caused much consternation regarding the survival of private wildlife ranching. This radical uncertainty can be tied to the "classical agrarian question," which has implications for the nation-state in relation to equitable land distribution meant especially to cater to the poor (Moyo et al. 2013, p. 93; see also Moraes 2012). As private landowners, wildlife ranchers are fully aware of the political importance of land, and given the experience of landowners in Zimbabwe, there is fear of widespread land dispossession without compensation. In the case of Zimbabwe, this "radical shift in agrarian property rights" has popularized the persistent agrarian question in a unique fashion (Moyo and Yeros 2005a, p. 3; Moyo et al. 2013). As alluded to earlier in the chapter, wildlife ranchers have been directly affected by land reform there, as a number of their farms have been claimed and transferred to land beneficiaries, through either labor tenant claims or restitution claims.

The South African state's struggle to control wildlife ranching developments on private land has made strengthening the institutions dealing with this sector all the more important. At the national level, a policy dialogue has been initiated through the Wildlife Forum (Snijders 2012). This is essential in facilitating the development of a clear policy on game farming, as the position of game farming at the intersection of agriculture, environment, tourism, and rural development (incorporating land reform) must be addressed. As the Wildlife Forum's policy dialogue shows, some actors in the government aim to develop an integrated approach to solving the rift between the reconfigured departments dealing with agriculture, forestry and fisheries, and environmental affairs. Aligning departmental functions and regulations along national, provincial, and local lines will bring policy consistency and pave the way for better implementation. This also requires the identification and addressing of the discrepancies caused by the implementation of governance arrangements, resulting, for instance, in a clear and consistent game farming policy that will increase stability in the wildlife ranching sector.

9.6   Toward a Different Environmental Policy

Horizon threats should not be addressed in silos; rather, due to the intricate connection of ecosystem constituents, they should be solved holistically (Bonebrake et al. 2019). Increased cooperation and institutional capacity across spatial and temporal scales would also help address the trend of increasing illegal wildlife trade in endangered species, flowing from the south to the north (Bonebrake et al. 2019). It is critical to acknowledge that some of the current challenges in wildlife conservation stem from the history of disenfranchisement through colonialism, which alienated African people from land and natural resources (Kamuti 2017, p. 48). Thus, there is a need for people-centered policies – in other words, policies based on and driven by the need to recognize the people who own, live with, and are affected by wildlife. Rather than try to please the clients who are detached from the day-to-day experiences and challenges faced by the rightful owners of wildlife resources, these policies emphasize the rights of these owners, who should be these resources' primary beneficiaries.

For example, since the 1990s, the Namibian state has enacted policies for ceding natural resource property rights to indigenous communities by adopting and amending relevant statutes, therefore enhancing community-based natural resource management (or CBNRM) (Boudreaux and Nelson 2011). In Namibia, the Nature Conservation Amendment Act of 1996 was promulgated to provide a framework for enhancing local conservancies (Bollig and Schwieger 2014). The country is well credited for overhauling its regulatory system in a way that sought to give local communities power over their natural resources, including wildlife (Bollig and Schwieger 2014). This was done to foster long-lasting natural resource governance in the countryside, to empower people at the local level to determine the use of their resources in their own right, and to make sure that they directly benefit from resources within their environs (Bollig and Schwieger 2014). In Zimbabwe, implementation of CBNRM was pioneered via the Communal Areas Management Program for Indigenous Resources (CAMPFIRE). Each of these programs has had its fair share of challenges, but there is a need to leverage these successes to continue restoring the dignity of African people as they work toward their reconciliation with the environment. South Africa, too, is bringing previously disadvantaged communities into the wildlife ranching sector through various initiatives under the auspices of the land reform program. The recommendation for all these countries is to increase the proportion of land that is under the majority population so that they can be brought into the mainstream economy, while framing and practicing conservation in a way that coincides with their identity and values.

Since various stakeholders have different views of what constitutes conservation, perhaps it is worth looking at the philosophy of the Indigenous and Community Conservation Areas (ICCA) Consortium, based on the Global Environment Facility's (GEF) 2018–2022 program's concept of inclusive conservation (Farvar et al. 2018). Their idea of inclusive conservation places indigenous communities at the forefront of conservation efforts, by empowering them to take charge of natural resources in a way that is geared toward "satisfying [the] livelihoods and vibrant cultural and biological diversity on our planet" (Farvar et al. 2018, p. 2). This also allows indigenous groups to use their inherited customs and traditions, thereby ensuring that they will use their agency to resist conservation values that are contrary to their way of deriving meaning from nature. In this way, communities themselves will embrace external parties to include them in their conservation efforts and not the other way around.

Community conservancies have spread the benefits of wildlife conservation from the select few private property owners to numerous community members. However, this success is based on the same *modus operandi* of entrepreneurial models, which of course place an economic value on wildlife. It is likewise argued that the distribution of income from communal conservancies – despite preventing the concentration of wealth in the hands of a few powerful individuals – does not happen in a fair and impartial way (Bollig and Schwieger 2014).

9.7 Conclusion

The steady growth of wildlife ranching in southern Africa goes hand in hand with the growing worldwide trend of wildlife commodification (Brooks et al. 2011; Snijders 2012; Spierenburg and Brooks 2014). International capital has played a critical role in this process, using the rhetoric of biodiversity conservation and sustainable development within a neoliberal framework. In the context of game farming, there is a shift of power away from producers to consumers, as game farmers have been gearing their products and services to meet the demands of the (international) market. Similarly, landowners market their products and services by networking with regular, local, and overseas clients. This is indicative of the global reach of the local wildlife industry, where farmers establish a niche market for high-paying clients. Local and external capital is entrenching itself through the private wildlife sector at a time when there is pressure to redistribute resources and transform the South African economy. The growth of the wildlife sector thus acts as a new frontier for capital investment in the developing world under conditions of a persistent agrarian question (Moyo and Yeros 2005b; Spierenburg and Brooks 2014).

Wildlife ranching is, to a large extent, a business that is now being presented as conservation. This is not to say that there is no conservation taking place on wildlife ranches; although some scientists question the conservation value of certain activities on wildlife ranches, there is no need to totally discount the contribution of private farming to biodiversity conservation. However, by solely characterizing the private wildlife ranching sector as conservation, its capital and profit-seeking imperatives are masked. Indeed, though it first emerged as a sector for wildlife and biodiversity conservation, wildlife ranching has drastically changed over time. On the other hand, "changes in the state of an ecosystem generate surprises for scientists and great uncertainties for stakeholders and policymakers" (Carpenter and Gunderson 2001, p. 456). In other words, conservation programs cause social changes, which in return impact the same environments that people seek to nurture (Miller et al. 2012). The wildlife economy is now in vogue in contemporary conservation narratives, as the wildlife value chain is now viewed as a significant influence on the broader economy (Kamuti 2015; Scales 2015). South Africa, for its part, is at an advanced stage in adopting a broad biodiversity economic strategy. Wildlife ranching constitutes a conduit between capitalism and nature under the guise of conservation (Büscher and Fletcher 2019). Thus, the development of private wildlife ranching is part of a continual process of redefinition of both nature and wilderness, in conjunction with capitalism and neoliberalism.

Notes

1 The southern African region here is loosely taken to be represented by the countries that are currently in the Southern African Development Community but in this chapter I have picked South Africa, Zimbabwe, Namibia, and Botswana which have some commonality and an intricate history of Anglo-Dutch influence that set the basis of state formation and the evolution of the wildlife ranching industry under discussion.

2 Canned hunting is understood as "where animals are shot in enclosures with no chance of escape" (Lindsey et al. 2009, p. 100). In "put and take" hunting, animals are first raised *in situ* and then released into the wild for a certain period before they are hunted.

References

Aliber, M. and Cousins, B. (2013). Livelihoods after land reform in South Africa. *Journal of Agrarian Change* 13 (1): 140–165.

Aylward, B. (2003). The actual and potential contribution of nature tourism in Zululand. In: *Nature Tourism, Conservation, and Development in KwaZulu-Natal, South Africa* (eds. B. Aylward and E. Lutz), 3–40. Washington: World Bank.

Bale, R. (2019). Botswana lifts ban on elephant hunting. *National Geographic*. www. nationalgeographic.com/animals/2019/05/botswana-lifts-ban-on-elephant-hunting/?fbclid= IwAR3U0fHPxn5DPUpwX0yquUAw3-IgXMpV88o14pHE9t4_AGVbwgK1DB41a0w

Bernstein, H. (2013). Commercial agriculture in South Africa since 1994: 'natural, simply capitalism'. *Journal of Agrarian Change* 13 (1): 23–46.

Bollig, M. and Schwieger, D.A.M. (2014). Fragmentation, cooperation and power: institutional dynamics in natural resource governance in North-Western Namibia. *Human Ecology* 42 (2): 167–181.

Bond, P., Dada, R., and Erion, G. (eds.) (2009). *Climate Change, Carbon Trading and Civil Society*. Pietermaritzburg: University of KwaZulu-Natal Press.

Bonebrake, T.C., Guo, F., Dingle, C. et al. (2019). Integrating proximal and horizon threats to biodiversity for conservation. *Trends in Ecology & Evolution* 34: 781–788. https://doi. org/10.1016/j.tree.2019.04.001.

Borrini-Feyerabend, G. and Hill, R. (2015). Governance for the conservation of nature. In: *Protected Area Governance and Management* (eds. G.L. Worboys, M. Lockwood, A. Kothari, et al.), 169–206. Canberra: ANU Press.

Boudreaux, K. and Nelson, F. (2011). Community conservation in Namibia: empowering the poor with property rights. *Economic Affairs* 31 (2): 17–24.

Brink, M., Cameron, M., Coetzee, K. et al. (2011). Sustainable management through improved governance in the game industry. *South African Journal of Wildlife Research* 4 (11): 110–119.

Brooks, S., Spierenburg, M., van Brakel, L. et al. (2011). Creating a commodified wilderness: tourism, private game farming, and 'third nature' landscapes in KwaZulu-Natal. *Tijdschrift Voor Economische en Sociale Geografie [Journal of Economic and Social Geography]* 102 (3): 260–274.

Brooks, S., Spierenburg, M., and Wels, H. (2012). The organisation of hypocrisy? Juxtaposing tourists and farm dwellers in game farming in South Africa. In: *African Hosts and Their Guests* (eds. W. van Beek and A. Schmidt), 201–224. Rochester: James Currey.

Browne-Nuñez, C. and Jonker, S.A. (2008). Attitudes toward wildlife and conservation across Africa: a review of survey research. *Human Dimensions of Wildlife* 13: 47–70.

Büscher, B. and Fletcher, R. (2014). Accumulation by conservation. *New Political Economy* 20 (2): 273–298.

Büscher, B. and Fletcher, R. (2019). Towards convivial conservation. *Conservation and Society* 17 (3): 283–296.

Cadman, M. (2009). *Lions in Captivity and Lion Hunting in South Africa – An Update: A Report Commissioned by the National Council of SPCA*. Alberton: National Council of SPCA.

Carpenter, S.R. and Gunderson, L.H. (2001). Coping with collapse: ecological and social dynamics in ecosystem management. *BioScience* 51 (6): 451–457.

Carruthers, J. (2008). "Wilding the farm or farming the wild"? The evolution of scientific game ranching in South Africa from the 1960s to the present. *Transactions of the Royal Society of South Africa* 63 (2): 160–181.

Castree, N. (2003). Commodifying what nature? *Progress in Human Geography* 27 (3): 273–297.

Child, B. (2009). Private conservation in southern Africa: practice and emerging principles. In: *Evolution and Innovation in Wildlife Conservation: Parks and Game Ranches to Transfrontier Conservation Areas* (eds. H. Suich and B. Child), 103–111. London: Earthscan.

Cloete, P.C., Taljaard, P.R., and Grove, B. (2007). A comparative economic case study of switching from cattle farming to game ranching in the northern Cape Province. *South African Journal of Wildlife Research* 37 (1): 71–78.

Cousins, J.A., Saddler, J.P., and Evans, J. (2008). Exploring the role of private wildlife ranching as a conservation tool in South Africa: stakeholders perspectives. *Ecology and Society* 13 (2) art. 43. www.ecologyandsociety.org/vol13/iss2/art43.

Cousins, J.A., Saddler, J.P., and Evans, J. (2010). The challenge of regulating private wildlife ranches for conservation in South Africa. *Ecology and Society* 15 (2) art. 28. www.ecologyandsociety.org/vol15/iss2/art28.

Crookes, D.J. and Milner-Gulland, E.J. (2006). Wildlife and economic policies affecting the bushmeat trade: a framework for analysis. *South African Journal of Wildlife Research* 36 (2): 159–165.

Department of Environmental Affairs and Tourism. (2005). Panel of Experts on Professional and Recreational Hunting in South Africa. Report to the Minister of Environmental Affairs and Tourism. Pretoria: Department of Environmental Affairs and Tourism.

Drutschinin, A., Casado-Asensio, J., Corfee-Morlot, J. & D. Roe. (2015). Biodiversity and Development Co-operation. OECD Working Paper 21. Paris: Organization for Economic and Development Co-operation.

Duffy, R., Massé, F., Smidt, E. et al. (2019). Why we must question the militarisation of conservation. *Biological Conservation* 232: 66–73.

Duminy, A. and Guest, B. (1989). *Natal and Zululand from Earliest Times to 1910: A New History*. Pietermaritzburg: University of Natal Press & Shuter and Shooter.

Farvar, M. T., Borrini-Feyerabend, G., Campese, J., Jaeger, T., Jonas, H., & Stevens, S. (2018). Whose 'Inclusive Conservation'? Policy Brief of the ICCA Consortium no. 5. Tehran: ICCA Consortium and Cenesta.

Fraga, J. (2006). Local perspectives in conservation politics: the case of the Ría Lagartos biosphere reserve, Yucatán, México. *Landscape and Urban Planning* 74: 285–295.

Gewald, J., Spierenburg, M., and Wels, H. (2019). Introduction: people, animals, morality, and marginality: reconfiguring wildlife conservation in southern Africa. In: *People, Animals, Morality, and Marginality: Reconfiguring Wildlife Conservation in Southern Africa* (eds. J. Gewald, M. Spierenburg and H. Wels), 1–21. Leiden: Koninklijke Brill NV.

Guy, J. (1980). Ecological factors in the rise of Shaka and the Zulu Kingdom. In: *Economy and Society in Pre-Industrial South Africa* (eds. S. Marks and A. Atmore), 102–119. Essex: Longman.

Higgins-Desbiolles, F. (2008). Justice tourism and alternative globalisation. *Journal of Sustainable Tourism* 16 (3): 345–364.

Hill, K.A. (1991). Zimbabwe's wildlife conservation regime: rural farmers and the state. *Human Ecology* 19 (1): 19–34.

Kamuti, T. (2014). The fractured state in the governance of private game farming: the case of KwaZulu-Natal, South Africa. *Journal for Contemporary African Studies* 32 (2): 190–206.

Kamuti, T. (2015). A critique of the discourse of the green economy approach in the wildlife ranching sector in South Africa. *Africa Insight* 45 (1): 146–168.

Kamuti, T. (2017). The changing geography of wildlife conservation: perspectives on private game farming in contemporary KwaZulu-Natal province. *New Contree: A Journal of Historical and Human Sciences for Southern Africa* 79: 39–64.

Kamuti, T. (2018a). Intricacies of game farming and outstanding land restitution claims in the Gongolo area of KwaZulu-Natal, South Africa. In: *Land Reform Revisited: Democracy, State Making and Agrarian Transformation in Post-Apartheid South Africa* (eds. B. Brandt and G. Mkodzongi), 124–148. Leiden: Brill NV.

Kamuti, T. (2018b). The Radical Challenge of Illegal Hunting on Privately Owned Wildlife Ranches in KwaZulu-Natal, South Africa. In: Proceedings of the Biennial Conference of the Society of South African Geographers, pp. 412–427. Bloemfontein: University of the Free State.

Kamuti, T. (2019). "If it pays it stays": the lobby for private wildlife ranching in South Africa. In: *People, Animals, Morality, and Marginality: Reconfiguring Wildlife Conservation in Southern Africa* (eds. J. Gewald, M. Spierenburg and H. Wels), 167–188. Leiden: Koninklijke Brill NV.

Kariuki, S. (2009). Agrarian Reform, Rural Development and Governance in Africa: A Case of Eastern and Southern Africa. Policy Brief 59. Johannesburg: Centre for Policy Studies.

Lindsey, P.A., Romanach, S.S., and Davies-Mostert, H.T. (2009). The importance of conservancies for enhancing the value of game ranch land for land mammal conservation in southern Africa. *Journal of Zoology* 277: 99–105.

Mace, G.M. (2014). Whose conservation? Changes in the perception and goals of nature conservation require a solid scientific basis. *Science* 345 (6204): 1558–1560.

Marsh, T.L. (2004). The economics of conserving wildlife and natural areas. *Wildlife Society Bulletin* 32 (1): 299–300.

McGranahan, D.A. (2008). Managing private, commercial rangelands for agricultural production and wildlife diversity in Namibia and Zambia. *Biodiversity Conservation* 17: 1965–1977.

Miller, B.W., Caplow, S.C., and Leslie, P.W. (2012). Feedbacks between conservation and social-ecological systems. *Conservation Biology* 26 (2): 218–227.

Minteer, B.A. and Miller, T.R. (2011). The new conservation debate: ethical foundations, strategic trade-offs, and policy opportunities. *Biological Conservation* 144: 945–947.

Moraes, R.C. (2012). Reclaiming the land, reclaiming the nation: adjacent or twin questions? *Agrarian South: Journal of Political Economy* 1 (1): 65–83.

Morrison, S.A. (2014). A framework for conservation in a human-dominated world. *Conservation Biology* 29 (3): 960–964.

Moyo, S. and Yeros, P. (2005a). *Reclaiming the Land: The Resurgence of Rural Movements in Africa, Asia and Latin America*. London: Zed Books.

Moyo, S. and Yeros, P. (2005b). The resurgence of rural movements under neoliberalism. In: *Reclaiming the Land: The Resurgence of Rural Movements in Africa, Asia and Latin America* (eds. S. Moyo and P. Yeros), 8–64. London: Zed Books.

Moyo, S., Jha, P., and Yeros, P. (2013). The classical agrarian question: myth, reality and relevance today. *Agrarian South: Journal of Political Economy* 2 (1): 93–119.

Muir-Leresche, K., & Nelson, R.H. (2000). Private property rights to wildlife: The southern African experiment. https://cei.org/studies-issue-analysis/private-property-rights-wildlife-southern-africa-experiment

Naughton-Treves, L. and Sanderson, S. (1995). Property, politics and wildlife conservation. *World Conservation* 23 (8): 1265–1275.

Nel, A. (2014). Sequestering Market Environmentalism: Geographies of Carbon Forestry and Unevenness in Uganda. Unpublished PhD thesis. Dunedin: University of Otago.

Scales, I.R. (2015). Paying for nature: what every conservationist should know about political economy. *Oryx* 49 (2): 226–231.

Sikor, T., Martin, A., Fischer, J., and He, J. (2014). Toward an empirical analysis of justice in ecosystem governance. *Conservation Letters* 7 (6): 524–532.

Smith, N. & Wilson, S. L. (2002). Changing Land Use Trends in the Thicket Biome: Pastoralism to Game Farming. Report Number 38. Port Elizabeth: Nelson Mandela University Terrestrial Ecology Research Unit.

Snijders, D. (2012). Wild property and its boundaries – on wildlife policy and rural consequences in South Africa. *Journal of Peasant Studies* 39 (2): 503–520.

Spierenburg, M. and Brooks, S. (2014). Private game farming and its social consequences in post-apartheid South Africa: contestations over wildlife, property and agrarian futures. *Journal of Contemporary African Studies* 32 (2): 151–172.

Steele, N. (1979). *Bushlife of a Game Ranger*. Cape Town: T.V. Bulpin Publications.

Sukhdev, P., Wittmer, H., and Miller, D. (2014). The economics of ecosystems and biodiversity (TEEB): challenges and responses. In: *Nature in the Balance: The Economics of Biodiversity* (eds. D. Helm and C. Hepburn), 135–152. Oxford: Oxford University Press.

Sullivan, S. (2013). Banking nature? The spectacular financialisation of environmental conservation. *Antipode* 45 (1): 198–217.

Torgerson, D. (2007). Expanding the green public sphere: post-colonial connections. *Environmental Politics* 15 (5): 713–730.

Wels, H. (2015). *Securing Wilderness Landscapes in South Africa: Nick Steele, Private Wildlife Conservancies and Saving Rhinos*. Leiden: Koninklijke Brill NV.

Section III

Energy, Emissions, and the Economy

10

Climate Change Regulations and Accounting Practices

Optimization for Emission-Intensive Publicly Traded Firms

Carol Pomare[1] and David H. Lont[2]

[1] *Ron Joyce Center for Business Studies, Mount Allison University, Sackville, New Brunswick, Canada*
[2] *University of Otago, Dunedin, New Zealand*

10.1   Introduction

Assessing transition risks and stranded asset risks (i.e., risks related to the transition to a more sustainable economy) requires a credible reporting system that captures its environmental, social, and governance impacts. Attempts by accounting academics to capture such information have been articulated through the lens of corporate social responsibility (CSR). Indeed, the triple bottom line and integrated reporting initiatives at the level of publicly traded firms have gained mainstream support in recent years with a focus on shared green values. However, these initiatives also were subject to a risk of greenwashing (i.e., the idea that some may claim to be green as a reputation management strategy without changing their actual practices), since one's ability to measure, report, and verify disclosures in relation to environmental, social, and governance impacts is not sufficiently developed. The triple bottom line and integrated reporting approach was designed to capture a wider range of forward risks and did include information on climate change. Despite its promise, it is not entirely clear if adequate models exist to incorporate transition and stranded asset risks.

A muted transition that proves inadequate to mitigate the effect of global warming is a threat to the stability of capital markets. Transitioning to a low-carbon economy is being made possible by technological breakthroughs in both fossil fuel use (i.e., increased use of cheap lower-carbon fuels, such as natural gas or a potential for carbon-capture technology) and the development of alternative energy sources that are increasingly becoming economically viable. However, if the resources to mitigate transition and stranded asset risks are inappropriately used, the new low-carbon economy may fail to maximize wealth, creating an alternative threat to capital markets.

Complicating the problem of economic efficiency are externalities, compounded by the impact of geopolitical or other counterforces that may distort the above economic and technological response. A move to new technology is part of a wider revolution that sees investors increasingly favoring knowledge-based firms, whose assets are primarily intangible, over fossil fuel-reliant firms. Such innovations are positive with regard to

Environmental Policy: An Economic Perspective, First Edition. Edited by Thomas Walker, Northrop Sprung-Much, and Sherif Goubran.
© 2020 John Wiley & Sons Ltd. Published 2020 by John Wiley & Sons Ltd.

mitigating potential economic shocks but have other consequences. For example, automation threatens many jobs and therefore creates a significant risk to labor and capital markets' stability. However, the self-driven technology also allows for much greater utilization of vehicles and road networks and therefore helps to lower carbon emissions related to transportation. Low or no emission vehicles also reduce carbon emissions and accident rates, providing significant economic and health benefits.

Academic research shows that emission-intensive publicly traded firms are affected by a sustainability-oriented economic system that takes disclosed and/or undisclosed greenhouse gas (GHG) emissions, as well as carbon taxes and cap-and-trade programs, into consideration when valuing emission-intensive companies. It is believed that such incremental changes toward sustainability mitigate the adverse impact of transition risks and the likelihood of stranded assets or a nonsustainable financial system. However, more research is needed to explore whether standard-setters and regulators should require a more substantial adjustment of accounting practices for emission-intensive companies to reflect their potential for transition and physical climate risk and for the fossil fuel industries (in particular, the risk their assets may become stranded).

This chapter explores the impact of changes in regulations and in accounting practices on the valuation of emission-intensive publicly traded firms. For more details, please refer to Pomare (2018a,b, 2019).

10.2 Emission-Intensive Publicly Traded Firms

A move toward a sustainability-oriented energy framework and a reporting system that captures environmental and social impacts, as articulated through the lens of CSR at the level of publicly traded firms, has gained more support in recent years with a focus on shared green values and greenwashing (Lyon and Maxwell 2011). For more details, please refer to Pomare (2018a,b, 2019).

Hawley (2015) believes that when thinking about short-term and long-term sustainability, an efficient market hypothesis (EMH) (i.e., the idea that capital markets are efficient in valuing publicly traded firms) is an inadequate model. Indeed, the drivers of transition and stranded assets risks principally include: (i) policy, legal, and political risks; (ii) technological development risks; and (iii) reputational risks (which in turn are linked to policy, legal, political and technological risks); and (iv) market and economic risks versus market and economic opportunities (Hawley 2015). These drivers interact in highly complex ways, which call for complex risk lenses.

As part of the complex risk lenses described above, greenwashing is what a company does when it focuses on claiming to be green through advertising, marketing, and stakeholder reporting, instead of implementing business practices that actually minimize its environmental impacts (Lyon and Maxwell 2011). Lyon and Maxwell (2011) believe that greenwashing is like a persuasion game, in which (i) one player has verifiable CSR information that he/she can either disclose or not disclose (i.e., moral hazard); and (ii) the other has little ability to verify the CSR information (i.e., information asymmetry).

As a deterrent against greenwashing, some argue for activist, legal, and/or regulatory sanctions. Lyon and Maxwell (2011) believe that one may develop an economic model of greenwashing in which an emission-intensive company discloses environmental information and activist audits or possibly the company may be penalized for disclosing positive (but not negative and realistic) environmental self-related information. However, Lyon and Maxwell (2011) also believe that greater activist pressure would not only deter greenwashing (i.e. positive outcome), but also lead to emission-intensive companies disclosing less about their environmental performance overall (i.e. negative outcome). In a related vein, others question the green narrative when it comes to a field dominated by multinational corporations (Tregidga et al. 2018).

Cooper (2016) discusses the recent growth of socially responsible investing (SRI) and its possible legal ramifications. Cooper (2016) argues that there may be legal ramifications for a company that misrepresents its environmental and social practices, even if such practices do not affect the expected future cash flows of the company, the company's cost of capital, or the price of the company's stock. The author adds that the Security Exchange Commission (SEC) Rule 10b-5 provided a private right for action for security fraud, including fraud related to greenwashing. However, SEC Rule 10b-5 requires an investor to sustain an economic loss as a result of a company's material misrepresentation for the right for action to be recognized (Cooper 2016). Cooper's reasoning extends to the idea that if SRI cannot affect a company's future cash flows and cost of capital, SRI cannot affect a company's value and stock price. Consequently, the SEC Rule 10b-5 cannot apply in most situations of greenwashing, since the economic loss element of the claim cannot be satisfied. Cooper (2016) adds that financial professionals make investment decisions based on companies' social and environmental representations, which greenwashing companies create mostly to gain reputational benefits in the eyes of investors and/or creditors. According to Cooper (2016), there should be a legal remedy for green misrepresentations, even when there is no economic loss (e.g., drop in share price). Indeed, independently of any economic harm, there may be moral harm in being misled into supporting environmental practices counter to one's beliefs.

According to Hawley (2015), there is a fiduciary obligation and associated need for audited climate change-related information, as investors and/or creditors demand high-quality information to counter the information asymmetry and moral hazard possibly related to greenwashing. Ballou et al. (2012) argue that accountants possess the ability to analyze and communicate information on sustainability. More specifically, they discuss how accountants can foster strategic integration through three facets of their accounting practice: (i) risk identification and measurement; (ii) financial reporting; and (iii) independent review and/or assurance. The authors use a survey of 178 corporate responsibility officers to explore how accountants add value to sustainability-related initiatives and they observe that these three areas of accounting practice (risk identification and measurement, financial reporting, and independent review/assurance) contribute to the strategic integration of sustainability initiatives. Accountants possess expertise in risk identification from experience with business risk auditing and risk management systems (Ballou et al. 2012), what some consider as the basis of strategic integration. However, Ballou et al. (2012) also observe that accounting professionals are rarely involved in sustainability initiatives. This finding suggests that increased involvement from accountants would provide significant benefits to companies and their stakeholders for sustainability-related insights.

In summary, a move toward a sustainability-oriented framework and a reporting system that captures the environmental, social, and governance impacts of publicly traded firms is gaining support despite greenwashing concerns. However, according to Hawley (2015), the question of how a greener economy may be affected by stranded assets and transition risks is an interesting test of the market efficiency theory. Indeed, there seems to be a lag in response to the transition and stranded asset risk analyses of fossil fuels, for example (Hawley 2015). In a related vein, Griffin et al. (2019) show that there is a tendency to underestimate physical risks associated with climate risks (i.e., financial assets underpricing climate risks). As such, capital markets may have underpriced carbon risks, much like capital markets may have underpriced subprime mortgage risks prior to the 2008 financial crisis.

10.3 Climate Change-Related Regulations and Their Impact

A question commonly asked by the public in relation to the concepts of stranded assets and a nonsustainable financial system is: "Are we not moving towards a more sustainability-oriented financial system already?" For more details, please refer to Pomare (2018a,b, 2019).

According to Ratnatunga et al. (2011), emission trading schemes (ETSs) have been established, as governments are coping with public pressure to foster an economically and socially sustainable economy. First, the Kyoto Protocol outlined various strategies to reduce global carbon emissions, including a proposal to establish international ETS, with trade carbon credits in carbon credit markets. The American Clean Energy and Security Act was then passed and proposed a variant to the cap-and-trade approach to reducing carbon emissions. Second, the Copenhagen Climate Summit attempted to establish a global climate framework. However, the outcome was reportedly a political accord that failed to achieve consensus (Ratnatunga et al. 2011). The 2015 Paris Agreement has intensified reciprocal influences between the financial world and issues around climate change (Stanley 2015). However, Dolsak and Prakash (2016) also discuss the political risks related to recent changes in governments in the US and other countries, with consequences that still need to be fully assessed.

Academic research shows that the stock price and equity value of emission-intensive companies reflect the present and/or future carbon taxes and cap-and-trade programs with mixed patterns of results. First, research shows that financial systems are taking the impact of present and/or future regulations in terms of GHG emissions (e.g., carbon taxes and cap-and-trade programs) into consideration when valuing emission-intensive companies, with a mixed pattern of results. In this context, some discuss the fact that previous empirical evidence provided mixed results on the relationship existing between corporate environmental performance and the level of environmental disclosures. Matsumura et al. (2014), as well as Griffin et al. (2017), showed that the link between levels of GHG emissions and stock valuation was negative in the US. However, Griffin et al. (2019) observed that this link between GHG emission and stock valuation is not always negative, as demonstrated in the case of Canada, where high levels of subsidies for emission-intensive sectors may be responsible for a positive overall relation. Clarkson et al. (2011) observed that companies with a higher pollution propensity disclosed more environmental information, which was interpreted as related to issues of reliability of voluntary environmental disclosures in an

Australian context (i.e., potentially related to greenwashing). As a consequence, Clarkson et al. (2011) advocated for the need for both enhanced mandatory reporting requirements and improved enforcement at a facility level. Second, carbon taxes and cap-and-trade programs share several major advantages over alternative policies: (i) both policies reduce GHG emissions by rewarding the lowest-cost emissions reduction; (ii) both policies encourage companies to develop new low-carbon technologies; and (iii) both policies generate government revenue, assuming that emission allowances are auctioned under cap-and-trade to be used against climate change.

The European Union pioneered the development of the ETS. Clarkson et al. (2015) show that ETSs are value relevant for emission-intensive companies in the European Union. However, according to the authors, carbon emissions affect the companies' valuation only to the extent that a company: (i) exceeds its carbon allowances under a cap-and-trade system; and (ii) is unable to pass on carbon-related compliance costs to consumers and end-users. As such, the companies' carbon allowances were not necessarily associated with valuations, but the allocation shortfalls were negatively associated with valuations. Clarkson et al. (2015) also found that the negative association between the companies' values and the carbon emission shortfalls is mitigated for companies with better carbon performance relative to their industry peers and for companies in less competitive industry sectors.

Chapple et al. (2013), using an Ohlson valuation model and an event study model, also provided insights into the capital market's assessment of the magnitude of the economic impact of a proposed ETS, as reflected in an Australian market capitalization system. They show that the future carbon permit price (i.e., in 2008, the Australian government announced its intention to introduce a national ETS by 2015) could be estimated, with the cost of GHG emissions on equity value going from AU$17 per ton to AU$26 per ton of carbon dioxide emitted, or from AU$15 to AU$74 per ton of carbon dioxide emitted, depending on the carbon emission profile of the selected Australian companies. Griffin and Sun (2013) examine the impact of the GHG emission allowances granted under California's cap-and-trade program on the balance sheets and income statements of the S&P 500 companies, as well as the significance of a cap-and-trade program for the US companies' financial statements. They showed that the average S&P 500 company's balance sheet and net income were to be adversely affected under several different accounting treatments for emission allowances, with the greatest impact being for emission-intensive companies.

Importantly, references that discuss the differences between these schemes do exist. Goulder and Schein (2013) discuss the concepts of a carbon tax, a "pure" cap-and-trade system, and a "hybrid" option (i.e., a cap-and-trade system with a price ceiling and/or price floor). They show that these approaches have different impacts, including some dimensions that have received little attention in previous publications. Although no one option dominates the others, the key finding of Goulder and Schein (2013) is that exogenous emissions pricing (whether through a carbon tax or through the hybrid option) has a number of important attractions over pure cap and trade. Beyond helping prevent price volatility and reducing expected policy errors in the face of uncertainties, exogenous pricing apparently helps avoid problematic interactions with other climate policies and potential wealth transfers to oil-exporting countries (Goulder and Schein 2013). Carl and Fedor (2016) investigate the current use of public revenues which are generated through both carbon taxes and cap-and-trade systems. They note that more than $28.3 billion in government "carbon revenues" are collected each year in 40 countries and another 16 states or provinces around the world. Of those revenues, they state that: (i) 27% (i.e., $7.8 billion) are reportedly used to subsidize "green" spending in energy efficiency or renewable energy; (ii) 26% (i.e., $7.4 billion) reportedly go toward state general funds; and (iii) 36% (i.e., $10.1 billion) are reportedly returned to corporate or individual taxpayers through paired tax cuts or direct rebates (Carl and Fedor 2016). Drawing from an empirical data set, Carl and Fedor (2016) also identify various trends in these systems' use of "carbon revenues" (i.e., in terms of the total revenues collected annually per capita in each jurisdiction) and offer qualitative observations on carbon policy designs. They argue that cap-and-trade systems (i.e., $6.57 billion in total public revenue) earmark a larger share of revenues for "green" spending (i.e., 70%), while carbon tax systems (i.e., $21.7 billion) more commonly refund revenues or otherwise direct them toward government general funds (i.e., 72% of revenues).

Also, Griffin (2018) argues a carbon tax or cap and trade may not actually work as well as intended. For example, it may shield energy companies from prior environmental damage. Griffin (2018) also cautions that a carbon price set too low may be insufficient to change behavior. Further, he suggests countries with carbon markets have not seen the expected reduction in emissions. His preference is for full disclosure of carbon risk to be made and for wider market forces to apply. This would include companies that have failed to meet their prior liabilities being held to account for their share of the liability.

10.4 Climate Change-Related Accounting Framework

There may be a gap between public opinion, academic research, and policymakers (Fahey and Pralle 2016). Carbon risks are related to a series of valuation, measurement, and financial reporting issues that are being scrutinized by accountants and/or auditors. For more details, please refer to Pomare (2018a,b, 2019).

In this context, the Carbon Tracker Initiative (2014) proposes to revisit the way emission-intensive companies are valued, including the accounting treatment of fossil energy reserves. The initiative argues that corporate disclosure of carbon risks has improved markedly over the past decade, but that some material amounts of risks are still hidden from reports issued by emission-intensive companies. The Carbon Tracker's argument is that, for emission-intensive companies, the challenge is the scale of their operational emissions and the realistic assessment of emissions associated with the product locked into their reserves because of regulations in terms of GHG emissions (Carbon Tracker 2014). Also, there has been little discussion in the literature on the valuation and reporting of assets capable of producing and using carbon credits, particularly when these assets are generated internally by the organization rather than acquired (Ratnatunga et al. 2011). Therefore, the Carbon Tracker (2014) recommends that regulators: (i) "require reporting of fossil fuel reserves and potential CO_2 emissions by listed companies and those applying for listing"; (ii) "aggregate and publish the levels of reserves and emissions using appropriate accounting guidelines"; (iii) "assess the systemic risks posed to capital markets and wider economic prosperity through the overhang of unburnable carbon"; and (iv) "ensure financial stability measures are in place to prevent a carbon bubble bursting."

According to Simnett et al. (2009a), concern over climate change and the need to limit GHG emissions have motivated officials to consider more stringent environmental standards. More specifically, there are three major issues that are being scrutinized by accountants: (i) the valuation and reporting of GHG emission permits; (ii) the measurement of tangible and intangible assets capable of creating GHG emissions; and (iii) the reporting of how a company is meeting its environmental and social responsibilities (Jones and Belkaoui 2009).

With regard to the first issue (the valuation and reporting of GHG emission permits), there are three main accounting issues that have been addressed in prior discussions: (i) how should the permits be valued at the grant date and over time, and when should permits be reported in the income statement?; (ii) how should the liability be valued over time and when should it be reported in the income statement?; (iii) how should the grant liability be recorded? (Jones and Belkaoui 2009). With regard to the second issue (measurement of tangible and intangible assets capable of creating GHG emissions), there has been little discussion on the valuation and reporting of assets capable of producing and using carbon credits. However, we will review the model presented by Ratnatunga et al. (2011). With regard to the third issue (reporting of environmental and social responsibilities), we will review the position of auditing practitioners and the literature on voluntary disclosures for internal versus external purposes.

10.4.1 Valuation and Reporting of GHG Emission Permits

Although standards continue to evolve, having a clear understanding of today's expectations and producing reliable information around sustainable performance may be priorities for emission-intensive companies in the US, for example (Ernst and Young 2016). For more details, please refer to Pomare (2018a,b, 2019).

The Clean Air Act (CAA) in the US established a cap-and-trade system for sulfur dioxide (SO_2) and nitrous oxide (NO_x) emissions of electric power producers. According to Ernst and Young (2016), many companies may need to account for related cap-and-trade activities associated with the program without specific accounting guidance. Many companies may also need to disclose the impact of climate change and the regulation of GHG emissions in their annual report or Form 10-K. Certain emission-intensive companies in the US were required to participate in the Environmental Protection Agency (EPA) and other carbon emissions programs for several years. However, varying accounting practices have emerged (Ernst and Young 2016): (i) under an inventory accounting model, emissions credits may be measured at a weighted-average cost; (ii) under the liability and gain recognition accounting model, the entity may not record an obligation to deliver emissions credits to the regulatory agency until the actual level of emissions for a given period may exceed the credits held on the balance sheet; (iii) under a vintage year swap accounting model related to US cap-and-trade programs, each emission credit or allowance may have a vintage year designation that is indicative of the first year an allowance may be used; (iv) under a derivative accounting model, companies may enter into forward contracts, swaps, and/or options pertaining to emissions credits used; and (v) under an intangible asset accounting model, companies may measure emissions credits or allowances issued to them and acquired in the open market at cost (Ernst and Young 2016).

Importantly, the Financial Accounting Standard Board (FASB) (standards-setter for US generally accepted accounting principles – GAAP) and the International Accounting Standard Board (IASB) (i.e. standards-setter for the International Financial Reporting Standards – IFRS) have been working on accounting for ETS since 2007. For more details and updates, please refer to Ernst and Young (2016).

10.4.2 Measurement of Tangible and Intangibles Assets Creating GHG Emissions

With regard to the second issue, there has been little discussion in the literature of the valuation and reporting of assets capable of producing and using carbon credits, particularly when these assets are generated internally by the emission-intensive company rather than acquired. For more details, please refer to Pomare (2018a,b, 2019).

According to Ratnatunga et al. (2011), the cap-and-trade approach remained the preferred political and economic approach for reducing carbon emissions in many countries. As such, there are two types of carbon credits that may be recorded by accountants: (i) the carbon credits issued by governments (similar to a ration card); and (ii) the carbon credits created by an emission-intensive company internally, to be sold in an ETS or used as an offset to reduce carbon liability. The first type of carbon credit originates as governments issue credits or sell them at a grant-date price (Ratnatunga et al. 2011). The second type of credit originates as governments may permit some emission-intensive companies to issue their own credits, if these companies document some carbon sequestration strategies during a given period (Ratnatunga et al. 2011).

The credits internally generated require a different accounting treatment to those issued by the government, as the underlying assets that have the capability of creating such credits are tangible or intangible (Ratnatunga et al. 2011). More specifically, the authors argue that a special case should be made for the valuation and recording of carbon sequestration assets, that is, with the Environmental Capability Enhancing Asset (ECEA) and Capability Economic Value of Intangible and Tangible Assets (CEVITA) models.

First, they propose an ECEA model for valuing an emission-intensive company's noncurrent carbon sequestration and emission capabilities. According to the authors, an ECEA is the total intangible capacity of an entity to produce carbon credits. Ratnatunga et al. (2011) propose obtaining these coefficients through a consensus from experts within the organization or external experts. Second, they propose CEVITA, a technique that not only makes valuations more relevant and contextual but also shows how tangible and intangible asset combinations provide true capability values. CEVITA, an index-based valuation approach, uses some amount of professional judgment to provide valuations and performance measurements. Ratnatunga et al. (2011) believe that this approach provides a potentially useful basis for the valuation of assets having carbon emission and sequestration capabilities for all types of carbon credits.

10.4.3 Reporting of Environmental and Social Responsibilities

With regard to the third issue, this can be explored through GHG emission assurance engagements. According to the International Standard Board on Assurance Engagement (ISAE), with increasing attention given to the link between GHG emissions and climate change, many emission-intensive companies are quantifying their GHG emissions for internal management purposes, and an increasing number are preparing GHG statements for external management purposes as well. For more details, please refer to Pomare (2018a,b, 2019).

In terms of internal factors related to GHG emission assurance engagements, Hartmann et al. (2013) observe that carbon reduction programs and corporate emissions reporting have expanded across emission-intensive companies in response to climate change. This development is driven by internal institutional demands and value-creation considerations. The consequences of these developments for management accounting and control (MAC) are not clear, despite anecdotal evidence that suggests an increasing effort to incorporate carbon accounting into traditional internal decision and reporting processes (e.g., activity-based costing strategies related to environmental activities). The reason for such a lack of clarity is reportedly the absence of academic debate on GHG emission assurance engagements from the MAC perspective. Hartmann et al. (2013) seek to stimulate academic debate by identifying key theoretical and empirical shortcomings and by outlining some directions for future studies on internal strategies for carbon accounting. These directions are inspired by more established MAC research streams to guide MAC research in the emerging field of internal carbon accounting.

In terms of external factors related to GHG emission assurance engagements, Green and Zhou (2013) conduct a detailed examination of the existing external assurance practices for carbon emission disclosures with an international sample of 3008 companies across 43 countries between 2006 and 2008. The demand for assurance services was found to be mainly from companies in Europe and companies from emission-intensive industries, with a majority hiring a specialist assurer (Green and Zhou 2013). Using interview data, Martinov-Bennie and Hoffman (2012) explore the perception of greenhouse gas and energy (GGE) audits within the context of the legislated Audit Determination in Australia. They provide evidence of the accounting profession's perceived influence on the views and methodologies adopted in the Audit Determination. Their results also indicated that accounting firms are perceived to be the ones gaining from an increasing share of the evolving sustainability assurance markets in Australia.

As a consequence, disclosures may be: (i) published as a standalone document; (ii) included as part of a broader sustainability report or an emission-intensive company's annual report; or (iii) made to support inclusion in a "carbon register." Simnett et al. (2009b) discuss the types of disclosures that may be assured and outline the issues involved in developing an international assurance standard on GHG emissions disclosures for Asia-Pacific, North America, and Europe. First, the Climate Disclosure Standards Board (CDSB), a non-profit organization working to provide material information for investors and financial markets through the integration of climate change-related information into mainstream financial reporting, acts as a forum for collaboration on how existing standards and practices can be used to link financial and climate change-related information using its Climate Change Reporting Framework (Climate Disclosure Standards Board 2016). Second, at its December 2007 meeting, the International Auditing and Assurance Standards Board (IAASB) reportedly approved a project to consider the development of a standard aimed at promoting trust and confidence in external disclosures of GHG emissions, including disclosures required under ETS. The IAASB later approved assurance engagements on greenhouse gas statements. Simnett et al. (2009a) also argue that the development of a new international assurance standard on GHG disclosures is an appropriate response by the auditing and assurance profession to meet challenges in terms of GHG emission assurance engagements. Ge et al. (2017) further argue that expanding the use of the IAASB's assurance standards to other assurance providers would be advisable. More specifically, Ge et al. (2017) provide an analysis of the reporting of ethical and quality control requirements that may need to be expanded in line with the IAASB's assurance standards.

10.5 Conclusion

Academic research shows that financial systems are becoming more sustainability oriented. Disclosed and/or undisclosed GHG emissions, as well as carbon taxes and cap-and-trade programs, are now taken into consideration when valuing emission-intensive companies. These changes toward a more sustainability-oriented financial system progressively help mitigate the adverse impact of transition risks while also better defining the risks associated with stranded assets. A failure to manage these risks is a threat to the financial system, but the effectiveness of carbon prices also needs to be monitored to ensure the markets work as intended.

Recent political commitments to reduce emissions require countries and firms to deliver a significant reduction in carbon emissions. Monitoring progress based on quality information is crucial if we are to develop a more sustainability-oriented financial system that mitigates transition risks and better manages stranded asset risk. Not correctly pricing carbon decreases the ability of capital markets to move toward a more sustainability-oriented financial system. If markets underprice carbon risks, as much as they underpriced the subprime mortgage and other financial instruments risks prior to the 2008 financial crisis, then the consequences to the financial system and real economy could be even more damaging.

Acknowledgments

We would like to thank the Chartered Professional Accountant (CPA) and the Canadian Academic Accounting Association (CAAA) for their joint financial support of the research project through a Financial Accounting Research Grant. We would like to thank the Marjorie Young Bell Fund and the J.E.A. Crake Foundation.

References

Ballou, B., Casey, R.J., Grenier, J.H., and Heitger, D.L. (2012). Exploring the strategic integration of sustainability initiatives: opportunities for accounting research. *Accounting Horizons* 26 (2): 265–288.

Carbon Tracker. (2014). The carbon tracker initiative. www.carbontracker.org

Carl, J. and Fedor, D. (2016). Tracking global carbon revenues: a survey of carbon taxes versus cap-and-trade in the real world. *Energy Policy* 96: 50–77.

Chapple, L., Clarkson, P.M., and Gold, D.L. (2013). Capital market effects of the proposed ETS. *Abacus* 49: 1–33.

Clarkson, P.M., Overell, M.B., and Chapple, L. (2011). Environmental reporting and its relation to corporate environmental performance. *Abacus* 47: 27–35.

Clarkson, P.M., Li, Y., Pinnuck, P., and Richardson, G. (2015). The valuation relevance of greenhouse gas emissions under the European Union carbon emissions trading scheme. *European Accounting Review* 24 (3): 551–580.

Climate Disclosure Standards Board (2016). About the Climate Disclosure Standards Board. www.cdsb.net/

Cooper, C.B. (2016). Rule 10b-5 at the intersection of greenwash and Green investment: the problem of economic loss. *Environmental Affairs Literature Review* 405: 234–237.

Dolsak, N. & Prakash, A. (2016). The US environmental movement needs a new message. The conversation. http://theconversation.com/the-us-environmental-movementneeds-a-new-message-70247

Ernst & Young (2016) Accounting guidance for emissions programs. www.ey.com/us/en/industries/oil---gas/carbon-market-readiness---4---accounting-guidancefor-emissions-programs

Fahey, B.K. and Pralle, S.B. (2016). Governing complexity: recent developments in environmental politics and policy. *Policy Studies Journal* 44: 28–49.

Ge, Q., Simnett, R., & Zhou, S. (2017). Expanding the Use of the IAASB's Assurance Standards to Other Assurance Providers: An Analysis of the Reporting of Underpinning Ethical and Quality Control Requirements. Available at SSRN 2837397.

Goulder, L.H. and Schein, A.R. (2013). Carbon taxes versus cap and trade: a critical review. *Climate Change Economics* 4 (3): 1–28.

Green, W. and Zhou, S. (2013). An international examination of assurance practices on carbon emissions disclosures. *Australian Accounting Review* 23: 54–66.

Griffin, P. (2018) Taxing carbon may sound like a good idea but does it work? http://theconversation.com/taxing-carbon-may-sound-like-a-goodidea-but-does-it-work-104871

Griffin, P. and Sun, Y. (2013). Cap-and-trade emission allowances and US companies' balance sheets. *Sustainability Accounting, Management and Policy Journal* 4 (1): 7–31.

Griffin, P., Lont, D., and Sun, Y. (2017). The relevance to investors of greenhouse gas emission disclosures. *Contemporary Accounting Research* 34: 1265–1297.

Griffin, P., Lont, D. & Pomare, C (2019). The curious case of Canadian corporate emission valuation. Proceedings of the Accounting & Finance Association of Australia and New Zealand (AFAANZ) Annual Conference. www.afaanzconference.com

Griffin, P., Lont, D., and Lubberink, M. (2019). Extreme high surface temperature events and equity-related physical climate risk. *Weather and Climate Extremes* 26: 100220.

Hartmann, F., Perego, P., and Young, A. (2013). Carbon accounting. *Abacus* 49: 539–563.

Hawley, J. (2015). Carbon risks and investment implications. https://www.insight360.io/carbon-risk-implications

Jones, S. and Belkaoui, A. (2009). *Financial Accounting Theory*. Australia: Engage Learning.

Lyon, T.P. and Maxwell, J.W. (2011). Greenwash: corporate environmental disclosure under threat of audit. *Journal of Economics and Management Strategy* 20: 3–41.

Martinov-Bennie, N. and Hoffman, R. (2012). Greenhouse gas and energy audits under the newly legislated Australian audit determination: perceptions of initial impact. *Australian Accounting Review* 22: 195–207.

Matsumura, E., Prakash, R., and Vera-Muñoz, S. (2014). Firm-value effects of carbon emissions and carbon disclosures. *Accounting Review* 89 (2): 695–724.

Morgan Stanley. (2015). *The Investors' Guide to Climate Change.* www.theatlantic.com/sponsored/morgan-stanley/the-investors-guide-to-climate-change/696/

Pomare, C. (2018a). Responsible investing and environmental economics. In: *Handbook of Engaged Sustainability* (eds. S. Dhiman and J. Marques). Heidelberg: Springer.

Pomare, C. (2018b). Socially Responsible Investing. Continuing Education Course and Presentation at the 2018 Annual Conference of the Canadian Institute of Financial Planning, Continuing Education for Certified Financial Planners (CFP), Halifax, Canada.

Pomare, C. (2019). United Nations (UN) Sustainable Development Goals (SDG) and accountability framework. In: *Encyclopedia of the UN Sustainable Development Goals: Partnerships for the Goals* (eds. W.L. Filho, P.G. Ozuyar, P.J. Pace, et al.). New York: Springer.

Ratnatunga, J., Jones, S., and Balachandran, K.R. (2011). The valuation and reporting of organizational capability in carbon emissions management. *Accounting Horizons* 25 (1): 127147.

Simnett, R., Nugent, M., and Huggins, A.L. (2009a). Developing an international assurance standard on greenhouse gas statements. *Accounting Horizons* 23 (4): 347–363.

Simnett, R., Vanstraelen, A., and Chua, W.F. (2009b). Assurance on sustainability reports: an international comparison. *Accounting Review* 84 (3): 937–967.

Tregidga, H., Milne, M.J., and Kearins, K. (2018). Ramping up resistance: corporate sustainable development and academic research. *Business & Society* 57 (2): 292–334.

11

The Economic Aspects of Renewable Energy Policies in Developing Countries

An Overview of the Brazilian Wind Power Industry

Elia E. Cia Alves[1] and Andrea Q. Steiner[2]

[1] *Federal University of Paraíba, Joao Pessoa, Paraíba, Brazil*
[2] *Federal University of Pernambuco, Recife, Pernambuco, Brazil*

11.1   Introduction

The use of renewable energy[1] sources has increased significantly in the global energy mix; this can be seen in the doubling of renewable electricity capacity between 2005 and 2015 compared to the previous decade (IEA 2017). Several countries started promoting public policies to encourage renewable energy. In 2005, 43 countries – most of them developed – had adopted at least one type of policy or goal linked to renewable energy. By 2014, this had increased to 138 countries (REN21 2015).

This shift can be attributed to three major events. First, concerns about the volatility of oil prices; after the first oil price shock, which generated a worldwide recession, energy became an essential part of strategic economic growth. Second, countries' quest for energy self-sufficiency; in addition to rising oil prices, countries did not wish to become dependent on a small number of oil-producing countries (Maugeri 2010). Finally, since the creation of the UN Framework Convention on Climate Change (UNFCCC) in 1992, there have been rising concerns around climate change and the environmental effects of carbon emissions, as well as an emphasis on environmental and climate problems in international agendas driving policy adoption (Falkner 2013).

The international community has become more concerned about greenhouse gases (GHGs) and related environmental impacts. In 2013, for example, energy generation and use accounted for 34% of global GHG emissions, and the largest volume of emissions is directly related to electricity (IPCC 2014). Furthermore, most of the energy produced in the world came from nonrenewable resources: 81.2% from fossil sources and 4.7% from nuclear sources. Only a small part came from renewable sources: 10.2% from biofuels, 2.4% from hydroelectric plants, and 1.4% from wind, solar and other renewables (IEA 2016).

Studies on climate change policies link the global diffusion of renewable energy policies (REPs) to several elements, yet there is still a considerable gap in research. Most studies

Environmental Policy: An Economic Perspective, First Edition. Edited by Thomas Walker, Northrop Sprung-Much, and Sherif Goubran.
© 2020 John Wiley & Sons Ltd. Published 2020 by John Wiley & Sons Ltd.

have focused on developed countries while developing nations account for an increasing share of emissions: 70% of carbon emissions produced by the energy sector worldwide in 2010 (IPCC 2014). The literature has also partially ignored the economic aspects of REP diffusion. In many cases, renewables have been unable to compete on an equal footing with other energy sources. Renewable energy sources have high capital costs and many countries have heavily subsidized fossil fuels, thus diminishing renewable sector competitiveness. REPs, in turn, help level the playing field for these comparatively new and costly technologies. Understanding the dynamic interplay of energy markets, technologies, and policies has never been more critical – policies adopted today are crucial for investment decisions and long-term trends.

We aim to help diminish this gap in the literature by answering the following question: what are the main economic drivers of REP diffusion?

In section 11.2, we define our object – REPs – and its global diffusion. In section 11.3, we review the literature on the drivers of REP adoption and present potential gaps. In section 11.4, we map six potential economic aspects to explain environmental policies adoption: (i) income, (ii) energy prices, (iii) financing, (iv) trade, (v) foreign direct investment (FDI), and (vi) lobbying, arising from a case study regarding the economic aspects of wind energy policy adoption in Brazil. We focus on the case of Brazilian wind sector regulation because it does not fit properly with traditional explanations for REPs adoption: energy security or climate change concerns. The Brazilian energy matrix is one of the cleanest in the world, with more than 40% of energy consumption coming from renewable sources (BEN 2018). We highlight the effects of the 2008 international crisis and reveal potential causal mechanisms not foreseen by the existing literature.

In our final considerations (section 11.5) we argue that the quest for energy cannot be analyzed solely from the standpoint of states' strategic decisions. One should also consider that the pervading process of political and economic liberalization since the 1990s has helped increase the number of actors involved in decision making. International interdependence became a key factor to explain the increasing number of REPs worldwide.

Thus, theoretical approaches must cover the multitude of domestic and international actors influencing states' policy choices. To tackle this, we apply a multifaceted theoretical and methodological apparatus. We build on policy diffusion and international political economy literature and shed light on the case of the Brazilian wind sector through the analysis of official reports, the review of specialized literature and through semi-structured interviews with specialists from different private, government, academic, and civil society sectors.

11.2 Renewable Energy Incentive Policies: A Space–Time Spread

Policies to support renewable energy can be divided into three categories: technological (including R&D), industrial (protection to domestic industry), and market regulation policies (Dutra 2007; Podcameni 2014).

Traditionally, REPs have focused mainly on market regulation. Policies may target different aspects of market equilibrium: quantity (quotas, such as renewable portfolio standards – RPS), price (establishing differentiated tariffs, such as feed-in tariffs – FITs) or auctions.[2] FITs and quotas became the two types of market policies that predominated during the 1990s (REN21 2016).

Feed-in tariffs, still popular to date, offer a guaranteed fixed payment rate for renewable energy (or fixed premiums) for a certain period (approximately 15–20 years), therefore encouraging producers to sell to the grid by allowing them to assume lower risks with a fixed rate and a long-term agreement (REN21 2016). Although by 2003, only 18 countries – all European – had implemented FITs (Busch and Jörgens 2005), by 2015 it became one of the most popular measures in the world.

Initially implemented in Germany and Denmark, its operation assumes that all the energy produced by the generator is injected into the grid (Sovacool 2013). Germany, alongside Denmark, has pioneered the adoption of premium rates (a variant of FITs) and its experience has become a reference for several countries that have adopted policies to support the wind industry. The country has developed a strong industrial base of wind equipment and has introduced major technological innovations in this segment. In 2013, it had the third largest installed capacity of wind energy in the world (IRENA-GWEC 2013).

Renewable investments started in the midst of the oil crises that culminated in profound questions about energy choices. Initially, the country had sought alternatives in the nuclear and coal sectors but these sources began to be increasingly questioned by public opinion. In 1980, the German Parliament recommended prioritizing energy efficiency measures and encouraging renewable energies. In this context, several demonstrative projects in the areas of wind and photovoltaics began to emerge from R&D projects (MME 2009b).

The 1986 Chernobyl accident fueled public opposition to nuclear power. Social democrats and the Green Party, which had a significant presence in the country, demanded the closure of nuclear plants and the formation of a commission to analyze political alternatives. In 1989, the concept of payment for cost coverage from electricity generation from renewable sources, a root of the feed-in tariff mechanism, was approved (Camillo 2013). In 1990, the German Parliament approved the Electricity Feed-in Law, providing for a long-term contract for power generators from renewable sources. Incentives soon began to expand markets and strengthen the political power of the wind industry: subsidies on technology purchase and capital cost, focusing expansion toward developing countries and technical advice on wind projects (IRENA-GWEC 2013; Podcameni 2014).

As a result, the industry grew stronger despite attacks by opposition sectors. With the consolidation of the market over two decades, and with the support of civil society and political coalitions, in 2000 the government launched a Renewable Energy Sources Code, reforming and broadening the scope of FITs. In 2008, a new reform was implemented due to the constraints caused by the financial crisis, but the system's operation remained guaranteed (Bjork et al. 2011). With the maturation of the wind power industry and the consolidation of Denmark and Germany's technological leadership in the international market, foreign financing programs were eliminated, highlighting export credit guarantees to finance both wind farms, as new production units abroad (Camillo 2013).

Along with Denmark and Germany, the US has also pioneered the development of a strong wind energy market. Like Germany, in response to the 1973 oil crisis, the government began implementing research support programs. In 1978, market incentives, such as tax credits and an instrument that was the embryo of FITs in the country, were implemented (IEA 2004; Salino 2011). An electricity market for independent producers was created through the Public Utility Regulatory Policies Act (PURPA), which required utility companies to purchase energy from independent producers at a certain price (Azuela et al. 2014).

Since the 1980s, the country has used intermittent production tax credits to promote the development of wind farms. This inconsistency was due to phases of oil price reduction, with consequent reduction of incentives to renewables, making it difficult to expand the sector systematically. Only in 1989, with the Gulf War, concerns about energy security, fuel price volatility and the reduction of the cost of new wind technologies, did the US government usher in a new phase of wind energy policy (Podcameni 2014).

Over the following decades, the country adopted different types of financial measures, such as the granting of accelerated depreciation, to encourage further development of renewable energy (Bjork et al. 2011). In 1992, the Renewable Energy Production Tax Credit, a 10-year tax credit, was implemented for renewable energy producers and private utilities – a measure that proved essential for the industrial development of the wind sector. Nonetheless, the most relevant incentive policies occurred within state governments. Many are now offering financial incentives such as property tax exemption, green credits, state-produced equipment credits, state tax exemptions, and production tax credits (Chandler 2009; Lyon and Yin 2010; Matisoff and Edwards 2014). Still, the most robust instrument that gained national support was the quota system.

The RPS, as the US quota system is known, has proliferated at the state and international levels. What began as an idea first described in detail in the pages of the *Electricity Journal* has emerged as an important policy instrument for encouraging renewables (Wiser et al. 2007). Quotas mandate that a percentage of the power supply must come from a renewable energy source. In other words, quotas impose on energy suppliers the obligation to obtain part of their power supply from renewable sources. Another example is the renewable portfolio obligations (RPO) in the United Kingdom (REN21 2016).

The most important precursor proposal in this line was the Alternative Energy Law, implemented in the state of Iowa (USA) in 1983, although discussions about a detailed RPS project took place in California in 1995 (Azuela et al. 2014). Following adoption, the measure spread rapidly and was interpreted as a promarket instrument as it did not require an explicit allocation of government funding. Indeed, the RPS combine policy regulatory strategies and market mechanisms, a hallmark of US environmental and energy policy innovations.

In 2009, the US Congress tried to make the measure a national law, facing strong resistance against the legislation. Even so, in the same year, the country expanded its promotion of industrial policies, increasing tax credits for the wind, solar, and geothermal sectors under the American Recovery and Reinvestment Act (ARRA) (Bjork et al. 2011).

In the 2000s, auctioning mechanisms became popular among developing countries, although it had previously failed during the 1990s in the United Kingdom (Del Río and Linares 2014). The growing interest in auctions sky-rocketed due to the long-term financial stability that contracts establish, but, at the same time, imposing ceiling prices on suppliers. To attract investment in generation, energy auctions for long-term energy contracts (15 and 30 years) were created to direct energy contracting by distribution utilities. This scheme aims at reducing risks for investors, while the auction by least price stimulates economic efficiency and in principle gives correct signals for system expansion cost through competition (Melo et al. 2010).

According to Azuela et al. (2014), auctions could represent the "best of both worlds" of FITs and RPOs, providing stable revenue guarantees for investors (similar to the FIT mechanism) while at the same time ensuring that the renewable generation target will be met precisely (similar to an RPO). As a result, despite the high transaction costs, the increased competition reduced energy prices (IRENA and CEM 2015). Notably, the option for auctions grew at a faster pace between 2010 and 2015 in developing countries and, as far as we are concerned, literature on REPs diffusion has not tackled this quest in depth.

11.3 The Drivers of REP Adoption: Advances and Gaps in the Literature on REPs Diffusion

In order to understand the process discussed above, many hypotheses are presented. In general, the literature surveyed suggested separate causal mechanisms for international or national drivers. Although simplistic, given the difficulty in differentiating between domestic and international phenomena, such a distinction can provide important insights into the trigger effects that lead countries to adopt certain policies. In short, the main finding regarding this classification, under a comparative strategy analysis, is that while domestic factors play a major role in REPs adoption in developed countries, drivers from the international arena, which encompasses the role of intergovernmental organizations (IOs), civil society movements, private groups and pioneer countries, are determinant to REPs diffusion toward developing countries (Cia Alves et al. 2019).

Besides that, the literature points to a broad set of environmental policy-related forces covering social, political, and environmental aspects of REPs diffusion. Regarding the first, several works shed light on the pressure from international organizations (Kern et al. 2005; Holzinger et al. 2008; Jenner et al. 2012; Biesenbender and Tosun 2014; Massey et al. 2014; Alizada 2018; Cia Alves et al. 2019) and nongovernmental organizations (NGOs) (Kern et al. 2005; Massey et al. 2014). Concerning political factors, Stadelmann and Castro (2014) and Cia Alves et al. (2019) covered energy security issues; Fankhauser et al. (2014) considered national political regimes and several works analyzed the influence of being part of an international regime, such as the Kyoto Protocol (Nicolli and Vona 2012; Biesenbender and Tosun 2014; Stadelmann and Castro 2014; Cia Alves et al. 2019). Baldwin et al. (2019) contribute to the literature on emulation driven by coercive donor–recipient aid relationships. In relation to environmental factors, the level of CO_2 emissions (Kern et al. 2005; Holzinger et al. 2008; Biesenbender and Tosun 2014) or natural resources availability (Jenner et al. 2012; Stadelmann and Castro 2014) are important.

Nevertheless, we identified two important gaps in the literature: samples analyzed and economic dimensions. The samples analyzed had a limited representation of developing countries, with contributions from Fankhauser et al. (2014), Stadelmann and Castro (2014), Baldwin et al. (2019), Alizada (2018), and Cia Alves et al. (2019), noting that most of this research considering developing countries is more recent.

As for the second gap, we noticed the importance of in-depth exploration of the economic dimensions of REPs diffusion. Regarding this, the literature analyzed investigates the importance of the economic dimension on similar grounds to other vectors and we consider that economics possibly is, in many aspects, the ultimate explanation to REP adoption. As Oates and Portney (2003) suggest, economics plays a growing role in both the setting of standards for environmental quality and the design of regulatory measures. We highlight that the existing literature had already addressed, unsystematically, economic aspects of policy diffusion. Falkner (2014) highlights the importance of strengthening the analysis of linkages between environmental and economic cooperation in order to tackle complex global problems, such as climate change. As an example, he comments on the UN environment summits, stressing the importance of integrating global economic and environmental policy.

Therefore, we looked for the main economic vectors of REPs diffusion discussed in the literature, in order to address the Brazilian case for wind power. The main factors identified were: (i) income – appearing directly or indirectly in all mentioned works; (ii) energy prices (Jenner et al. 2012); (iii) financing (Massey et al. 2014; Stadelmann and Castro 2014); (iv) trade (Holzinger et al. 2008; Biesenbender and Tosun 2014; Alizada 2018); (v) FDI (Biesenbender and Tosun 2014; Alizada 2018), and (vi) lobbying (Holzinger et al. 2008; Jenner et al. 2012; Biesenbender and Tosun 2014; Nicolli and Vona 2019).

11.3.1 Income

Several studies find a correlation between measures of economic wealth and environmental policy adoption (Kern et al. 2005; Holzinger et al. 2008; Nicolli and Vona 2012). Apergis and Danuletiu (2014) report strong evidence of the interdependence between renewable energy consumption and economic growth, indicating that renewable energy is important for economic growth and likewise economic growth encourages the use of more renewable energy. According to Sadorsky (2009), increased economic growth and demand for energy in emerging economies are creating an opportunity for these countries to increase their usage of renewable energy. Although they are not dealing specifically with policies, they argue that the presence of causality provides an avenue to the implementation of policies that enhance the development of the renewable energy sector.

The basic mechanism behind this variable's influence is that the higher the income of a country, the greater the demand for energy and, possibly, the higher the demand for renewable sources. Furthermore, the preferences of consumers in richer societies tend to change toward more environmentally friendly alternatives.

The argument could also be supported by the logic of the environmental Kuznets curve,[3] which suggests that the richer the country is, the more it can afford renewable technologies. It is reasonable to consider that richer countries have more financial capacity to invest in capital-intensive renewable energy technologies (Stadelmann and Castro 2014). Consequently, countries with higher per capita income have a greater ability to invest in technologies that are initially offered at prohibitive costs, since they can take advantage of their leadership in innovation (Dunning 2001).

11.3.2 Energy Prices

Another consideration before implementing renewable energy is the price of alternative energy sources. Apergis (2019) argues that oil prices are the most volatile among all commodity prices and affect real and financial economies. In that context, Maugeri (2010) claims that, for years, low oil prices have discouraged research and expansion into renewable energy resources and dissuaded technological advances in the search for energy efficiency (Maugeri 2010).

The oil market structure has few producers with a limited and finite resource. Oil is becoming scarcer, and as its scarcity increases so does its price. Due to a substitution effect, it is expected that the increase in oil prices will encourage governments, businesses, and households to reduce their demand for oil by buying more efficient products and seeking out alternative sources. This theoretical prediction implies a direct relationship between the increase in oil prices and the increase in the demand for renewable energy.

From this perspective, the energy choices of countries are significantly influenced by oil prices. The rise in energy input prices affects the costs of the global industry, promoting investments in alternatives for the energy sector. Besides that, oil prices also connect politically to REPs adoption, once governments seek energy security by overcoming dependency on exhaustible sources of energy and on oil-exporting countries, which are usually politically unstable (Sovacool and Mukherjee 2011).

In fact, as shown in Figure 11.1, the rise in oil prices of the early 2000s was followed by the spread of REPs. Although we are not claiming causality, it is reasonable to expect that the greater the oil prices, the more reason governments may have to incentivize other energy technologies.

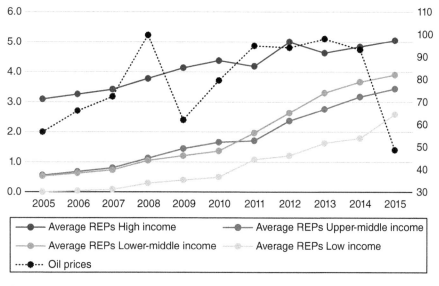

Figure 11.1 Oil prices and renewable energy policy adoption. *Source:* Authors' elaboration based on publicly available data from EIA (n.d.) and REN21 (2005, 2007, 2009, 2010, 2011, 2012, 2013, 2014, 2015, 2016). Oil price information from oil spot price FOB (dollars per barrel).

11.3.3 Financing

Massey et al. (2014) discuss the high costs of adaptation projects and that financial support from international grants or funds can serve as a variable for adaptation policy uptake. Stadelmann and Castro (2014) also highlight this issue and test whether developing country governments may have reacted to the financial opportunity provided by the Clean Development Mechanism (CDM) and to RE-related capacity building under development and environmental finance initiatives.

Not dealing specifically with REPs diffusion but from a broader perspective of renewable energy financing, Newell (2011) comments that the landscape of energy finance faces a series of challenges regarding sourcing of and access to finance, and the predictability of flow, which the 2008 financial crisis had made more uncertain. From the domestic perspective, the author also suggests that with the resurgence of a state capitalism model in the energy sector, especially in those countries that will increasingly dominate energy production and consumption such as China and India, public institutions are heavily involved in governing public energy finance. In addition, they largely control the underlying infrastructure investments that affect opportunities in the energy sector. In spite of that, Newell points out that governments account for less than 15% of global economy-wide investment, which evidences the huge gap in renewable energy projects funding.

Therefore, Newell (2011) also reinforces the idea presented before, suggesting that multilateral institutions such as the World Bank and International Monetary Fund (IMF) have a key role to play in energy financing, as do regional development banks and bilateral donors. In this context, Newell claims that those organizations maintain direct channels of influence over the states that created them and to whom they lend. The power they wield is clearly a function of which states fund and support them, representing an important vector on energy choices of developing countries.

Through seemingly neutral policy advice and technical assistance, embedded ideas may influence energy governance. Nakhooda (2011) discusses this issue, arguing that multilateral development banks have played a significant role in influencing institutions and instruments that should govern the energy sector. According to the author, moments of crisis provide opportunities to press for far-reaching reforms, as was seen with the promotion of electricity privatization in the wake of the East Asian financial crisis, as part of broader packages of macroeconomic reform (Nakhooda 2011).

Globally, there are at least 20 different bilateral and multilateral funds for climate change (World Bank 2010). One of the most important credit lines is the Climate Investment Funds, created in 2008 to support the implementation of country-led programs and investments on climate mitigation. Climate Investment Funds provide financing to these programs and investments through the provision of concessional loans (World Bank 2011).

11.3.4 Trade

Trade openness appears in some of the work presented above (Stadelmann and Castro 2014; Cia Alves et al. 2019) and in one of the most important publications on policy diffusion (Simmons et al. 2008).

The role of trade in diffusion studies is twofold. On one hand, there is policy diffusion by regulatory competition. Policies are supposed to be diffused between pairs of countries that have strong ties, such as having high bilateral trade, due to a competition mechanism (Prakash and Potoski 2006; Baccini et al. 2013). Saikawa (2013) states that adoption by an importing country puts pressure on the exporting country to adopt the policy as well. Trade liberalization facilitates the movement and exchange of goods and services, promoting competitive pressure on country regulation. From the point of view of technological competition, not encouraging the renewable sector can signal asset misallocation. As the consumption and production of tradable goods and services involve the efficient use of energy, trade liberalization influences total energy demand, including renewable energy. Therefore, an exporting country's adoption of REPs gives it a competitive advantage.

On the other hand, trade is generally related to the flexibility of environmental policies, consequently minimizing the chances of REPs adoption, due to the costs related to their implementation. Oates and Portney (2003) relate trade liberalization to the increase in political interaction of governments with various interest groups in the setting of environmental standards and the choice of regulatory instruments, which may negatively influence REPs adoption.

11.3.5 FDI

Similar to trade liberalization, FDI might also have a positive impact on REP diffusion, once investors can promote regulatory pressure domestically. Elkins et al. (2006) address the role of Germany and the United States in encouraging countries to join bilateral investment agreements, which implies possible domestic policy adaptation, in face of the need for a country to increase its attractiveness compared to its competitors.

Dechezleprêtre et al. (2012) include FDI as one factor that promotes international diffusion of climate-friendly technologies using detailed patent data from 96 countries for the period 1995–2007. They find that restrictions on both international trade and FDI hinder the diffusion of climate-friendly technologies, including renewable energy.

Clapp and Meckling (2013) point to several ways in which multinationals can influence the adoption of environmental policies, from lobbying to influencing the drafting of laws. In general, private action in politics is carried out in overlapping and complex ways, through formal and informal interventions, and at global and local scales. These same authors note that the role played by certain private groups is usually linked to the performance of business councils. The influence of such actors on political decisions is also related to their performance and use of the technology in question. This topic is especially relevant to the wind sector, as it is shaped by a few large, global companies. Multinationals that own property rights to certain technologies, for example, can use them to encourage policies that increase their market position. This has been documented by Falkner (2003), when studying the role of private actors in environmental governance, and by Levy and Newell (2005), on climate change-related policies. In addition, it is possible to identify indirect ways of influencing these groups on the formulation of environment-related policies, such as through the language of official documents, as suggested by Clapp (2005) and Cox (2012).

While Baldwin et al. (2019) pay special attention to foreign aid, Newell (2011) notes that overseas development assistance (ODA) is less than 10% of the size of FDI flows in the energy sector. Therefore, the author claims that many governments, through processes of power sector liberalization and energy reform, have relinquished at least some control over the provision of energy policy.

11.3.6 Lobbying

Classic models of international political economy emphasize the multiple lobbies competing for specific sectoral policies. Grossman and Helpman (1994) introduced a new era of intensive research on the political economy of trade and cited special interest groups as one of the main reasons for the lack of free trade. The principle behind their work is that special-interest groups make political contributions in order to influence a government's choice of policy.

The amount of influence a special group exerts depends primarily on information (lobbying) and on financial contributions to political campaigns. In the first case, the purpose is to persuade politicians to support certain policies, although some authors argue that only a small portion of lobbying takes this form in real politics (Rodrik 1995). In addition, although civil society movements can also be considered a kind of lobby (Keck and Sikkink 1998), private corporations play a more significant role. Their advantages include access to material resources, expertise, and personal connections with government officials (Falkner 2003; Vormedal 2008).

In the case of energy-related issues, the traditional providers of the energy sector – the so-called brown lobby[4] – generally support less stringent policies in relation to the adoption of renewable technologies. Falkner (2013) points out that Germany is an international leader on issues such as climate change, particularly after the Green Party formed a coalition government with the Social Democrats in 1998. On the other hand, environmental groups and companies selling renewable technologies seek more policies to promote this niche (the green lobby). Nesta et al. (2014) also reinforce lobbying as having important effects on the adoption of environmental policies. They consider that the recent liberalization of energy markets favors the adoption of REPs and conclude that the more liberalized the sector, the lower the influence of the brown lobby.

Fredriksson et al. (2004), Nicolli and Vona (2012), and Falkner (2013) document some examples of interest group action in environmental policy international diffusion. Faced with strong pressure on domestic regulation and international competition from countries with low environmental standards, some business groups in Germany have opted for a strategy of encouraging regulatory exportation, with the aim of creating a level playing field or gaining an early advantage. According to Falkner (2013, p. 256):

> It is the competitive dynamic of an increasingly global marketplace that has led certain business groups to join forces with environmental groups ... Once industry groups start calling for regulatory export to extend domestic regulation to their competitors, a so-called "Baptists-and-bootleggers-coalition" becomes the key driving force behind the government's foreign environmental policy. Similar patterns of domestically driven attempts to internationalize national environmental regulations have been observed in a number of contexts ...

Officially, Germany has prioritized international development activities that promote the adoption and implementation of REP recipient countries under the label of the fight against climate change (GIZ 2012). This might be due to two reasons: first, to ensure that their competitors operate under similar regulatory constraints and costs, and second, to pressure countries to adopt more environmentally friendly regulations. According to Drezner (2008), the United States and the EU, in particular, have actively sought to spread their regulatory models worldwide, which has set off processes of regulatory competition.

Overall, we consider that all the aspects mentioned above have contributed to the adoption of wind energy policies in Brazil. In the next section, we discuss each one in the specific case of the Brazilian wind sector.

11.4   The Economic Drivers of Wind Energy Policy Adoption: A Look to the Brazilian Case

This section focuses on REP adoption in Brazil, particularly the wind sector. The promotion of REPs in Brazil is based mainly on two formats: a government-driven model, similar to FITs, and a market-driven model, based on a system of auctions (Azuela and Barroso 2012). The two models that supported the growth of wind energy in the Brazilian market for almost two decades were the Program to Incentive Alternative Sources of Electricity (Proinfa), introduced in 2002, and exclusive regulated auctions, promoted annually between December 2009 and 2015. As a result of both efforts, in December 2016 Brazil had more than 10 GW of installed wind energy, equivalent to approximately 7% of the total (Abeeolica n.d.).

Although Brazil is recognized for its great renewable energy potential, it had to overcome two challenges to develop its wind industry: (i) high generation costs and (ii) lack of manufacturing companies in the country.

As shown in Table 11.1, cost for renewable energy has been significantly reduced since the introduction of auctions in Brazil. Although in 2009 the price of energy produced by the parks originally contracted by Proinfa varied between 247 and 280 R$/MWh, depending on the capacity factor of the park, in the same year, the price negotiated by exclusive auctions reached 135 R$/MWh. In 2011, a new milestone was achieved: the megawatt-hour price fell to 102 R$/MWh, a price similar to energy generated by hydroelectric plants (Azevedo et al. 2012). As a result, wind energy has become more competitive compared to other sources.

In 2004, the country had only one manufacturer of wind turbines. In 2013 due to the emergence of a relatively stable market, there were 11 manufacturers. Eventually, over 100 related businesses flourished in response to the development of a productive and effective supply chain (MCTI 2014).

Brazil employed specific regulations to ensure these results in the installation and marketing of its wind industry. In 2005, the country had implemented only two instruments: feed-in tariffs (by Proinfa) and public investment. By 2015, Brazil had employed five different types of REPs: auctions, net metering, tax credits for production, or investment, tax reductions for consumption. and public investment or financing.

Table 11.1 Energy auctions that contracted wind energy, 2009–2015 (values in *reals[a]*)

Date	Auction	Contracted installed power(MW)	Number of projects contracted	Average sales price (reals per MWh)	Average initial prices (reals per MWh)
12/14/09	02°LER (reserve auctions)	1805.7	71	R$148.33	R$189
8/26/10	03°LER and 02°LFA (alternative sources actions)	2047.8	70	R$134.13	NA
8/17/11	12°LEN (new power auction)	1066.9	44	R$99.38	NA
8/18/11	04°LER	861.1	34	R$99.58	R$146
12/20/11	13°LEN	976.5	39	R$105.53	NA
12/14/12	15°LEN	281.9	10	R$87.98	R$110
8/23/13	05°LER	1505.2	66	R$110.42	R$117
11/18/13	17°LEN	867.6	39	R$124.45	R$125
12/13/13	18°LEN	2337.8	97	R$119.08	R$120
6/6/14	19°LEN	550.0	21	R$130.00	R$131
10/31/14	06°LER	769.1	31	R$142.31	R$144
11/28/14	20°LEN	925.9	36	R$136.05	R$137
4/27/15	03°LFA	ND	3	R$177.47	ND
8/21/15	22°LEN	538.8	19	R$181.09	ND
11/13/15	08°LER	548.2	20	R$203.30	R$213

[a] Throughout the period, the Brazilian real's value averaged 2–3 US dollars.
ND, no data.
Source: Authors' elaboration based on publicly available data from EPE (n.d.).

Which drivers explain this evolution? As mentioned in the previous section, our goal here is to shed light on the economic aspects of this process. Therefore, after a brief institutional-historic contextualization, we follow the framework presented previously and discuss each of the six economic aspects of REP adoption in Brazil.

First, it may be helpful to examine the history of Proinfa and its evolution to understand what shaped Brazil's growth. Until 1995, the Brazilian power sector was owned by the federal and state governments. In 1996, the Brazilian government contracted a new model of organization that proposed the privatization of power utility companies under a project entitled RE-SEB (restructuring the Brazilian electricity system, in Portuguese). By the turn of the century, the Brazilian power sector had undergone drastic changes (Dutra 2007).

The new model separated generation, transmission, distribution, and commercialization services. Energy generation and trading became competitive activities where prices were defined by the market. Several institutions were established to promote higher reliability of the system, lower tariffs, universal access to energy, and regulatory stability (Leite 2009).

Technical literature suggests that the main driver for Brazil to start promoting renewable energy was the 2001 energy crisis, which resulted from low investments in the sector and a severe drought that reduced water levels at hydroelectric dams. The solution then was to ration energy supplies for eight months, in large part because the nation relied on such dams for 88% of generating capacity (Dutra 2007; IRENA-GWEC 2013). At the same time, the government launched the Emergency Wind Energy Program (Proeólica), which targeted 1050 MW of installed capacity from wind power by 2023. According to Dutra (2007), Proeólica did not reach the set goal of installed capacity due to the lack of short-term regulations for its applications and the resulting lack of interest from investors. Nevertheless, a new program was launched in 2002, the above-mentioned Proinfa. The program originally had two phases but only the first was operationalized. The first phase expected to implement 144 plants, totaling 3299.40 MW of new installed capacity from wind power plants, biomass, and small hydroelectric plants (SHP).

With this structure, the Brazilian choice of REPs was a hybrid model that encompassed feed-in tariffs and quotas (Dutra 2007). The program only started in 2004, after additional legislation was implemented.[5] Proinfa was instrumental in developing the market for alternative sources in Brazil but it also faced problems that caused serious delays in schedule. According to Dutra and Szklo (2008), the main challenge for Proinfa was the limited financial capacity of entrepreneurs. An industrial park was needed in order to establish the renewable energy source. The limited funds led to delays in financing operations to develop the park. Additionally, lack of funds affected the ability to review projects, which made it difficult to guarantee that the industrial park met the program's requirements and criteria.

In March 2004, before the program was implemented, the Brazilian government modified the power sector model. Supply was ensured through a series of measures, including the promotion of auctions. The model designed the coexistence of two energy contracting environments: a regulated one (under the acronym, in Portuguese, ACR), where distributors would buy energy from a pool supplied by public bids, by the lowest tariff, and a liberalized one (known as the ACL), in which consumers could freely negotiate bilateral contracts with generators and energy traders. The distributors would serve around 85% of the final demand for energy and would be responsible for forecasting demand, being obliged to contract 100% of their long-term needs within the ACR (Dutra 2007).

This new arrangement changed the rules for the commercialization of energy in Brazil and all sources started to compete with each other (reflecting the operation of Proinfa). With low tariffs as one of its pillars, the new model restricted the participation of new renewable energy sources, thus delaying the development of the sector in Brazil (Dutra 2007).

Confidence around the long-term continuity of Proinfa was threatened by the 2004 regulatory changes and by political uncertainty regarding energy goals, especially in the wind generation sector. In the second edition of Proinfa, in 2007, the government proposed a direct dispute with other energy sources in an alternative energy auction market. Wind farms, initially registered in the 2008 New Energy Auctions, declined to participate as they considered the competition unfair. Once again, renewable resources were at a disadvantage because of their higher cost. Consequently, no wind power was contracted in the 1st Alternative Sources Auction – LFAs/2007, nor in the 1st Reserve Energy Auction – LER/2008. Instead, the auctions were dominated by SHPs, thermoelectric, and biomass.

The wind sector demanded changes in auctions criteria. The articulation was consolidated in 2008, with the strengthening of the Brazilian Wind Energy Association (ABEEólica). The goal was greater interaction between businesses and government (GWEC 2011). This is examined more in detail in section 11.4. In May 2009, the Minister of Mines and Energy approved guidelines for the promotion of the first exclusive auction within the power generation sector. In December 2009, the government promoted the second auction of power reserves, the 2nd LER, and the first exclusive to wind power, with the aim of expanding the wind energy source in Brazil (Minc 2014). This initiative changed the direction of public policies and encouraged alternative sources of energy to become more market driven (Salino 2011; Azuela et al. 2014).

The 2009 auction attracted a significant number of generation ventures. More than 339 projects were registered and 1806 MW of wind power was sold to start operations (see Table 11.1). According to declarations from the Ministry of Energy and Mines (MME 2009a), the main motivation for this auction was to guarantee the security of national supply. The prices negotiated in this first exclusive auction were substantially lower than the values established in the first phase of Proinfa, which made wind energy more competitive, to the point of encouraging new bidding processes.

Table 11.1 summarizes information regarding the 16 auctions – promoted from 2009 to 2015 – in which wind power participated and more than 15 GW were contracted in new projects. Now, we will analyze the specific economic aspects of the above-described process.

11.4.1 Income

As mentioned previously, literature relates how economic growth and the rise of income may affect electricity demand and consumption as the adoption of REPs.

Figure 11.2 shows the trajectory of gross domestic product (GDP) growth and the number of REPs adopted in the period 2005–2015. The average economic growth was 2.87% and Brazil went from two kinds of instruments to five within the period. We can see from Figure 11.2 that the number of REPs adopted increased significantly after 2011, remaining stable after 2013. Meanwhile, renewable electricity fell in the period, due to the decrease in participation of hydroelectricity in total output and electricity production from renewable sources (excluding hydro) increased, which may be possible due to the maturation of wind and solar plants negotiated in 2009.

11.4.2 Energy Prices

At the time of Proinfa's implementation, there were high costs associated with building large dams for hydro-energy, and high costs for importation of natural gas from Bolivia. Since both of these energy resources were highly priced, the government, private individuals, and the market opted for wind energy (Barroso Neto 2010).

Nevertheless, the redirection of the Brazilian government to promote auctions in 2009 at the expense of FITs was part of a broader movement that included several developing countries, such as China, Mexico, and India (Azuela and Barroso 2012; Azuela et al. 2014; Bayer et al. 2018b). For Azevedo et al. (2012), the 2009 wind wave in Brazil resulted from a specific situation triggered by the 2008 international financial crisis.

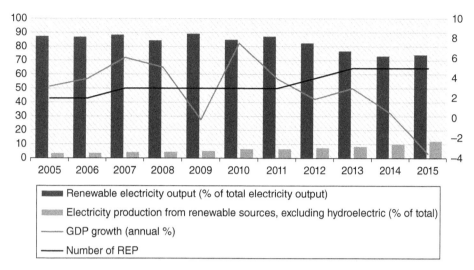

Figure 11.2 Brazilian economic growth (in percentage) and number of renewable energy policies (right axis) and renewable electricity output and production from renewables (in percentage, left axis). *Source:* Authors' elaboration based on publicly available data from World Bank (2019) and REN21 (2005, 2007, 2009, 2010, 2011, 2012, 2013, 2014, 2015, 2016).

Decisions on costs and competitiveness were significant in reshaping the industry's attractiveness. More importantly, changes in the priority of policymaking goals from effectiveness (increase in implantation) to efficiency (cost of the policy engine and impacts on supply costs) affected the political direction toward a preference for auctions. The increasing costs in advanced economies, followed by markets shrinking due to the economic crisis, contributed decisively to this change of focus toward developing countries.

11.4.3 Financing

Brazil launched Proinfa in order to promote alternative power sources. The idea behind Proinfa was to allow Brazil to obtain international financing for projects eligible for carbon credits under the CDM. In the context of the Global Environmental Facility (GEF), considering the 1991–2016 period, Brazil benefited from international financing for 12 projects, out of 435 exclusively related to renewable energy or energy efficiency (approximately 3% of them) (GEF n.d.).

Although it represents a relatively low proportion compared to the total number of projects, the value of the renewable energy and energetic efficiency projects represented 9.3% of the total loans. Encompassing climate change in general, Brazil has benefited from international financing in 34 of the 1558 projects (around 2% of all projects). In financial terms, this represented 6.4% of loans granted. The data indicate that Brazil has received an important volume of funds, and that international financing encouraged the development of the renewables sector in the country.

11.4.4 Trade

The mixed system of fixed quotas and guaranteed prices (FITs) adopted by Proinfa allowed the emergence of the first wind farms. Additionally, it helped improve the technology used for the farms (Barroso Neto 2010). When analyzing the most important partners related to the wind sector, Germany and the United States have an outstanding performance. These two countries accounted for at least 40% of the Brazilian imports of the items analyzed related to wind generation, according to Table 11.2. This trend was maintained in almost all periods from 2002 to 2016, except 2008–2010. Meanwhile, during that period, China was also gaining a prominent place with large investments in wind energy, and by expanding its companies in the global market (Melo et al. 2010).

During the 2008–2010 period, after the first exclusive auction, there was a 400% increase in imports of nacelle, an item of high added value for the fabrication of wind turbines. This reinforces the argument that migration to auctions, coupled with the international financial crisis, boosted the wind sector in Brazil (and once again we highlight the participation of the US and Germany). Figure 11.3 shows the trajectory of the increase in nacelle imports by country, revealing the importance of the most important trade partners to this industry: China, Germany, and the USA.

11.4.5 FDI

As discussed previously in the energy prices section, wind technology was becoming cheaper. Coupled with high interest rates in the Brazilian market, this attracted multinationals through its restructured regulatory framework, which included many incentives for international capital and few sanctions for delays.

Table 11.2 Value of Brazilian imports of items related to wind generation, by exporting country and period (in millions of US dollars), 2002–2016

Country	2002–2004	2005–2007	2008–2010	2011–2013	2014–2016
United States	$99.30	$174.20	$243.94	$1041.03	$672.97
Germany	$77.93	$234.18	$328.73	$794.13	$449.09
Spain	$7.96	$25.27	$158.02	$548.08	$437.99
China	$21.66	$30.92	$158.39	$448.53	$669.84
Italy	$61.58	$94.71	$147.77	$225.33	$134.18
Denmark	$1.23	$1.54	$1.51	$204.39	$31.76
India	$1.59	$30.03	$422.04	$157.54	$147.83
Others	$176.90	$210.74	$389.38	$546.44	$452.10

Source: Authors' elaboration based on publicly available data from MDIC (n.d.). The main components of a wind generator were consulted, including the following products, according to NCM code: 73082000 (steel tower), 94060099 (concrete tower), 85023100 (nacelle + housing), 84821090 (yaw bearing, bearing), 85016400 (generator), 84834010 (gearbox + cooling system + pitch gear or yaw drive), 85030090 (wind turbines). The countries were listed by decreasing value of the period from 2011 to 2013, years in which Brazil most imported the analyzed items.

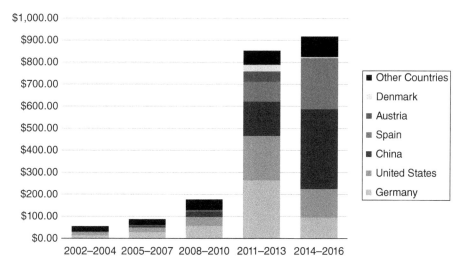

Figure 11.3 Value of nacelle imports by country and period (in millions of US dollars). *Source:* Authors' elaboration based on publicly available data from MDIC (n.d.). In the chart, we considered only imports of item 85 030 090 (wind turbines).

The growing activities of foreign groups in the domestic market can be seen as a sign of the reinforcement of REP adoption (GWEC 2011). Wobben Windpower, from the Enercon Group, began operations in Brazil in 1995, when it installed a factory next to the Brazilian company Tecsis. At the time, the German company exported part of the blades since the Brazilian wind market was virtually nonexistent (Camillo 2013).

During Proinfa, the construction of the first wind farms started; it was then that Wobben turned to the domestic market and expanded its business using the Brazilian local production of wind turbines (Podcameni 2014). Since it was a pioneer in Brazil, there was a significant effort by the company to structure the supply chain in the country and to train providers. Initially, higher technology components were imported and the equipment was designed abroad. As shown in Table 11.4, the Argentine company Impsa, the third company operating in the country, opened its first wind turbine factory in Brazil in 2008. In contrast to Wobben, it outsourced blades and towers as well as other parts requiring greater technological imput (Podcameni 2014).

These companies, however, worked in a small market with few growth prospects until 2008, when the international crisis led to the wind energy auctions in Brazil. As mentioned previously, the international financial crisis of 2008 was a relevant external factor that profoundly rearranged the strategic direction of the power sector in Brazil and other emerging countries (Salino 2011; IRENA-GWEC 2013; Melo 2013). It also had important consequences for the wind power industry in other parts of the world, since renewable energy investments in developed countries dropped (Maurer and Barroso 2011; Melo 2013).

Given the absence of orders in the major markets of the West and with their stocks full, equipment manufacturers had to look for alternatives, such as the promising markets of developing countries, and especially in the BRIC countries (Brazil, Russia, India, and China). China was a good alternative for these manufacturers as it had the fastest

Table 11.3 Main companies of the wind sector in the Brazilian wind market

Component	Company	Origin	Start of operations in Brazil
Wind turbines	Wobben (Enercon)	Germany	1995
	Suzlon	India	2006
	IMPSA	Argentina	2008
	Gamesa	Span	2011
	Alstom	USA	2011
	WEG	Brazil	2011
	Vestas	Denmark	2012
	Siemens	Germany	2013
	Acciona	Spain	2013
	GE Wind Energy	USA	2014
Blades	Tecsis	Brazil	1995
	Wobben (Enercon)	Germany	2002
	LM	Denmark	2013
	Aeris	Brazil	2013
Towers	Tecnomaq	Brazil	2006
	Intecnial	Brazil	2008
	Engebasa	Brazil	2009
	Piratininga	Brazil	2010
	RM eólica	Brazil	2010
	Wobben (Enercon)	Germany	2011
	Inneo Torres	Spain	2011

Source: Authors' elaboration based on publicly available data from Podcameni (2014).

growing wind energy market (REN21 2009, p. 9). However, this market was essentially supplied by local companies. European and American wind turbine manufacturers focused their sales in new markets such as South America. In this sense, international factors drove the promotion of exclusive wind energy auctions (Melo et al. 2010; Melo 2013; Podcameni 2014).

Therefore, in 2009, Brazil experienced the arrival of a large number of manufacturers interested in the Brazilian wind energy market. Table 11.3 shows the different foreign and local companies that brought their operations to Brazil; out of the 21 companies depicted, 14 entered the market in or after 2009. The firms' strategy was an aggressive price-penetration entry into the Brazilian market while at the same time, increased competition through bidding in the auction process made energy prices drop sharply (Melo 2013). The December 2009 auctions significantly increased power capacity and even encouraged the Brazilian private sector to venture into the wind market (Maurer and Barroso 2011).

11.4.6 Lobbying

Besides the traditional kind of lobby, other important forces behind Brazilian incentives for REP came from environmental activism and climate change-related issues widespread through international regimes. Brazil, along with many other developing countries, made clean energy development a central component of their Nationally Determined Contributions (NDCs) to the global effort to fight climate change, raising the need to adopt new RE policies, most likely by emulating policies already used in other countries (Baldwin et al. 2019). In this context, Brazilian commitment to reduce GHGs created a favorable political environment for the promotion of renewable energy sources (MME 2009a; Salino 2011).

It is important to highlight that the flexibilization of Brazilian power sector started with the 1988 Constitution. Article 175 states that "the provision of public services is the responsibility of the Public Power, in the form of Law, directly or under concession or permission, always through bidding." Nevertheless, until 1995, the Brazilian electricity sector was characterized as a hybrid state model, owned by the federal and state governments. Electricity generation was predominantly of hydraulic origin (95%), with a thermal component. In 1996, in the context of a wide liberalization process of Brazilian infrastructure sectors, the Brazilian government contracted a new model of organization of the electricity sector aimed at a restructuring of the entire sector, proposed by the project RE-SEB (restructuring of the Brazilian electricity sector) accompanied by privatization of energy utilities. Thus, based on a study by a consortium of British consultants, headed by Coopers & Lybrand and Latham & Watkins, the Brazilian electricity sector underwent drastic changes at the turn of the century (Dutra 2007).

As a consequence, there was the unbundling of generation, transmission, distribution, and trading companies (segment created in that period). Commercialization and generation have become competitive activities, with prices set by the market, signaling a latent process of liberalization with a great drop in the barriers of entry in this market. In this context, Aneel (National Electric Energy Agency, in 1996), MAE (Wholesale Energy Market) and ONS (National Electric System Operator) emerged. In 1997, the National Energy Policy Council (CNPE) was created. The proposed model would be based on higher system reliability, lower cost to consumers (low tariffs), universal access to energy, and regulatory stability (Leite 2009).

The Brazilian Wind Energy Association (Abeeólica) represents the wind energy industry in Brazil, including companies from the entire production chain. Abeeólica, founded in 2002, became part of the board of the Global Wind Energy Council (GWEC) in 2009, for a parallel event at COP15 (GWEC 2011). The organization's goal is to "promote power generation from wind as a complementary source of the national energy matrix, and to defend the consolidation and competitiveness of the wind sector, mainly through a government long-term program" (Abeeolica n.d.).

This last target has taken place through direct and indirect channels. Abeeólica representatives have been invited by the Chamber of Deputies on different occasions to present and defend the interests of the sector, as shown in Table 11.4.

On several occasions, Abeeólica declared itself in favor of Proinfa. According to GWEC (2011), the role of Abeeólica was essential to ensure exclusive auctions, after the previously

Table 11.4 Participation of Abeeólica in special commissions of the Chamber of Deputies

Date	Meeting	Commission
06/05/2014	0851/14	Commission of Mines and Energy
05/18/2010	0600/10	Commission of the Environment and Sustainable Development
10/13/2009	1755/09	Other events
03/24/2009	0141/09	Special Commission (PL 630/03) on Renewable Sources of Energy
02/03/2009	0007/09	Special Commission (PL 630/03) on Renewable Sources of Energy

Source: Authors' elaboration based on publicly available data from Brazilian Legislative Chamber (n.d.).

discussed 2007 and 2008 auctions, where renewable energies were unable to compete with lower-cost nonrenewable energy. The organization began to demand exclusive auctions, one for nonrenewable energy and one for renewable energy. This took place the following year, allowing renewable energy, including wind energy, to remain competitive (GWEC 2011).

As the sector developed in Brazil, the Association itself became stronger and the number of members went from 19 to 106 between 2008 and 2016 (Lacombe 2009). In 2009, the Association declared itself against the Brazilian Foreign Trade Chamber (Camex) decision to increase the rate of import taxes on wind turbines to 14%. The following year, after the success of the 2009 auction, Abeeólica advocated for exclusive annual auctions for wind power, and won that cause. That same year, Abeeólica proposed a tax exemption scheme for the wind sector, advocating the need for long-term policy instruments for the sector (some of these policies were eventually approved). In 2012 and 2013, it defended a technical reform in the bidding model by claiming that the current model prioritized price over quality (Melo 2012; Bahnemann 2013).

In June 2014, a law that would result in losses for the wind sector was suppressed in response to the mobilization of industry representatives. According to the president of Abeeólica at the time, Elbia Melo, action was taken jointly with parliamentarians to explain economic spillovers of particular political measures (Menna 2014; Moura 2014).

Thus, it is clear that the work of Abeeólica has helped promote renewable energy through the Brazilian regulatory framework by influencing governments and legislators to support the wind industry.

11.5 Final Considerations

How does economics influence politics? The answer to this question is not straightforward as it involves several aspects. Our goal was to identify the economic triggers behind the policy diffusion process for the Brazilian wind sector, between 2005 and 2015. We tried to shed light on some of the aspects that drove the adoption of two specific instruments related to climate change policies in Brazil.

In particular, this reveals that the expectations placed on international treaties (politics) as the traditional solution to the climate problem may be too high. Although climate change is a global collective problem that requires international coordination, we found that six main areas of economics can be just as important as international instruments in regard to policy adoption: (i) income, (ii) energy prices, (iii) financing, (iv) trade, (v) FDI, and (vi) lobbying.

In Brazil, all of those factors contributed to consolidate the wind market. The role of the business council, Abeeólica, was pivotal in systematizing companies' demands in policy changes. The shift toward more market-oriented policies, as auctions, made it possible to increase competition, promoting significant decreases in wind energy prices and, at the same time, giving investors some assurance in relation to the sustainability of wind sector in Brazil.

Overall, the Brazilian case is interesting because it has become a model for other developing countries. In this sense, there is also space for future research, following studies focusing on the Brazilian case and the evolution of wind energy prices, rates of project completion (Bayer 2018; Bayer et al. 2018a, 2018b), and the effect of potentialities and fragilities of REPs in Brazil (Aquila et al. 2017).

Notes

1 According to the US Energy Information Agency (EIA) and REN21, renewable energy includes geothermal, solar, and hydroelectric energy, as well as wind power and biomass.
2 Although auctions are not policies *per se,* they are mechanisms that encourage activity from a competitive structure and therefore are ordinarily considered as a type of policy (Maurer and Barroso 2011, p. 80).
3 The controversial idea of the environmental Kuznets curve is assigned to a 1991 article authored by economists Gene Grossman and Alan Krueger, who found a relationship in the form of an inverted U when investigating the relationship between air quality and economic growth (Stern 2004).
4 The brown lobby represents the sectors that rely heavily on energy from fossil fuels, such as the oil and coal industries (UNEP 2011). It is measured as a proxy for the barriers to entry to the energy sector, based on information from the OECD's Indicators of Product Market Regulation database (electricity and gas sectors). The index is a rank (0–6), where 6 is the maximum anticompetitive regulation. Therefore, the higher the score, the stronger the brown lobby and the lower the expectation of implementing policies for renewable energy.
5 Proinfa's legal framework is spread among several legal provisions, such as Law n. 10 438/02, amended by laws 10 762/03 and n. 11 075/04, Law n. 11 943/2009; decrees n. 5025/04 and n. 5882/06; MME ordinances n. 45/04, n.452/05, n. 86/07 and n. 263/07; ANEEL resolutions 56, 57, 62, 65, 127, 287 and 250; CAMEX Resolution n. 07/07.

References

Abeeolica (n.d.). Nosso setor. http://abeeolica.org.br/energia-eolica-o-setor

Alizada, K. (2018). Rethinking the diffusion of renewable energy policies: a global assessment of feed-in tariffs and renewable portfolio standards. *Energy Research & Social Science* 44: 346–361.

Apergis, N. (2019). The role of oil price volatility in the real and financial economy. In: *Routledge Handbook of Energy Economics* (eds. U. Soytaş and R. Sarı), 355. Abingdon: Routledge.

Apergis, N. and Danuletiu, D. (2014). Renewable energy and economic growth: evidence from the sign of panel long-run causality. *International Journal of Energy Economics and Policy* 4 (4): 578–587.

Aquila, G., Pamplona, E., de Queiroz, A. et al. (2017). An overview of incentive policies for the expansion of renewable energy generation in electricity power systems and the Brazilian experience. *Renewable and Sustainable Energy Reviews* 70: 1090–1098.

Azevedo, R., Seiceira, D., Carvalho, M., and Pereira, F. (2012). Comunicação: Brazil Windpower 2012. *Revista do BNDES* 38: 219–228.

Azuela, G. and Barroso, L. (2012). *Design and Performance of Policy Instruments to Promote the Development of Renewable Energy: Emerging Experience in Selected Developing Countries*, World Bank Study. Washington, DC: World Bank.

Azuela, G. E., Barroso, L., Khanna, A., et al. (2014). Performance of Renewable Energy Auctions: Experience in Brazil, China and India. Policy Research Working Paper 7062. Washington, DC: World Bank.

Baccini, L., Lenzi, V., and Thurner, P.W. (2013). Global energy governance: trade, infrastructure, and the diffusion of international organizations. *International Interactions* 39 (2): 192–216.

Bahnemann, W. (2013). Abeeólica: restrição para eólicas pode reduzir oferta. www. jornaldebeltrao.com.br/noticia/109506/abeeolica--restricao-para-eolicas-pode-reduzir-oferta

Baldwin, E., Carley, S., and Nicholson-Crotty, S. (2019). Why do countries emulate each other's' policies? A global study of renewable energy policy diffusion. *World Development* 120: 29–45.

Barroso Neto, L. (2010). Implementation of Renewable Energy Market Development Policies in Brazil. Consultant Report. Washington, DC: World Bank.

Bayer, B. (2018). Experience with auctions for wind power in Brazil. *Renewable and Sustainable Energy Reviews* 81 (2): 2644–2658.

Bayer, B., Berthold, L., and de Freitas, B.M.R. (2018a). The Brazilian experience with auctions for wind power: an assessment of project delays and potential mitigation measures. *Energy Policy* 122: 97–117.

Bayer, B., Schauble, D., and Ferrari, M. (2018b). International experiences with tender procedures for renewable energy – a comparison of current developments in Brazil, France, Italy and South Africa. *Renewable and Sustainable Energy Reviews* 95: 305–327.

BEN (2018). Balanço Energético Nacional. Ministério de Minas e Energia. www.epe.gov.br/pt/publicacoes-dados-abertos/publicacoes/balanco-energetico-nacional-2018

Biesenbender, S. and Tosun, J. (2014). Domestic politics and the diffusion of international policy innovations. *Global Environmental Change* 29: 424–433.

Bjork, I., Connors, C., Welch, T. et al. (2011). *Encouraging Renewable Energy Development: A Handbook for International Energy Regulators*. Washington, DC: United States Agency for International Development (USAID), the National Association of Regulatory Utility Commissioners (NARUC) & Pierce Atwood LLP.

Brazilian Legislative Chamber (n.d.). Speeches in Commissions. www2.camara.leg.br

Busch, P. and Jörgens, H. (2005). The international sources of policy convergence: explaining the spread of environmental policy innovations. *Journal of European Public Policy* 12 (5): 860–884.

Camillo, E.V. (2013). As políticas de inovação da indústria de energia eólica: uma análise do caso brasileiro com base no estudo de experiências internacionais. Tese de Doutorado. Instituto de Geociências, Unicamp. Campinas, SP.

Chandler, J. (2009). Trendy solutions: why do states adopt sustainable energy portfolio standards? *Energy Policy* 37 (8): 3274–3281.

Cia Alves, E.E., Steiner, A., de Almeida Medeiros, M., and da Silva, M.E.A. (2019). From a breeze to the four winds: a panel analysis of the international diffusion of renewable energy incentive policies (2005–2015). *Energy Policy* 125: 317–329.

Clapp, J. (2005). Global environmental governance for corporate responsibility and accountability. *Global Environmental Politics* 5 (3): 23–34.

Clapp, J. and Meckling, J. (2013). Business as a global actor. In: *The Handbook of Global Climate and Environment Policy* (ed. R. Falkner), 286–303. New York: Wiley.

Cox, R. (2012). *Environmental Communication and the Public Sphere*. London: Sage Publications.

Dechezleprêtre, A., Glachant, M., and Meniere, Y. (2012). What drives the international transfer of climate change mitigation technologies? Empirical evidence from patent data. *Environmental and Resource Economics* 54 (2): 161–178.

Del Río, P. and Linares, P. (2014). Back to the future? Rethinking auctions for renewable electricity support. *Renewable and Sustainable Energy Reviews* 35: 42–56.

Drezner, D.W. (2008). *All Politics is Global: Explaining International Regulatory Regimes*. Princeton: Princeton University Press.

Dutra, R. M. (2007). Propostas de políticas específicas para energia eólica no Brasil após a primeira fase do PROINFA. Rio de Janeiro: COOP-PE/UFRJ.

Dutra, R.M. and Szklo, A.S. (2008). Incentive policies for promoting wind power production in Brazil: scenarios for the alternative energy sources incentive program (PROINFA) under the New Brazilian electric power sector regulation. *Renewable Energy* 33 (1): 65–76.

Dunning, J.H. (2001). The eclectic (OLI) paradigm of international production: past, present and future. *International Journal of the Economics of Business* 8 (2): 173–190.

EIA (n.d.). Oil prices information. www.eia.gov/dnav/pet/pet_pri_spt_s1_a.htm

Elkins, Z., Guzman, A.T., and Simmons, B.A. (2006). Competing for capital: the diffusion of bilateral investment treaties, 1960–2000. *International Organization* 60 (4): 811–846.

EPE (n.d.). Leilões de energia. www.epe.gov.br/pt/leiloes-de-energia/leiloes

Falkner, R. (2003). Private environmental governance and international relations: exploring the links. *Global Environmental Politics* 3 (2): 72–87.

Falkner, R. (2013). *The Handbook of Global Climate and Environment Policy*. Chichester: Wiley.

Falkner, R. (2014). Global environmental politics and energy: mapping the research agenda. *Energy Research & Social Science* 1: 188–197.

Fankhauser, S., Gennaioli, C., Collins, M. (2014). Domestic dynamics and international influence: What explains the passage of climate change legislation? Centre for Climate Change Economics and Policy Working Paper 156. www.lse.ac.uk/GranthamInstitute/wp-content/uploads/2014/05/Wp156-Domestic-dynamics-and-international-influence-what-explains-the-passage-of-climate-change-legislation.pdf

Fredriksson, P.G., Vollebergh, H.R.J., and Dijkgraaf, E. (2004). Corruption and energy efficiency in OECD countries: theory and evidence. *Journal of Environmental Economics and Management* 47 (2): 207–231.

GEF (n.d.) Projects. www.thegef.org/projects

GIZ (2012). Legal Frameworks for Renewable Energy: Policy Analysis for 15 Developing and Emerging Countries. www.icafrica.org/fileadmin/documents/Knowledge/GIZ/Legal%20Frameworks%20for%20Renewable%20Energy.pdf

Grossman, G.M. and Helpman, E. (1994). Protection for sale. *American Economic Review* 84 (4): 833–850.

GWEC (2011). Análise do marco regulatório para geração eólica no Brasil. http://gwec.net/wp-content/uploads/2012/06/2ANALISE_DO_MARCO_REGULATORIO_PARA_GERACAO_EOLICA_NO_BRASIL.pdf

Holzinger, K., Knill, C., and Sommerer, T. (2008). Environmental policy convergence: the impact of international harmonization, transnational communication and regulatory competition. *International Organization* 62 (4): 553–587.

IEA (2004). *Renewable Energy – Market and Policy Trends in IEA Countries*. Paris: International Energy Agency.

IEA (2016). *Key World Energy Trends: Excerpt From World Energy Balances*. Paris: International Energy Agency.

IEA (2017). *Renewables*. Paris: International Energy Agency.

IPCC (2014). Climate Change 2014: Synthesis Report. IPCC: Geneva.

IRENA; CEM (2015). *Renewable Energy Auctions: A Guide to Design*. IRENA: Abu Dhabi.

IRENA-GWEC (2013). *30 Years of Policies for Wind Energy. Lessons from 12 Wind Energy Markets*. Abu Dhabi: IRENA.

Jenner, S., Chan, G., and Frankenberger, R. (2012). What drives states to support renewable energy? *Energy Journal* 33 (2): 1–12.

Keck, M. and Sikkink, K. (1998). *Activists Beyond Borders*. Ithaca: Cornell University Press.

Kern, K., Jorgens, H., and Janicke, M. (2005). The diffusion of environmental policy innovations: a contribution to the globalisation of environmental policy. Discussion Paper FS II 302. Berlin: Social Science Research Center.

Lacombe, F. (2009). ABEEólica critica aumento de imposto de importação para aerogeradores. www.nuca.ie.ufrj.br/blogs/gesel-ufrj/index.php?/archives/2819-ABEEolica-critica-aumento-de-imposto-de-importaco-para-aerogeradores.html

Leite, A.D. (2009). *Energy in Brazil: Towards a Renewable Energy Dominated System*. London: Routledge.

Levy, D.L. and Newell, P.J. (eds.) (2005). *The Business of Global Environmental Governance*. Cambridge: MIT Press.

Lyon, T.P. and Yin, H. (2010). Why do states adopt renewable portfolio standards? An empirical investigation. *Energy Journal* 31 (3): 131–155.

Massey, E., Biesbroek, R., Huitema, D., and Jordan, A. (2014). Climate policy innovation: the adoption and diffusion of adaptation policies across Europe. *Global Environmental Change* 29: 434–443.

Matisoff, D.C. and Edwards, J. (2014). Kindred spirits or intergovernmental competition? The innovation and diffusion of energy policies in the American states (1990–2008). *Environmental Politics* 23 (5): 795–817.

Maugeri, L. (2010). *Beyond the Age of Oil: The Myths, Realities, and Future of Fossil Fuels and Their Alternatives*. Santa Barbara: Praeger.

Maurer, L. and Barroso, L. (2011). *Electricity Auctions: An Overview of Efficient Practices.* Washington, DC: World Bank.

MCTI. (2014). Agência Brasileira de Desenvolvimento Industrial (ABDI) e Ministério do Desenvolvimento, Indústria e Comércio Exterior, Brasília.

MDIC (n.d.). Ministério da Indústria, Comércio Exterior e Serviços. http://comexstat.mdic.gov.br/pt/home

Melo, E. (2012). Modelo dos leilões deve ser revisitado. https://luizabrito67.blogspot.com/2012/04/elbia-melo-da-abeeolica-modelo-dos.html

Melo, E. (2013). Fonte eólica de energia: aspectos de inserção, tecnologia e competitividade. *Estudos Avançados* 27 (77): 125–142.

Melo, E., Neves, E. M. A., Pazzini, L. H. A., and Ogawa, K. (2010). An evaluation of the regulation of incentives for alternative electricity sources in Brazil. Presented at 8th Academic Conference in Association with UK Energy Research Centre, St John's College Oxford, England, September 22–23.

Menna, V. (2014) Desconto que beneficia eólicas deve ser mantido. www.tribunadonorte.com.br/noticia/desconto-que-beneficia-eolicas-deve-ser-mantido/284411

Minc, C. (2014) Minc no Executivo. Revista Mais Ambiente. https://carlosminc.files.wordpress.com/2014/09/revista_mais_ambiente_17_set_fio-compressed.pdf

Ministério de Minas e Energia (MME) (2009a) Programa de Incentivo às Fontes Alternativas de Energia (Proinfa), Brasília, DF. www.aneel.gov.br/proinfa

Ministério de Minas e Energia (MME) (2009b). Relatório do Grupo de Trabalho em Sistemas Fotovoltaicos – GT-GDSF. MME, Brasília.

Moura, R. (2014). A indústria eólica obteve uma vitória extraordinária. Entrevista com Élbia Melo, presidente executiva da Abeeólica. www.tribunadonorte.com.br/noticia/a-a-indaostria-ea-lica-obteve-uma-vita-ria-extraordina-riaa/295435

Nakhooda, S. (2011). Asia, the multilateral development banks and energy governance. *Global Policy* 2: 120–132.

Nesta, L., Vona, F., and Nicolli, F. (2014). Environmental policies, competition and innovation in renewable energy. *Journal of Environmental Economics and Management* 67 (3): 396–411.

Newell, P. (2011). The governance of energy finance: the public, the private and the hybrid. *Global Politics* 2: 94–105.

Nicolli, F., and Vona, F. (2012). The evolution of renewable energy policy in OECD countries:aggregate indicators and determinants. Working Paper No. 51. Fondazione Eni Enrico Mattei, Rome.

Nicolli, F. and Vona, F. (2019). Energy market liberalization and renewable energy policies in OECD countries. *Energy Policy* 128: 853–867.

Oates, W.E. and Portney, P.R. (2003). The political economy of environmental policy. In: *Handbook of Environmental Economics*, vol. 1 (eds. K.G. Mäler and J.R. Vincent), 325–354. Amsterdam: North-Holland and Elsevier Science.

Podcameni, M.B. (2014). *Sistemas de inovação e energia eólica: a experiência brasileira.* Instituto de Economia, UFRJ: Tese de Doutorado.

Prakash, A. and Potoski, M. (2006). Racing to the bottom? Trade, environmental governance, and ISO 14001. *American Journal of Political Science* 50 (2): 350–364.

REN21 (2005). Renewables 2005. Global Status Report. REN21 Secretariat, Paris.

REN21 (2007). Renewables 2007. Global Status Report. REN21 Secretariat, Paris.

REN21 (2009). Renewables 2009. Global Status Report. REN21 Secretariat, Paris.

REN21(2010). Renewables 2010. Global Status Report. REN21 Secretariat, Paris.

REN21 (2011). Renewables 2011. Global Status Report. REN21 Secretariat, Paris.

REN21 (2012). Renewables 2012. Global Status Report. REN21 Secretariat, Paris.

REN21 (2013). Renewables 2013. Global Status Report. REN21 Secretariat, Paris.

REN21 (2014). Renewables 2014. Global Status Report. REN21 Secretariat, Paris.

REN21 (2015). Renewables 2015. Global Status Report. REN21 Secretariat, Paris.

REN21 (2016). Renewables 2016. Global Status Report. REN21 Secretariat, Paris.

Rodrik, D. (1995). Political economy of trade policy. In: *Handbook of International Economics*, vol. 3 (eds. G. Grossman and K. Rogoff), 1457–1494. Amsterdam: Elsevier.

Sadorsky, P. (2009). Renewable energy consumption and income in emerging economies. *Energy Policy* 37 (10): 4021–4028.

Saikawa, E. (2013). Policy diffusion of emission standards is there a race to the top? *World Politics* 65 (1): 1–33.

Salino, P. J. (2011). Energia eólica no Brasil: Uma comparação do PROINFA e dos novos leilões. Doctoral Thesis, UFRJ.

Simmons, B.A., Dobbin, F., and Garrett, G. (2008). *The Global Diffusion of Markets and Democracy*. Cambridge: Cambridge University Press.

Sovacool, B.K. (2013). Energy policymaking in Denmark: implications for global energy security and sustainability. *Energy Policy* 61: 829–839.

Sovacool, B.K. and Mukherjee, I. (2011). Conceptualizing and measuring energy security: a synthesized approach. *Energy* 36 (8): 5343–5355.

Stadelmann, M. and Castro, P. (2014). Climate policy innovation in the south–domestic and international determinants of renewable energy policies in developing and emerging countries. *Global Environmental Change* 29: 413–423.

Stern, D.I. (2004). The rise and fall of the environmental Kuznets curve. *World Development* 32 (8): 1419–1439.

UNEP (2011). *Towards a Green Economy: Pathways to Sustainable Development and Poverty Eradication*. Paris: United Nations Environment Programme.

Vormedal, I. (2008). The influence of business and industry NGOs in the negotiation of the Kyoto mechanisms: the case of carbon capture and storage in the CDM. *Global Environmental Politics* 8 (4): 36–65.

Wiser, R., Namovicz, C., Gielecki, M., and Smith, R. (2007). The experience with renewable portfolio standards in the United States. *Electricity Journal* 20 (4): 8–20.

World Bank (2010). *World Development Report 2010: Development and Climate Change*. Washington, DC: World Bank.

World Bank (2011). Promotion of new clean energy technologies and the World Bank Group. In: *Background Paper for the World Bank Group Energy Sector Strategy*. Washington, DC: World Bank.

World Bank (2019). World Bank indicators. https://data.worldbank.org

12

Ontario's Energy Transition

A Successful Case of a Green Jobs Strategy?

Bruno Arcand

University of Quebec in Montreal, Montreal, Quebec, Canada

12.1   Introduction

In 2009, in the context of the 2008 financial crisis, the Liberal government introduced the Green Energy and Green Economy Act (GEGEA) to align its need for economic growth with that of a cleaner environment. The objective was to make Ontario a "leader" in the green economy (Office of the Premier 2009). It promised to develop renewable industries to reduce fossil fuel consumption and combat climate change. Additionally, one of the main objectives of the GEGEA was to create 50 000 green jobs in three years (Ontario Ministry of Energy and Infrastructure 2009). This target was important for an economic reason (tackling unemployment during a recession) and for a political reason (an opportunity to gain political support for the green transition from voters).

Ten years after the introduction of the GEGEA, the findings surrounding the employment objective and whether it has been attained are still debated in the literature. While several studies suggest that the GEGEA has generated a significant number of green jobs (Association of Power Producers in Ontario [APPrO] 2015; Cosbey et al. 2017), others argue that the GEGEA and, in particular, the high electricity rates which occurred while the policy was in place are responsible for destroying more jobs in Ontario than it created, and for reducing the competitiveness of several industries (Auditor General of Ontario [AGO] 2011; McKitrick and Aliakbari 2017).

Academic and private associations and governmental studies give important insights into the economic benefits and costs of the GEGEA on the labor market, income, and industries. It is mainly private firms, such as ClearSky Advisors and Compass Renewable Energy Consulting (CREC), and private actors with vested interests, such as the APPrO (2015), who have calculated the number of green jobs in Ontario. Only two academic papers, Pollin and Garrett-Peltier (2009) and Böhringer et al. (2012), have presented an

Environmental Policy: An Economic Perspective, First Edition. Edited by Thomas Walker, Northrop Sprung-Much, and Sherif Goubran.

analysis on the topic. Each of these papers had a very different methodology and conclusion. Additionally, they did not use explicit criteria for what a successful green jobs strategy is. It is therefore difficult to draw policy lessons from the Ontario experience from these analyses. This chapter aims to close this gap and examine it under a new light by asking the following question: is the GEGEA a successful case of a green jobs strategy?

12.2 Methodology

To address this question, the chapter uses the theoretical framework of the Green Industrial Policy (GIP), defined as: "any government measure aimed to accelerate the structural transformation towards a low-carbon and resource-efficient economy in ways that also enable productivity enhancements in the economy" (Altenburg and Rodrik 2017, p. 12). Like traditional industrial policies, the GIP is based on the idea that the market alone is unable to produce green growth (Rodrik 2014). In addition to the "market failures" that prevent economic development from aligning with a green trajectory, the existence of a "path dependency," due for example to existing energy infrastructures (pipelines, refineries, and service stations), favors fossil fuels over renewable energies (Unruh 2000). State intervention is therefore necessary to create opportunities for green industries, to penalize "brown" ones and ultimately to orient the economy toward a sustainable path (Lütkenhorst et al. 2014).

Since the government of Ontario used industrial policy instruments – such as feed-in tariffs (FIT) and domestic content requirements – to stimulate green industries and jobs as well as to decrease dependence on fossil energies, it is suitable to analyze the GEGEA as a GIP.

The literature distinguishes three main and interrelated criteria for a successful GIP (Pegels et al. 2018; Rodrik 2014).

1) *Efficacy* (Pegels et al. 2018). This refers to the achievement of the objectives set by the state. Because outcomes of public policies are hard to predict (Lütkenhorst et al. 2014), the state needs to fix short- and long-term objectives, and monitor them, to guide economic development toward a sustainable path. To be effective in this data production, GIPs require an institutional capacity from the government to collect information and influence the behavior and practices of private actors (Schneider and Maxwell 1997).

2) *Efficiency* (Pegels et al. 2018). GIPs are based on the idea that it is possible to combine economic growth with environmental sustainability, but because green policies can generate significant economic costs in the short term, the role of the state should be less about avoiding those costs and more about minimizing their impact on the economy. Efficiency thus refers to how successful the government was in minimizing the costs of the GIP.

3) *Accountability* (Rodrik 2014). It is the state's responsibility to communicate how and why the policy is needed. Additionally, it is also part of its mandate to give civil society and private actors the means to evaluate its impacts. By being transparent in its successes and failures, the state ensures the credibility and legitimacy of its interventions (Rodrik 2014, p. 488).

These criteria will allow this chapter to evaluate whether or not the policy was a success in terms of the number of green jobs it created, as well as a number of aspects related to its governance, more specifically regarding policy making and monitoring. To do so, the chapter first defines the concept of green jobs, as well as the main indicators used to measure them. In the second part, a metaanalysis is used to evaluate different methodologies used to estimate green job creation in Ontario and assess whether or not the objective set by the GEGEA was met. Then, the chapter compares job losses before and after the introduction of the GEGEA, using available data from Statistics Canada, to evaluate the relationship between a higher electricity rate in Ontario and job losses. In conclusion, the GIP criterion will be used to identify successful and failed aspects of the GEGEA, as well as policy lessons that need to be drawn from this experience in terms of green jobs strategy.

12.3   What is a Green Jobs Strategy?

The transition to a green economy – that is, "low carbon, resource efficient and socially inclusive economy" (United Nations Environment Program [UNEP] 2011, p. 16) – requires structural transformations that will affect labor markets and incomes (International Labour Organization 2011). Such a green transition can produce substantial economic and environmental benefits but in the short term, it can also bring negative externalities on polluting industries, such as a possible decrease in competitiveness and employment (Esposito et al. 2017). According to the GIP authors, because negative and positive effects vary across different actors and industries, the state needs to put policies in place for the "losers" of the energy transition (Altenburg and Rodrik 2017). Promoting the creation of green jobs is, therefore, an effective way for the government to reap environmental, economic, and political benefits from this structural transformation.

On the analytical level, however, studying the evolution of green jobs involves at least two important challenges. First, the definition of green jobs is often confused at the analytical level (Furchtgott-Roth 2012), which makes their monitoring difficult (Davis 2013). In Canada, for example, the definition of green employment varies across jurisdictions (MacCallum 2015). Besides the absolute number of jobs created, the International Labour Organization (ILO) uses the quality of jobs as a definitional criterion (ILO 2012). The blurriness of the concept makes it more difficult to compare between jurisdictions and to analyze the effectiveness of strategies to develop a green economy.

The second problem is that the objective of creating a green economy does not only entail creating green jobs, but also greening existing ones (Gülen 2012). Evaluating the "greening" process of industries requires being able to distinguish green from nongreen sectors, which is often difficult (Katz et al. 2012). Esposito et al. (2017) give the example of the automotive sector: parts that are assembled for an electric vehicle can be considered green goods, while the same parts used for a gasoline vehicle will be considered "nongreen" goods.

Despite those challenges, there are two indicators usually used to measure green jobs. The first one, "direct" green jobs, refers to jobs created in green industries. It includes temporary positions, such as construction and project management jobs, as well as operations-based positions, technicians, business and administration jobs (Earley and Mabee 2011). The second indicator, "indirect" green jobs, refers to the employment effect of green industries on related sectors, even if they do not themselves produce green goods or services. The development of wind farms, for example, stimulates production and employment within industries of intermediate goods such as lumber, steel, glass, and transportation (Pollin and Garrett-Peltier 2009). In Ontario, the addition of a domestic content requirement within the GEGEA was motivated by a concern to stimulate the economic activity and employment of intermediate sectors related to the green industries, also called "industrial clusters."

Understanding the social transformation that green jobs bring requires a broad definition of the concept (Esposito et al. 2017). For this reason, the chapter uses the ILO definition of green jobs, which "includes those in sectors that produce green goods and services as well as occupations in green processes that are environmentally favourable" (as cited in Esposito et al. 2017, p. 53). In addition to evaluating the progression of direct and indirect green jobs, it is necessary to assess the number of jobs lost. This makes it possible to compare the number of jobs lost with those created and understand the possible negative impacts of environmental measures on the employment level of certain sectors.

12.4   Measuring Green Jobs in Ontario

In addition to the general challenges of evaluating green jobs, there are four major difficulties specific to the Ontario case which create methodological challenges in monitoring green jobs. First, the government did not give a clear definition of what a green job is (Task Force on Competitiveness, Productivity and Economic Progress 2010). Second, the objective of creating 50 000 green jobs does not specify if this corresponds to the gross (i.e., total number of jobs created by a policy) or net number of jobs (i.e., total number of jobs created minus jobs destroyed by a policy) (Task Force on Competitiveness, Productivity and Economic Progress 2010). Third, the lack of transparency in the government's methodology to estimate how 50 000 green jobs would be created makes it difficult to evaluate the data or compare it with other results (Pollin and Garrett-Peltier 2009; Task Force on Competitiveness, Productivity and Economic Progress 2010). Fourth, the government did not adjust its statistical categories to include renewable energy as an industry.

Notwithstanding these important challenges, at the public debates in Ontario, differing numbers were mentioned to estimate the number of green jobs. In its 2011 review, the government estimated that roughly 30 000 jobs related to renewable energy were created between 2009 and 2012 in sectors such as "construction, installation, energy auditing, operations and maintenance, engineering, consulting, manufacturing, finance, IT and software" (AGO 2013, p. 315). The Ministry of Energy states that the estimation "relied on standard Ontario government methodology, including standard investment and job multipliers" (AGO 2011, p. 119). However, the lack of details on the parameters of the job multiplier makes it difficult to replicate the analysis. Moreover, the study being limited to the period between 2009 and 2012 requires the use of other analyses to cover the entire period of the GEGEA in Ontario. For this reason, the chapter refers to three different models used to assess the creation of green jobs in Ontario (see Table 12.1).

Table 12.1 Different estimations of green jobs in Ontario

Authors	Models	Time period	Jobs estimation (direct/indirect)
Government of Ontario	"Standard" jobs multiplier	2009–2012	30 000 between 2009 and 2012
Pollin and Garrett-Peltier (2009)	I/O	Investments scenarios on 10 years	The scenario of 47.1 billion: 90 000 jobs per year The scenario of 18.6 billion: 35 000 jobs per year
ClearSky Advisor (2011)	I/O (JEDI)	2011–2018	80 328 in the wind industry
CREC 2018	I/O (JEDI)	2006–2019	49 300 in the wind industry
APPrO (2015)	Employment ratios	2009–2014	88 842 in wind, 91 972 in solar and 21 378 in biomass between 2009 and 2014
Böhringer et al. (2012)	Computable general equilibrium	2011–2018	12 400 direct jobs, but 1.97 jobs lost for each new one created

I/O, input/output; JEDI, Jobs and Economic Development Impact.

12.4.1 Input/Output Models

First, Pollin and Garrett-Peltier (2009), ClearSky Advisors (2011), and Compass Renewable Energy Consulting (CREC) (2015, 2018) use a static *input/output table* (I/O) to estimate the creation of direct and indirect green jobs, as well as the effects of the GEGEA on Ontario's economy. An I/O table measures the interaction of flows of goods and services between industries associated with a certain amount of production. Although this method is often used by governments to estimate the impacts of economic activities of industries, it requires specific information for each industry that is not available for renewable energy in Canada.

For this reason, based on information from various Ministries of the United States and Canada, Pollin and Garrett-Peltier (2009) created "synthetic" green industries to estimate inputs needed by green industries to produce renewable energy. This allowed them to develop a multiplier of jobs for each type of renewable energy and to estimate the direct and indirect jobs needed to produce a certain quantity of energy. The study was done the same year as the implementation of the GEGEA, therefore, they used investment scenarios to project the employment effect of investment in renewable energy production. According to this methodology, each CAD 1.86 billion invested will create 27 068 direct and indirect jobs in Ontario, in industries such as solar, onshore wind, bioenergy, waste energy recycling, and hydroelectric as well as conservation and demand management (Pollin and Garrett-Peltier 2009, p. 12).

The analysis completed by ClearSky Advisors (2011) and CREC (2015, 2018) focuses on green jobs in the wind industry using the Jobs and Economic Development Impact (JEDI) model, which is an I/O model used by state decision makers and public utility commissions. To adapt the model to the specific economic characteristics of Ontario, ClearSky

Advisors conducted 43 in-depth interviews with different industry stakeholders to understand the wind energy sectors in Ontario (ClearSky Advisors 2011) The CREC (2015) integrated "historic and forecast installations, capital costs and domestic content obligations" of the province (CREC 2015, p. 3). According to these studies, between 2011 and 2018, an anticipated new installed capacity of 5.6 GW and an investment level of CAD 16.4 billion would create 80 328 direct and indirect green jobs in Ontario. Using the real level of investment (CAD 12.5 billion) and installed capacity (5552 MW) available in Ontario in 2018, the CREC (2018) estimated that 49 300 direct and indirect green jobs in wind industries have been created between 2006 and 2019. Of this total, 35 400 jobs were created between 2013 and 2019. Even though it is more conservative, the estimate produced by the CREC appears to be roughly consistent with the one produced by ClearSky Advisors.

12.4.2 Employment Ratio

The APPrO (2015) used Pollin and Garrett-Peltier's (2009) job multiplier but based it on the total value of new generation investments. It calculated by multiplying the total capital costs of production with the new installed capacity of electricity. According to this *employment ratio* approach, 88 842 direct and indirect jobs were estimated in Ontario's wind industry, 91 972 in the solar industry and 21 378 in biomass, for a total of 202 192 green jobs between 2009 and 2014. While this method is often used in the literature because it provides a simple way to assess employment impacts of the renewable energy production (see Jenniches 2018), its accuracy in reflecting renewable energy production structures depends on the quality of the multiplier used. The magnitude of the jobs multiplier depends on various information, such as regional characteristics, size of industries, and economic growth (Rutovitz and Atherton 2009). This augments the importance of using a job multiplier at the regional level and within a specific period of time (Moreno and López 2008) to reflect the production conditions of different types of industries and products (Llera-Sastresa et al. 2010).

This was not done in the APPrO analysis of the Ontario case (2015). Since the job multiplier is grounded on renewable industry based on the US and Canada experiences before 2009, this raises questions regarding the reliability of the job multiplier to analyze the Ontario case between 2009 and 2014. In fact, since the number of jobs estimated by the APPrO is almost twice that of the CREC, over a period of time that is less than half as long, we suspect that the multiplier does not accurately estimate the number of jobs in the wind industry in Ontario. This uncertainty about the reliability of the job multiplier used by the APPrO does not necessarily mean that its estimation overstates all the data. Ontario's solar industry, for example, is mainly composed of small businesses that distribute electricity through local distributors. Based on interviews and regional analyses to evaluate the job multiplier in the Kingston, Ontario, area, MacCallum (2015) stated that "smaller projects very likely employ more people per project than very large developments" (MacCallum 2015, p. 32). The scope of the analysis is, however, too limited to be generalized to the economy as a whole since solar energy is overrepresented in this region (MacCallum 2015).

12.4.3 Limits of an I/O and Employment Ratio Analysis

There are also theoretical limits associated with an employment ratio and an I/O analysis. First, most of the job multipliers used are stable over time, which means they do not take into account the efficiency gains of industries (Jenniches 2018). Second, employment ratio and I/O give a "snapshot" of the jobs created during a short period of time (ILO 2013); long-term implications of green policies on the market employment are neglected. Using highly aggregated data tends to underestimate the quality as well as the durability of the jobs created. In Ontario, the government estimated that 75% of the green jobs created between 2009 and 2014 were short-term jobs and would last only between one and three years (AGO 2011, p. 117). If temporary jobs are not necessarily negative, since they can be an effective strategy in a recession to stimulate economic activity and reduce unemployment, the duration of jobs created must be specified to ensure the development of a green economy, which relies on permanent green jobs.

12.4.4 Computable General Equilibrium

These critiques show the limits of analyses only focused on green jobs creation and the need to assess the long-term implications of green policies on the labor market more holistically. The study by Böhringer et al. (2012) is, in this sense, useful. They used a computable general equilibrium (CGE) model, which, essentially, simulates the effect of price changes on the rest of the economy (Bowen and Kuralbayeva 2015), to evaluate the long-term impacts of higher electricity prices on the employment market. In doing so, they argued that higher electricity prices reduce real wages and discourage workers, and thus increase unemployment. This relationship between a decrease in real wages and a rise in unemployment is known as the "wage curve" (Blanchflower and Oswald 1990). In Ontario's case, the authors stated that the GEGEA has been able to create roughly 12 400 green jobs – including only direct ones – but that for each job created, 1.97 jobs were lost in other sectors of the economy.

12.4.5 Selecting the "Right" Economic Model

Each economic model has its specific theoretical assumptions. For example, in the study by Böhringer et al. (2012), jobs are considered to be a social cost, which means that a decrease in real wages discourages workers from giving up their leisure time to work (Bowen and Kuralbayeva 2015, p. 12). On the contrary, in input/output and employment ratio models, a reduction in real wages will have little effect on employment. These abstract considerations can be illustrated by the two visions presented on the domestic content requirement in Ontario. Pollin and Garrett-Peltier (2009) considered the policy to be a benefit for the economy since it stimulates local jobs creation and economic activity by creating industrial clusters between green and nongreen industries. On the other hand, Böhringer et al. (2012) saw this as an additional cost – estimated at around 30% – to the Ontario economy as it exacerbates the "distortion" created by the increase in electricity costs to several sectors of the economy (Böhringer et al. 2012, p. 24).

Political and ideological considerations may influence the choice of one model over another, but it is important to recognize the different contexts in which a model should be favored. As we can observe, from 2009 employees stayed at their positions, even with a wage cut, as job opportunities were low. Therefore, the assumption that individuals will leave their jobs because of a decline in their real wages seems implausible in a recessionary economy. Because of that, Böhringer et al.'s (2012) evaluation probably overestimates the economic impact of a rise in electricity rates on the production and employment loss of industries. On the other hand, in a context of full employment, where job opportunities are numerous and it is relatively easy for a worker to change jobs, the assumption of the wage curve appears more plausible. In this sense, the challenge to estimate green jobs creation is not just about developing the "best" model, but also about choosing the right one according to the right context (Rodrik 2015).

Using an I/O model in Ontario seems justified, but the high number of methods using various timelines make a comparison between green jobs estimates difficult. The analysis by the government of Ontario and the CREC suggests that the GEGEA has resulted in significant green job creation, but it remains difficult to evaluate the efficacy of the policy since it is unclear whether or not the initial objective of 50 000 green jobs was achieved in three years.

The next section evaluates Ontario's labor market to assess the other side of a green jobs policy: the potential negative effects on economic growth. To do so, this chapter provides an analysis of the manufacturing sector in Ontario to estimate the impact of the GEGEA green jobs strategy on job losses.

12.5 Impact of the GEGEA on Ontario's Manufacturing Sector

A strategy to promote green industries and green jobs and phase out others, such as the one adopted in Ontario, can have multiple and sometimes unpredictable effects on the economy. For policymakers, anticipating and managing those negative effects is a key issue to ensure the political and economic sustainability of a green transition. To do so, they need to evaluate the impact, including job losses, that green policies may have by targeting polluting industries or causing negative externalities such as a rise in electricity prices. In Ontario, the rise in electricity prices raised concerns about significant job losses (AGO 2011). That was particularly true for jobs in many manufacturing industries which, due to their high electricity consumption, are very sensitive to changes in electricity rates.

Estimating job losses related to green policies is challenging for at least two reasons. First, various factors can influence job losses in the economy (globalization, innovation, monetary policy), so making a clear-cut distinction between job losses caused by green policies and other factors is very difficult. Second, employment alone gives an incomplete picture of the economic health of a sector, as job losses do not always equal declining industries. Indeed, it is possible for a company to increase its production while reducing its employees through technological progress. Taking these efficiency gains into account requires an overview of the manufacturing sector that also considers the output.

While various models have been used in the literature to estimate job losses associated with green policies, the main critique made against the GEGEA focused on job losses in the manufacturing industries (McKitrick and Aliakbari 2017). As such, this chapter also focuses on that sector. Since the rise in electricity prices occurred while the policy was in place, the performance of the manufacturing sector before and after the introduction of the GEGEA will be evaluated. Indeed, the average monthly electricity bill for consumers (adjusted for inflation) increased by a total of 19% between 2006 and 2017, with a decrease of roughly 9% from 2006 to 2009 (Ontario Energy Board [OEB], retrieved from Environmental Commissioner of Ontario 2018, p. 113).

To understand if this increase in electricity prices paid by consumers harmed employment in Ontario's manufacturing sector, the chapter evaluates the gross output – which measures the total value of industries sales – the number of jobs, and the investment in Ontario's manufacturing industries before and after the introduction of the GEGEA. A specific focus on electricity-intensive industries is also provided.

12.5.1   Output and Employment in the Manufacturing Industries

In Ontario, output and employment declined in a similar way between 2001 and 2009, but experienced divergent growth following the 2008 economic crisis (Figure 12.1). Most of the job losses in Ontario's manufacturing sector occurred before the GEGEA came into effect: employment declined by 27.22% between 2005 and 2009, while it increased by 0.54% between 2010 and 2018. Output has risen more quickly than employment: after a decrease of 25.64% between 2005 and 2009, output increased by 34.14% between 2009 and 2015. This faster increase in output compared to employment suggests an increase in industrial productivity.

However, using the output numbers might be deceiving since they include intermediate inputs – that is, the goods and services required for the production of one unit – which tends to inflate the performance of manufacturing production. An optimistic interpretation of these data assumes that most of the intermediate products used by manufacturing industries come from firms in Ontario, thereby stimulating national economic activity.

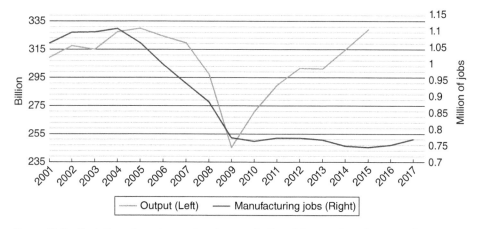

Figure 12.1   Evolution of employment and output in Ontario's manufacturing sector between 2001 and 2017. *Source:* Manufacturing jobs: Statistics Canada. Table: 14-10-0092-01 (CANSIM 282-0125) using the North American Industry Classification System (NAICS). Reproduced and distributed on an "as is" basis with the permission of Statistics Canada.

A more pessimistic interpretation would be that intermediate products are mainly imported, thus reducing the economic impact of the manufacturing sector on the Ontario economy.

McKitrick and Aliakbari (2017), in their study, used the output in gross domestic product (GDP) rather than gross output and argued that the performance of manufacturing production between 2010 and 2017 was weak because of higher electricity prices. While it is true that output in GDP never reached its 2005 level, this indicator is limited in assessing the manufacturing sector's performance for at least two reasons.

First, the output in GDP calculates the production in relative price, which means that the growth of the manufacturing sector is assessed with the growth of other industries, including the service industry. As an illustrative example, if the growth in the service industry is higher than that of manufacturing industries, the effect of manufacturing growth will be reduced in the data despite a rise in the production level (Baldwin and Macdonald 2009).

Second, the input in GDP assumes that the performance of the manufacturing sector is independent of the service sector and that their progress can be assessed by comparing them. But as pointed out by various authors, the line between manufacturing and services is not necessarily always clear, since there is strong interdependence between these two sectors (Baldwin and Macdonald 2009; Guerrieri and Meliciani 2005; Park 1989). For this reason, an increase in service industries is not necessarily at the expense of the manufacturing sector. The use of gross output overcomes, in part, this difficulty by tracking the gross volume of production in the manufacturing industries as well as the interrelationships between economic sectors.

12.5.2 Investments

Another facet of the health and productivity of an economic sector is investment. In Ontario, the recovery of investment has been slower than production. That is not necessarily surprising: investment is a bet on the future, and therefore depends on the anticipation and mood of investors. As shown in Table 12.2, investment was more stable than production between 2006 and 2008, hinting at investors' confidence. Since Ontario had been severely hit by the 2008 economic crisis due to its high economic dependence on US markets (Fortin 2018), this negatively affected investments. That being said, the level of investment in 2018 was CAD 7350.6 billion, which represents a level similar to that of 2006, which was CAD 7851.1 billion. Even this partial recovery in the level of investment seems to be more in line with the recovery in the level of gross output (Figure 12.1) than with the output in GDP (Figure 12.2).

Table 12.2 Evolution of capital expenditures (in CAD billion) in Ontario's manufacturing sector between 2006 and 2018

Year	2006	2008	2010	2012	2014	2016	2018
Investment	7851.8	7528.3	5096	5250.9	6046.3	6041.2	7350.6

Source: Statistics Canada. Reproduced and distributed on an "as is" basis with the permission of Statistics Canada.

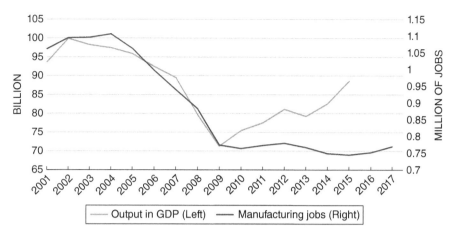

Figure 12.2 Evolution of output in GDP (chained [2012] CAD) and employment between 2001 and 2017. *Source:* Output in GDP: Statistics Canada. Table 36-10-0487-01. Gross domestic product (GDP) at basic prices; Employment: Statistics Canada. Table 14-10-0092-01. Employment by industry. Reproduced and distributed on an "as is" basis with the permission of Statistics Canada.

Evaluating the performance of the manufacturing industry is still a challenging task but comparing the period before and after the GEGEA allows us to make two observations. First, employment and output (in both measures) start declining before the rise of electricity bills to consumers. This suggests that other factors – for example, an overvaluation of the Canadian dollar (Beine et al. 2012), the emergence of Asian countries or the dependence on US markets (Sharpe 2015) – could be responsible for the decline in production and employment between 2005 and 2009. Second, the stability of employment and the rapid growth of the manufacturing sector between 2010 and 2017 suggest that the GEGEA has had a limited effect on job losses.

Comparing this performance with other Canadian provinces, McKitrick and Aliakbari (2017) argued that the increase in electricity prices in Ontario was responsible for the poor performance of manufacturing industries. While it is true that manufacturing industries in the Western provinces have grown more quickly than in Ontario, this is not necessarily due to differences in electricity rates. In fact, "interjurisdictional retail rate comparisons" have a limited explanatory scope since they underestimate significant institutional and market differences between electricity systems across Canadian provinces (Market Surveillance Administrator [MSA] 2017, p. 3). As indicated by the MSA: "Various factors influence differences in electricity prices between regions, including resource allocations, taxes, subsidies, and regulation (among others), some of which create indirect consumer costs are not accounted for by a simple comparison of all-in rates" (MSA 2017, p. 11). For these reasons, the MSA (2017) warns against using the report produced by Hydro-Quebec and used by McKitrick and Aliakbari (2017) to assess the performance of electricity markets between jurisdictions in Canada.

Another way to gain insight regarding the effects of a rise in electricity prices on employment is to evaluate the performance of the manufacturing industries that consume the most electricity.

12.5.3 Intensive Electricity Industries

According to the 2017 Industrial Consumption of Energy Survey, the largest consumers of electricity in Canada were the industries of "primary metal (38.9%), pulp and paper (22.4%), chemical (10.8%), food (5.5%), and wood (4.0%), which represented 81.6% of the total electricity consumed in the production process" (Statistics Canada 2018). As we can see in Figure 12.3, most of the decline in employment occurred before the introduction of the GEGEA. Job creation – where it has taken place – has developed at a slower pace than the job loss before the crisis. For example, after decreases of 2.27% and 16.92% between 2004 and 2009, food and chemical industries experienced respectively a 0.44% and 10.02% increase between 2010 and 2018. The trend for the wood and pulp and paper industries has been similar. After being relatively stable at the beginning of the 2000s, wood industries fell by 42.87% and pulp and paper industries by 39.99% between 2004 and 2009. Since 2010, the wood industry employment has grown by 8.56% while the pulp and paper industry has fallen by 11.23%. For its part, primary metal industries experienced a strong declining job trend as early as 2001, 44.84% between 2001 and 2009, and a modest decline of 0.84% between 2010 and 2018. We can see that most of the job losses came before the 2008 crisis and that employment, except for the pulp and paper industry, experienced growth or stable development between 2010 and 2018.

On the output side, Figure 12.4 shows that industries that performed well before the 2008 crisis – primary metal, basic chemical and meat (used here to illustrate the food industry) – also experienced a robust growth before 2009. In contrast, industries that declined between 2005 and 2009 (wood and pulp and paper industries) also experienced weak performance after 2009. When we compare Figures 12.3 and 12.4, it is clear that the rising output of meat, chemical, and primary metal industries (before 2008) was also accompanied by a decline in employment. This rise in productivity implies that many of the jobs lost

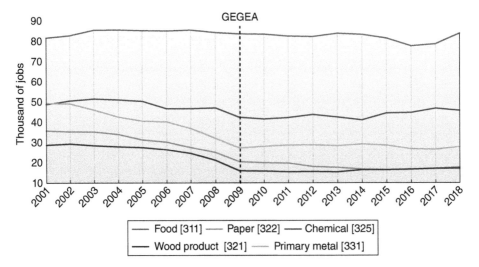

Figure 12.3 Evolution of employment of electricity-intensive manufacturing industries in Ontario between 2001 and 2018. *Source:* Statistics Canada. Table 14-10-0202-01, Employment by industry, annual. Reproduced and distributed on an "as is" basis with the permission of Statistics Canada.

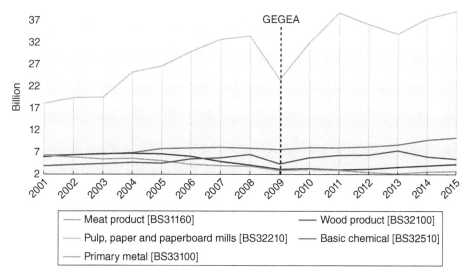

Figure 12.4 Evolution of the output of electricity-intensive manufacturing in Ontario between 2001 and 2015 in CAD. *Source:* Statistics Canada. Table 36-10-0488-01, Output, by sector and industry, provincial and territorial. Reproduced and distributed on an "as is" basis with the permission of Statistics Canada.

in those industries will not return. Only the pulp and paper and wood industries experienced a simultaneous decline in their production and employment levels. McKitrick and Aliakbari (2017) argued that the declining performance of the pulp and paper industry is directly caused by high electricity prices in Ontario, but this interpretation seems improbable. Between 2010 and 2017, the average effective price for high energy consumption industries actually decreased by 1.5% (OEB 2018, p. 12).

The reason is that, in 2011, the Ontario government introduced the Industrial Conservation Initiative (ICI) which created incentives for industries in Class A – that is, businesses with an average monthly peak demand of more than 500 KWh or less or equal to 1 MW[1] – to help them "reduce [their] consumption at critical peak demand times" (OEB 2018). Because the amount of Global Adjustment[2] paid by these industries was determined by their demand for electricity during the five highest hours of peak demand in a year,[3] energy efficiency has been an effective way of reducing electricity costs for large industries.

According to the OEB, the ICI program helped industries in Class A – about 1200 in 2017 compared to 200 in 2011 – reduce their electricity consumption during peak time by 42%, 33%, and 26% in 2016, 2013, and 2011 respectively. These reductions represent savings of up to CAD 4.91 billion (not adjusted for inflation) during this period (OEB 2018, p. 11). Due to the high electricity consumption of the pulp and paper industry, it is legitimate to assume that "these industries are likely to fall primarily in Class A" (Environmental Commissioner of Ontario 2018, p. 125). It is therefore improbable that a rise in electricity price is the main cause for its declining production. It should be noted, however, that the costs saved by industries in Class A have been transferred to small enterprises and residents.

In light of this analysis, two observations can be made. First, most of the job losses in manufacturing industries occurred before the 2008 crisis. Second, the data provided by Statistics Canada do not confirm that the increase in electricity prices is the main cause of the loss of manufacturing jobs in Ontario. This does not mean that the increase in electricity prices does not affect competitiveness or industrial production, but it does suggest that the ICI policy has been an effective response by the Ontario government to mitigate the effects of higher electricity prices on Class A consumers. However, the fact that the ICI did not provide specific measures for ratepayers and small businesses shows the limitations of this policy.

12.6 Discussion

In terms of the objective of creating 50 000 green jobs in three years, the estimation by the Ontario government suggests that a close figure (about 30 000) of the primary target of the policy has been achieved. However, the monitoring of green jobs has been limited to three years, and the method used for these estimates has not been made available. In addition to these restraints in the monitoring process, the absence of clarity concerning the nature and duration of the green jobs objective devalues this achievement. As other estimations of green jobs have been communicated in public debates in Ontario, the uncertainty about the reliability of the job multiplier used reduces the utility of these studies.

This ambiguity regarding the number of green jobs generated by the GEGEA makes the second criteria – efficiency – hard to evaluate in terms of job creation. On the other hand, analyzing job destruction in the manufacturing sector provides a more general picture of the situation in Ontario. This analysis illustrates the relatively good performance of the employment and growth of the manufacturing sector, compared to the 2005–2009 period, which suggests that the number of job losses between 2010 and 2017 has been limited. Of course, this is not necessarily attributed to the GEGEA, but the results suggest that the net creation of green jobs has been positive. The ICI program has proven to be an effective policy to help large industries maintain levels of growth and employment, and thus support the efficiency of green jobs creation by reducing job losses.

Concerning the accountability of the green jobs strategy, the lack of transparency in the data provided by the government of Ontario and the numerous studies using different timelines and methods make it difficult to assess the benefits of the GEGEA. This, in turn, decreases its social desirability.

On the other hand, Germany offers a contrasting experience in measuring green jobs. Unlike Ontario, the German Ministry of Environment, Nature Conservation and Nuclear Safety (BMU) has produced various reports assessing the long-term impact of the Renewable Energy Sources Act (EEG) on employment (BMU 2010, 2012). In those reports, the methodology and the scope of the measurement (gross and net employment) are provided as well as qualitative information (i.e., type of work and duration) on jobs created. In addition to this active green job monitoring by the state, an online and regularly updated database (BERFUFENET) provides valuable information on the classification of occupations in Germany (Paulus and Matthes 2013). Managed by the Federal Employment Agency (FEA) and using official data (Janser 2018), this database helped independent analysts develop indexes to measure the greening process of jobs at the national and regional levels (Janser 2018; Ulrich and Lehr 2014). As a result, there is a relatively strong consensus concerning both the gross and net number of green jobs created between 2010 and 2013 (BMU 2012; O'Sullivan et al. 2014; Ulrich and Lehr 2014). This, in turn, helps demonstrate the positive employment effect of the EEG in Germany (BMU 2012).

Progress in green employment policies has also been observed in the European Union. In addition to databases that help measure green employment, information concerning the labor market has been promoted to identify "skills bottlenecks" in the green economy (European Commission 2014, p. 5). The 2012 Commission Employment Package has accordingly listed some tools to stimulate "green employment, bridge skill gaps, labour shortages and anticipate change in human capital needs" (European Commission 2014, p. 2). The promotion of Sectoral Skills Councils, which bring together industries, government, social partners and researchers, aims to foster information gathering about the skills needed for a green economy (European Commission 2014). Such policies are lacking in Ontario since the government "does not collect or analyze regional information on labour force skills supply and demand to identify what jobs will have a shortage of skilled workers" (AGO 2016, p. 252). Several difficulties have, however, been encountered by the European employment strategy. According to the 2018 report by the European Centre for the Development of Vocational Training (CEDEFOD), connections between organizations and government concerning labor markets and skill policies remain "weak" (CEDEFOD 2019, p. 11). The differences in the definitions of a green skill and a green job between European countries also make a general evaluation of the European Union's strategy challenging (CEDPEFOD 2019).

12.6.1 Limitations

This chapter has three main limitations that need to be acknowledged. First, the models used have limitations that need to be reinforced by other indicators of the real economy in order to properly evaluate the creation of green jobs. Economic models, for example, often imply the same assumption: the mobility of workers between economic sectors. This means that workers can quickly change jobs, which makes them very responsive to changes in the labor market. This assumption may be true in the long run but seems improbable in the short term (Consoli et al. 2016). Second, the study assumes that the increase in electricity prices in Ontario was caused by green policies, such as the GEGEA. While several studies claim that the price of FIT-guaranteed contracts to renewable energy producers was too high (AGO 2015; Ontario Society of Professional Engineers 2016), an important refurbishment of supply capacities and a high level of debt of the Ontario power grid have also contributed to higher costs (Winfield and Dolter 2014). To better determine the impact of the GEGEA, there is a need to analyze the distribution of electricity system costs as well as the strategy adopted by the Ontario government to finance electricity projects. Third, if this analysis suggests that the ICI program has been effective in helping large industries manage the rise of electricity prices, the costs saved by those industries were transferred to small businesses and residents. This policy configuration raises questions about the fair allocation of the electricity system costs between taxpayers and its economic impact on small industries.

12.7 Conclusion

The structural changes needed to develop a green economy usually transform labor markets by creating and destroying jobs. Assessing the impact of the energy transition on employment is complex and there is no clear-cut answer to the question of whether or not Ontario's strategy was a success. According to the three objectives of a GIP, the Ontario experience represents mixed results.

Nonetheless, several lessons can be drawn from Ontario's experience concerning a green jobs strategy. First, a green jobs strategy requires more than just creating jobs. It involves reforming public statistics to incorporate the specificities of renewable energies, such as the number and type of jobs, and the skills that need to be promoted. Future green jobs strategies need to include green industries in transparent and independent public statistics. This reform will increase public confidence in the project and signal to private actors that the energy transition is permanent, that there is no turning back. Second, to increase the reliability of the estimation, assessment of green jobs must be done on an ongoing and long-term basis to account for the changes in the economy. This is particularly important for static economic models (i.e. I/O and employment ratio), since they assume price and output stability. Third, the government must anticipate the negative effects of the energy transition on polluting industries and those which are sensitive to changes in electricity prices to support their economic growth. Numerous factors can decrease the number of jobs in the economy, and governments need to assess the evolution of employment and the reasons for its variations, especially regarding green policies, to support the growth and productivity of its industries.

A shift toward sustainability brings growth opportunities but also challenges that are hard to predict. To ensure that workers benefit from it, states need to adapt labor policies to take into account the specificities of a green economy. This chapter highlights the need for public policies that support both green and brown industries throughout the energy transition. It also provides insights to policymakers about the importance of accurate and consistent information on the labor market to correctly assess green jobs. In Canada, stronger institutional capacities and a clear framework are needed to measure and promote green employment. In line with those observations, future green policies in Canada should focus more on the changes in the policy-making process that a successful green jobs strategy requires. Other counties which are in the energy transition process should ensure the simultaneous development of their environmental, economic, and labor policies, and apply successful measurement strategies such as those used in Germany and other countries of the EU.

Acknowledgments

I am grateful to Maya Jegen, Simon Langlois-Bertrand, and Cassandre Gratton for their valuable comments and suggestions.

Notes

1 This rate corresponds to the one established in 2011. In 2017, the rate to be eligible for the ICI was reduced to an average monthly peak of over 1 MW (OEB 2018).

2 The cost of electricity production that is not fully amortized by the price paid on the wholesale electricity market (OEB 2018).

3 "For example, if a Class A consumer was responsible for 1% of Ontario demand during the five peak demand hours in a 12-month period, they would pay 1% of the Global Adjustment in the ensuing 12-month period" (OEB 2018, p. 7).

References

Altenburg, T. and Rodrik, D. (2017). Green industrial policy: accelerating structural change towards wealthy green economies. In: *Green Industrial Policy. Concept, Policies, Country Experiences* (eds. T. Altenburg and C. Assmann), 1–20. Geneva, Bonn: UN Environment; German Development Institute (DIE).

Auditor General of Ontario (AGO). (2016). 2016 Annual report. www.auditor.on.ca/en/content/annualreports/arreports/en16/2016AR_v1_en_web.pdf

Association of Power Producers of Ontario (APPrO). (2015). The Value of Electricity to Ontario. www.appro.org/2-uncategorised/762-the-value-of-electricity-to-ontario.html

Auditor General of Ontario (AGO). (2011). Annual Report 2011. www.auditor.on.ca/en/content/annualreports/arreports/en11/2011ar_en.pdf

Auditor General of Ontario (AGO). (2013). Annual Report 2013. www.auditor.on.ca/en/content/annualreports/arreports/en13/2013ar_en_web.pdf

Auditor General of Ontario (AGO). (2015). Annual Report 2015. www.auditor.on.ca/en/content/annualreports/arreports/en15/2015AR_en_final.pdf

Baldwin, J.R. and Macdonald, R. (2009). *The Canadian Manufacturing Sector: Adapting to Challenges*. Toronto: Statistics Canada.

Beine, M., Bos, C., and Coulombe, S. (2012). Does the Canadian economy suffer from Dutch disease? *Resource and Energy Economics* 34: 468–492. https://doi.org/10.1016/j.reseneeco.2012.05.002.

Blanchflower, D. and Oswald, A.J. (1990). The wage curve. *Scandinavian Journal of Economics* 92 (2): 215–237.

Böhringer, C., Rivers, N.J., Rutherford, T.F., and Wigle, R. (2012). Green jobs and renewable electricity policies: employment impacts of Ontario's feed-in tariff. *BE Journal of Economic Analysis & Policy* 12 (1) www.degruyter.com/view/j/bejeap.2012.12.issue-1/1935-1682.3217/1935-1682.3217.xml.

Bowen, A. & Kuralbayeva, K. (2015). Looking for green jobs: The impact of green growth on employment. www.lse.ac.uk/GranthamInstitute/wp-content/uploads/2015/03/Looking-for-green-jobs_the-impact-of-green-growth-on-employment.pdf

Bundesministerium für Umwelt (BMU) (2010). Renewably employed! Short and long-term impacts of the expansion of renewable energy on the German labour market. Study by GWS

Osnabrück, DIW Berlin, DLR Stuttgart, Fraunhofer ISI Karlsruhe, ZSW Stuttgart. https://elib.dlr.de/65777/1/broschuere_erneuerbar_beschaeftigt_final.pdf

Bundesministerium für Umwelt (BMU) (2012). Renewably employed. Short- and long-term impacts of the expansion of renewable energy on the German labour market. https://www.solarify.eu/wp-content/uploads/2013/04/EE_beschaeftigt_BMU.pdf

CEDEFOP (2019). Skills for green jobs: 2018 update. European synthesis report. Luxembourg: Publications Office. http://data.europa.eu/doi/10.2801/750438

ClearSky Advisors Inc. (2011). The economic impacts of the wind energy sector in Ontario 2011–2018. https://canwea.ca/pdf/economic_impacts_wind_energy_ontario2011-2018.pdf

Compass Renewable Energy Consulting Inc. (CREC). (2015). Wind dividends: an analysis of the economic impacts from Ontario's wind procurements. https://canwea.ca/wp-content/uploads/2018/10/FINAL-CanWEA-Economic-Analysis-Report-Nov_25-2015_PUBLIC.pdf

Compass Renewable Energy Consulting Inc. (CREC). (2018). Wind dividends: 2018 update – an analysis of the economic impacts of Ontario's wind energy industry. https://canwea.ca/wp-content/uploads/2018/10/winddividends-ontarioeconomicimpacts-web.pdf

Consoli, D., Marin, G., Mazucchi, A., and Vona, F. (2016). Do green jobs differ from non-green jobs in terms of skills and human capital? *Research Policy* 45: 1046–1060. https://doi.org/10.1016/j.respol.2016.02.007.

Cosbey, A., Wooders, P., Bridle, R., and Casier, L. (2017). In with the good, out with the bad: phasing out polluting sectors as green industrial policy. In: *Green Industrial Policy. Concept, Policies, Country Experiences* (eds. D.T. Altenburg and C. Assmann), 69–86. Geneva/Bonn: UN Environment/German Development Institute.

Davis, G. (2013). Counting (green) jobs in Queensland's waste and recycling sector. *Waste Management & Research* 31 (9): 902–909. https://doi.org/10.1177/0734242X13487580.

Earley, S. & Mabee, W.E. (2011). The impact of bioenergy and biofuel policies on employment in Canada. Papers presented at Working in a Warming World, York University, Canada.

Environmental Commissioner of Ontario (ECO). 2018. Making connections: straight talk about electricity in Ontario. http://docs.assets.eco.on.ca/reports/energy/2018/Making-Connections.pdf

Esposito, M., Haider, A., Samaan, D., and Semmler, W. (2017). Enhancing job creation through green transformation. In: *Green Industrial Policy. Concept, Policies, Country Experiences* (eds. T. Altenburg and C. Assmann), 50–67. Geneva, Bonn: UN Environment; German Development Institute.

European Commission (2014). Green employment initiative: tapping into the job creation potential of the green economy. http://ec.europa.eu/transparency/regdoc/rep/1/2014/EN/1-2014-446-EN-F1-1.Pdf

Fortin, P. (2018). Qu'avons-nous appris depuis 10 ans? https://lactualite.com/lactualite-affaires/quavons-nous-appris-depuis10-ans

Furchtgott-Roth, D. (2012). The elusive and expensive green job. *Energy Economics* 34: S43–S52. https://doi.org/10.1016/j.eneco.2012.08.034.

Guerrieri, P. and Meliciani, V. (2005). Technology and international competitiveness: the interdependence between manufacturing and producer services. *Structural Change and Economic Dynamics* 16 (4): 489–502. https://doi.org/10.1016/j.strueco.2005.02.002.

Gülen, G. (2012). Defining, measuring and predicting green jobs. www.lsarc.ca/Predicting%20Green%20Jobs.pdf

International Labour Organization (ILO) (2011). *Towards a Green Economy: The Social Dimension*. Geneva: ILO.

International Labour Organization (ILO) (2012). *International Standard Classification of Occupations*. Geneva: ILO.

International Labour Organization (ILO) (2013). *Methodologies for assessing green jobs*. Geneva: ILO.

Janser, M. (2018). The greening of jobs in Germany – first evidence from a text mining-based index and employment register data. http://doku.iab.de/discussionpapers/2018/dp1418.pdf

Jenniches, S. (2018). Assessing the regional economic impacts of renewable energy sources – a literature review. *Renewable and Sustainable Energy Reviews* 93: 35–51. https://doi.org/10.1016/j.rser.2018.05.008.

Katz, J. (ed.). (2012). Emerging green jobs in Canada: insights for employment counsellors into the changing labour market and its potential for entry-level employment. https://ceric.ca/resource/emerging-green-jobs-in-canada/

Llera-Sastresa, E., Usón, A.A., Bribián, I.Z., and Scarpellini, S. (2010). Local impact of renewables on employment: assessment methodology and case study. *Renewable Sustainable Energy Reviews* 14 (2): 679–690. https://doi.org/10.1016/j.rser.2009.10.017.

Lütkenhorst, W., Altenburg, T., Pegels, A., and Vidican, G. (2014). *Green Industrial Policy: Managing Transformation Under Uncertainty*. Bonn: German Development Institute.

MacCallum, M.A. (2015). Employment associated with renewable and sustainable energy development in the Kingston region. Master's thesis, Queen's University. https://qspace.library.queensu.ca/bitstream/handle/1974/13921/MacCallum_A_Megan_201601_MA.pdf?sequence=1&isAllowed=y

Market Surveillance Administrator (MSA). (2017). Report to the Minister for the year ending December 31, 2016. https://resources.albertamsa.ca/uploads/pdf/Archive/00000-2017/2017-03-10%2MSA%20Report%20to%20Minister%202016.pdf

McKitrick, R. & Aliakbari, E. (2017). Rising electricity costs and declining employment in Ontario's manufacturing sector. www.fraserinstitute.org/sites/default/files/rising-electricity-costs-and-declining-employment-in-ontarios-manufacturing-sector.pdf

Moreno, B. and López, A.J. (2008). The effect of renewable energy on employment. The case of Asturias (Spain). *Renewable and Sustainable Energy Reviews* 12 (3): 732–751. https://doi.org/10.1016/j.rser.2006.10.011.

Office of the Premier. (2009). Green Energy Act will attract investment, create jobs. https://news.ontario.ca/opo/en/2009/09/green-energy-act-will-attract-investment-create-jobs.html

Ontario Energy Board (OEB). (2018). The Industrial Conservation Initiative: evaluating its impact and potential alternative approaches. www.oeb.ca/sites/default/files/msp-ICI-report-20181218.pdf

Ontario Ministry of Energy and Infrastructure. (2009). Green Energy Act. www.mei.gov.on.ca/en/energy/gea

Ontario Society of Professional Engineers. (2016). Ontario's energy dilemma. Reducing emissions at an affordable cost. www.ospe.on.ca/public/documents/advocacy/2016-ontario-energy-dilemma.pdf

O'Sullivan, M., Edler, D., Bickel, P., Lehr, U., Peter, F. & Sakowski, F. (2014). Gross employment from renewable energy in Germany in 2013 – a first estimate. www.bmwi.de/Redaktion/EN/

Publikationen/bruttobeschaeftigung-durch-erneuerbare-energien-in-deutschland-im-jahr-2013. pdf?__blob=publicationFile&v=1

Park, S. (1989). Linkages between industry and services and their implications for urban employment generation in developing countries. *Journal of Development Economics* 30 (2): 359–379. https://doi.org/10.1016/0304-3878(89)90009-6.

Paulus, W. & Matthes, B. (2013). The German classification of occupations 2010: structure, coding and conversion table. http://doku.iab.de/fdz/reporte/2013/MR_08-13_EN.pdf

Pegels, A., Vidican-Auktor, G., Lütkenhorst, W., and Altenburg, T. (2018). Politics of green energy policy. *Journal of Environment & Development* 27 (1): 26–45. https://doi. org/10.1177/1070496517747660.

Pollin R. & Garrett-Peltier, H. (2009). Building the green economy: employment effects of green energy investments for Ontario. www.peri.umass.edu/fileadmin/pdf/other_publication_types/ green_economics/Green_Economy_of_Ontario.PDF

Rodrik, D. (2014). Green industrial policy. *Oxford Review of Economic Policy* 30 (3): 469–491. https://doi.org/10.1093/oxrep/gru025.

Rodrik, D. (2015). *Economics Rules: The Rights and Wrongs of the Dismal Science*. New York: Norton.

Rutovitz, J. & Atherton, A. (2009). Energy sector jobs to 2030: a global analysis. https://opus. lib.uts.edu.au/bitstream/10453/35047/1/rutovitzatherton2009greenjobs.pdf

Schneider, B. and Maxwell, S. (1997). *Business and the State in Developing Countries*. Ithaca: Cornell University Press.

Sharpe, A. (2015). Ontario's productivity performance, 2000–2012: a detailed analysis. www. csls.ca/reports/csls2015-04.pdf

Statistics Canada. (2018). Energy consumption by the manufacturing sector, 2017. www150. statcan.gc.ca/n1/daily-quotidien/181030/dq181030e-eng.htm

Task Force on Competitiveness, Productivity and Economic Progress. (2010). Today's innovation, tomorrow's prosperity. www.competeprosper.ca/uploads/ICP_AR9_Final.pdf

Ulrich, P. & Lehr, U. (2014). Erneuerbar beschäftigt in den Bundesländern: Bericht zur aktualisierten Abschätzung der Bruttobeschäftigung 2013 in den Bundesländern. www. gws-os.com/discussionpapers/EE_besch%C3%A4ftigt_bl_2013.pdf

United Nations Environmental Program (UNEP) (2011). *Towards a Green Economy: Pathways to Sustainable Development and Poverty Eradication*. Geneva: UN Environmental Program.

Unruh, G.C. (2000). Understanding carbon lock-in. *Energy Policy* 28 (12): 817–830. https://doi. org/10.1016/S0301-4215(00)00070-7.

Winfield, M. and Dolter, B. (2014). Energy, economic and environmental discourses and their policy impact: the case of Ontario's green energy and green economy act. *Energy Policy* 68: 423–435. https://doi.org/10.1016/j.enpol.2014.01.039.

13

Ethical and Sustainable Investing and the Need for Carbon Neutrality

Quintin G. Rayer

P1 Investment Management, Exeter, UK

13.1 Introduction

Unsustainable human activities are responsible for the threats associated with climate change, including rising sea levels, extreme weather conditions, and flooding, and can result in the disruption of food and fresh water supplies, property damage, and fatalities (Stern 2006; National Academies of Sciences, Engineering, and Medicine 2016). Proponents of responsible investing argue that unsustainable behaviors are no longer an option. While environmentally focused investors often consider climate risks, sustainable investing is already used to encourage companies to improve their environmental, social, and governance practices, including monitoring and reducing their carbon emissions (Porritt 2001; Stern 2006). Ethical and sustainable fund managers also define investment policies for activities in these areas to direct resources away from firms with harmful practices and toward those providing solutions.

The criteria for sustainable and ethical investing often include, among other factors, the reduction of carbon emissions; however, very few ethical investors feel the need to directly target zero carbon emissions or net zero carbon emissions (NZCE), also called carbon neutrality, which are necessary for the stabilization of global warming (Matthews and Caldeira 2008; Rayer 2017a, 2018a). This chapter focuses on carbon neutrality rather than net negative emissions (i.e., net removal of greenhouse gases), and calls for ethical investors to explicitly require companies not only to report and reduce carbon emissions, but also target NZCE.

Companies that are significant carbon emitters will need to develop plans to transition to lower-carbon technologies before seeking carbon neutrality. Likewise, ethical investors considering these companies should scrutinize their investees' low-carbon transition strategies before supporting them. This approach would represent a significant step forward from the ethical investment community, moving well ahead of the current general practice which regards the reduction of emissions or the transition to low (but not zero) carbon

Environmental Policy: An Economic Perspective, First Edition. Edited by Thomas Walker, Northrop Sprung-Much, and Sherif Goubran.

technologies as sufficient. This applies particularly to the UK-based ethical fund management community, where there is little rigorous definition of ethical investing practices and policies. At present most companies are also largely unaware of the need to achieve NZCE, so the promotion of suitable policies within the investment community would represent a useful indicator in the wider policy landscape.

This chapter is comprised of five main sections. The first outlines the climate challenge in terms of carbon emissions and the United Nations Framework Convention on Climate Change (UN FCCC) global average temperature targets, describing the ease with which companies can externalize emissions, and the investor response (UN FCCC 2015). The second major section considers the current emphasis on carbon reduction, the reasons why it is likely to prove an insufficient measure, and the dangers of a Malthusian trap. The third section explores the role of carbon offsetting as a tool for addressing emissions, providing an example illustrating the transition to low (net zero) carbon as well as recommendations for its use. The fourth section considers the role of ethical investing in terms of carbon emissions and sustainability within the sustainable investing framework, and calls for NZCE investment policies. The section then considers other investment approaches that address global warming, including fossil engagement and divestment, the internet of things (IoT) and its role in monitoring emissions, as well as the uptake challenges of each approach. The fifth section provides recommended investment policies based on the material presented, with specific recommendations indexed within the text using the codes described in Table 13.1. Finally, the conclusion suggests some next steps for ethically minded investors, as well as an outline of further required developments and the challenges faced by investors seeking to support a move toward NZCE.

13.2   Carbon Emissions and Climate Challenges

Human activity is well established as the leading cause of global warming (Stern 2006), with cumulative carbon dioxide emissions as the primary driver, as supported by over 97% of peer-reviewed scientific papers on the subject (Cook et al. 2013). Current efforts to limit the impacts of global warming have focused on encouraging companies to report and reduce their emissions. However, this may prove insufficient to meet the UN FCCC intended aims of limiting the increase in global average temperatures to 1.5 °C above pre-industrial levels, or well below 2 °C (UN FCCC 2015; Climate Action Tracker n.d.-b). Indeed, the 2018 South Korea climate summit emphasized the need to contain warming within 1.5 °C (McGrath 2018). Current global warming stands at around 1.13 °C (October 2019), and is currently increasing at a rate of about 0.2 °C per decade – and possibly accelerating (Haustein et al. 2017; University of Oxford Environmental Change Institute and University of Leeds Priestley International Centre for Climate 2019). With the current emissions trajectory, the 1.5 °C threshold could be exceeded between 2030 and 2052 (IPCC 2018). Based on current policies, projections range between 3.1 and 3.5 °C by 2100 (Climate Action Tracker n.d.-b). These estimates are based on existing country-specific policy scenarios from the literature, governments, national independent research, or international sources (Climate Action Tracker n.d.-a).

Table 13.1   Summary policy recommendations

	NZCE – Z	IoT – I
Companies and organizations – C	• C-Z.1: Identify and implement NZCE strategies with time-bound targets. • C-Z.2: Monitor and report progress toward NZCE status in both relative (% reduction) and absolute terms (tonnes $CO_2[e]$). • C-Z.3: Monitor and report progress toward reducing emissions by 40% relative to business as usual by 2030, or other scientifically based targets as new evidence emerges. • C-Z.4: Monitor and report emissions reduction undertaken by offsetting, both in terms of proportion of emissions offset (% offset) and in absolute terms (tonnes $CO_2[e]$ offset). • C-Z.5: Report details of offsetting schemes used so that their quality and reliability can be independently verified. • C-Z.6: Confirm that the offsetting schemes are used for (a) only addressing residual emissions, or (b) as a temporary measure. • C-Z.7: Confirm that the offsetting schemes used meet the minimum requirements of a recognized external standard, such as PAS 2060. • C-Z.8: Confirm that the NZCE strategies used meet the minimum requirements of a recognized external standard such as PAS 2060. • C-Z.9: Disassociate with, and ensure no financial support for, individuals or organizations that actively promote a climate (anthropogenic global warming) denial agenda. • C-Z.10: Educate staff, employees, and stakeholders where possible on the consensus climate science view on global warming and the steps required to address it.	• C-I.1: Identify and implement IoT strategies to deliver energy use reductions and energy efficiency gains. • C-I.2: Identify and implement IoT strategies to deliver fossil-based energy use reductions and fossil-based energy use efficiency gains. • C-I.3: Monitor and report progress on energy use reductions and energy efficiency gains. • C-I.4: Monitor and report progress on reductions in fossil-based energy usage in both relative (% reduction) and absolute terms (kWh or equivalent). • C-I.5: Identify and implement IoT strategies to monitor carbon and greenhouse gas emissions from firm activities (scopes 1, 2, and 3). • C-I.6: Report comprehensive emissions data (scopes 1, 2, and 3), broken down by geographical location and activity types, to national bodies and climate science researchers, both on a historical (after the fact) basis and in real-time. • C-I.7: Where possible, collaborate with climate scientists in the collection and use of real-time climate-related data. • C-I.8: Report details of collaborations with climate scientists in the collection and use of real-time climate-related data. • C-I.9: Respect personal data privacy and abide by legal requirements on data collection and use in relation to private individuals.

(Continued)

Table 13.1 (Continued)

	NZCE – Z	IoT – I
Direct ethical investors – EI	• EI-Z.1: Engage with companies on policy objectives C-Z.1 through C-Z.10. • EI-Z.2: Underweight equity and bond investment in firms acting contrary to policy objectives C-Z.1 through C-Z.10. • EI-Z.3: Underweight equity and bond investment in firms not meeting policy objectives C-Z.1 through C-Z.10. • EI-Z.4: Avoid equity and bond investment in firms not meeting policy objectives C-Z.1 through C-Z.10. • EI-Z.5: Use fossil engagement or fossil divestment to encourage fossil fuel intensive firms to reduce their carbon-based activities and emissions (scopes 1, 2 and 3).	• EI-I.1: Engage with companies on policy objectives C-I.1 through C-I.9. • EI-I.2: Underweight equity and bond investment in firms acting contrary to policy objectives C-I.1 through C-I.9. • EI-I.3: Underweight equity and bond investment in firms not meeting policy objectives C-I.1 through C-I.9. • EI-I.4: Avoid equity and bond investment in firms not meeting policy objectives C-I.1 through C-I.9. • EI-I.5: Actively invest in equity and bonds of companies developing technologies and services that support policy objectives C-I.1 through C-I.9.
Ethical fund selectors – EF	• EF-Z.1: Engage with ethical fund managers on policy objectives EI-Z.1 through EI-Z.5. • EF-Z.2: Underweight investment in ethical fund managers not meeting policy objectives EI-Z.1 through EI-Z.5. • EF-Z.3: Avoid investment in ethical fund managers not meeting policy objectives EI-Z.1 through EI-Z.5.	• EF-I.1: Engage with ethical fund managers on policy objectives EI-I.1 through EI-I.5. • EF-I.2: Underweight investment in ethical fund managers not meeting policy objectives EI-I.1 through EI-I.5. • EF-I.3: Avoid investment in ethical fund managers not meeting policy objectives EI-I.1 through EI-I.5.

The current increase rate of atmospheric concentrations of CO_2 – and other greenhouse gases, such as methane and nitrous oxide – is unprecedented in the ice core records from the last 22 000 years. There is also evidence that atmospheric CO_2 concentration plays an important role in colder periods associated with glacial advancement, as well as periods of warmer climate (Masson-Delmotte et al. 2013). An analysis of long-term temperature data suggests that global warming could be sufficient to significantly disrupt modern-day agriculture, to the extent that it would result in significant food shortages. Global sea levels are also affected by climate change; estimated rates of sea level rises average between 2.8 and 3.6 mm/yr from 1993 to 2012, a rate that is unusual compared to estimates from the last two millennia. If these rates persist over long periods, it will have important consequences for densely populated, low-lying coastal regions, where even a small sea level rise can inundate large areas (Masson-Delmotte et al. 2013). For an example of the potential impact of climate risks in investment terms, see Rayer and Millar (2018a, b). As we can see, the risks associated with climate change are severe; indeed, global climatic change may be the single most important proximate agent of mass extinction (Stanley 1987).

Analyses suggest that global CO_2 emissions need to reduce by at least 40% by 2030 to attain the 1.5 °C Paris agreement goal (Millar et al. 2018) (C-Z.3). However, for global warming to stabilize at any level, it is necessary to reach net zero carbon emissions (Matthews and Caldeira 2008; Millar et al. 2018) (C-Z.1, C-Z.2). Consequently, this chapter focuses on carbon neutrality, rather than the net removal of greenhouse gasses, as even the more ambitious scenarios for global warming stabilization focus on the reduction of net emissions rather than net removal (see for example Meinshausen et al. 2011). Though removal technologies may have a part to play in controlling net emissions, given the risks of uncontrolled climate change, it would be unwise to rely exclusively on unproven technologies for this task (Steffen et al. 2018). As Article 3.3 of the UN FCCC states, "Parties should take precautionary measures to anticipate, prevent, or minimize the causes of climate change and mitigate its adverse effects," meaning it would be far safer to focus on reducing emissions rather than CO_2 removal of emissions at this time (United Nations 1992).

The UN FCCC goals already accept some of the irreversible consequences of global warming, recognizing that the risks and impacts of climate change can only be reduced, not eliminated. But unfortunately, current efforts may prove insufficient, particularly with slow uptake from international governments and the potential for political pressure to delay or disrupt progress.

13.3 The Problem with Companies

The ability of companies to externalize costs associated with carbon emissions is likely to inhibit progress (Porritt 2001). For example, a company's activities or products may release significant quantities of atmospheric CO_2. Though these practices are unsustainable, generally the responsible company will not make significant financial contributions toward the reduction of atmospheric CO_2, or include the cost of CO_2 removal in the price of finished products. Multinational companies can also avoid significant taxation and regulations regarding carbon emissions by outsourcing their factory production to other countries. The costs of adapting to climate change then fall to society, with poorer countries often suffering disproportionately more than affluent, industrialized countries, which have historically benefited the most from the carbon-generating industry. This is typical of externalized costs, where the company responsible for the emissions has not paid for the consequences. Generally, consumers do not pay a price that reflects the real costs of dealing with the CO_2 involved in the production, use, and disposal of goods. The company in question, therefore, has little incentive to reduce carbon emissions or develop a strategy for achieving carbon neutrality (Porritt 2001; Rayer 2017b).

13.4 The Investor Response

Sustainable investing with carbon emissions in mind is one way of encouraging companies to reduce emissions. Awareness of ethical investment appears to be increasing rapidly, with £16.3 billion assets under management in the UK ethical funds sector in November 2018, an increase of £1.3 billion since November 2017 (Investment Association 2018). However, ethical investors need to be aware that while encouraging companies to report and reduce carbon emissions is helpful, at the current rate this is unlikely to be sufficient to meet the UN FCCC goals. What is required is a definite movement from companies toward NZCE to achieve carbon neutrality (C-Z.1).

To move forward, the ethical investment community (including clients, wealth managers, and fund providers) needs to start influencing companies to develop NZCE strategies and selecting investments based on their progress toward developing and implementing these policies (Rayer 2017a, 2018a).

13.5 Carbon Reduction

Current efforts to limit climate change have primarily focused on reducing CO_2 emissions. Examples include the CDP (formerly known as the Carbon Disclosure Project) which encourages companies to report their carbon emissions, or the UNPRI (UN Principles for Responsible Investment n.d.) which supports the adoption of ESG (environmental, social and governance) principles (Krosinsky and Robins 2008; Krosinsky et al. 2012; Griffin 2017). These encourage the incorporation of ESG principles into the decision-making process, including policies, practices, disclosures, acceptance, implementation, and reports.

When accounting for carbon emissions, the Greenhouse Gas Protocol (GHG Protocol) has categorized emissions into three groups, or "scopes" (GHG Protocol n.d.). Scope 1 emissions, or direct emissions, originate from sources that are owned and controlled by a company, including for example fuel used by company vehicles. Indirect emissions are covered by scopes 2 and 3; scope 2 emissions result from energy used by a company, including electricity, steam, heating, and cooling, while scope 3 emissions cover all other indirect emissions arising due to company activities. Scope 3 emissions also include upstream and downstream value chain emissions, including those of suppliers and customers using their products (for more detail see Carbon Trust n.d.). While avoiding double-counting, all three scopes of carbon emissions should be considered. Although many carbon emissions may take place when customers use their purchased products, efforts to reduce their emissions can be significantly constrained by the range of available products. Hence companies are best placed to reduce emissions by designing products which generate less carbon throughout the production process.

13.6 Why Reduction Alone Is Not Enough

Carbon reduction reporting and reduction initiatives are useful steps in the right direction, but given anticipated population growth and desirable economic development in less developed countries, it is reasonable to question whether they are likely to be sufficient to meet the UN FCCC aims (Shragg 2015).

13.6.1 Malthusian Trap

To quote Malthus, "The power of population is indefinitely greater than the power in the earth to produce subsistence for man" (Malthus 1798). This is the danger of the Malthusian trap: as less developed countries modernize and an expanding proportion of a growing world population demands improved living standards, any reduction in carbon emissions is likely to be negated by either increased global population or increased economic activity and emissions per capita (Rayer 2017b). For example, if emissions are reduced by 10% per capita but the carbon-based economy increases by 10% per capita, overall emissions remain unchanged.

We may see evidence of the Malthusian trap in data from the UK government's Green Finance Strategy on UK and G7[1] gross domestic product (GDP) and emissions from 1990 to 2015 (HM Government 2019). While UK GDP has grown by 72% during this time, national emissions have fallen by −42%. For G7 countries, over the same period, GDP has grown by +65%, while emissions are nearly static at −5% (1990–2016). While emissions reductions have been made in some G7 countries (such as the UK), increases from others have resulted in little change in emissions overall.

The issue of population levels and the impact of population growth is explored in some detail in *Move Upstream, A Call to Solve Overpopulation* (Shragg 2015). Addressing overpopulation is identified as a key upstream issue, since more people consume more resources and produce more waste. Shragg argues that working on downstream issues such as saving the environment and feeding the hungry is noble but ineffective, since the primary problem is a growing population. However, addressing emissions from a population perspective can be a highly controversial and politically charged issue, since it challenges many accepted social and cultural norms. Interested readers are referred to Shragg (2015) and the arguments for and against are not explored in this chapter. However, if carbon emissions are thought of in "per capita" terms, it becomes clear that as population increases, emissions per capita must decrease proportionately to keep overall emissions stable. To avoid this, emissions per capita must drop to zero; at this point, population increases would not affect emissions (C-Z.1, C-Z.2). It should be noted, however, that carbon emissions and carbon neutrality are only one aspect of a larger challenge; ultimately, sustainability is required across all aspects of human activity, including the use of resources, both of minerals and living organisms.

If there is continued exponential GDP growth in the carbon-based economy, anything less than an exponential reduction in carbon emissions will make no difference (Allen 2016). A faster-growing world economy may permit more rapid investment in low-carbon technologies, making it cost-effective to achieve NZCE sooner; however, what matters for peak global warming, primarily determined by accumulated CO_2 emissions, is the total emissions used to achieve a given rate of economic growth (Allen 2016). Apart from the need for dramatic reductions in carbon emissions, this highlights the need for companies to reach a stage where they rapidly become net zero carbon emitters, whilst maintaining growth. Ethical investment policies can be used to help direct capital market flows toward the most progressive firms in this respect.

13.7 Transition and the Role of Carbon Offsetting

The scale of the challenge can be seen by considering energy production from fossil fuels, which remains a significant contributor to global warming. Some 833 Gt of CO_2 equivalents were emitted from 1988 to 2015 – slightly more than the quantity emitted from the start of the Industrial Revolution in 1751 to 1987 (Griffin 2017). The fossil fuel industry and its products made up 91% of global industrial greenhouse gas emissions in 2015, with 25 entities (both companies and state producers) accounting for 51% of global industrial emissions. Seven of these top 25 emitters were publicly owned companies which accounted for 2.9 Gt of scope 1 and 3 emissions in 2015. Using emissions data for 224 fossil fuel companies in 2015, public investor influence reached 20% of global industrial emissions (Griffin 2017). While many ethical investment strategies will exclude such companies, this highlights the need for investor engagement as well as a significant market shift away from fossil fuels. Major emitters will need to develop a transition plan to move away from carbon-intensive technologies. The following section illustrates how the transition to low-carbon technologies and the use of carbon offsetting could hypothetically help the UK achieve NZCE.

13.7.1 Electricity Generation in the UK

Historically, coal-fired power stations have generated much of the UK's electricity. More recently, much of that capacity has been replaced by CCGT (combined cycle gas turbine) generation, with a further progressive move into renewables, including wind power.

For the sake of discussion, we will consider all three types: coal, CCGT, and wind. Over a full life-cycle (including manufacturing, installation, maintenance, and decommissioning of a plant), coal generation emits around 1000 g CO_2e per kWh (grams of CO_2 equivalent per kilowatt-hour); CCGT emits about 500 g CO_2e per kWh; and wind, up to approximately 32 g CO_2e per kWh (Thomson and Harrison 2015). Consider a hypothetical company generating 1 TWh (terawatt-hour) of power annually. If coal-fuelled, it would be responsible for around 1 Mt of CO_2e emissions; replacing with CCGT would halve this to 0.5 Mt of CO_2e emissions per year, while wind power would only generate 0.032 Mt of CO_2e emissions.

The above outlines a possible transition from high- to low-emission technologies. However, wind-power generation is not carbon neutral, as greenhouse gases are emitted during manufacture, installation, maintenance, and decommissioning (Thomson and Harrison 2015). For an offshore windfarm, around 78% of the CO_2 emissions arise from manufacturing and installation, with 20% resulting from operation and maintenance and 1% from dismantling and disposal (Thomson and Harrison 2015).Thus, progression from coal, through CCGT to wind power gives a significant reduction in emissions but does not achieve NZCE.

Indeed, once emissions have been reduced through the transition to low-carbon technologies, the remaining emissions must be offset. This means that a company must participate financially in reducing atmospheric CO_2 to a degree considered equivalent to its own emissions. For example, the hypothetical generator company using wind power has residual emissions of 0.032 Mt ($0.032 \times 10^6 = 32\,000$ tonnes) of CO_2 that need offsetting. Using a notional price for offsetting from January 2019 of £10 per tonne (for more information see, for example, Carbon Footprint 2019), the resulting cost would be £320 000. Although it can be subject to considerable variability and tends to be higher in winter, a notional wholesale electricity price – meaning the price that electricity suppliers are receiving – might be around £50 per MWh (megawatt-hour) (Energy Solutions 2019). On this basis, the 1 TWh (1×10^{12}Wh) annual generation of the company would be worth around £50 million at wholesale market prices ($1 \times 10^{12} \times £50/10^6 = £50 \times 10^6$). The £320 000 cost of offsetting the residual carbon emissions then represents only 0.6% of the generated electricity's value, a level that would appear economically affordable.

This example is not intended to be a blueprint as to how the electricity generation sector could achieve NZCE, but merely gives a concrete example as to how a combination of technological transition and carbon offsetting could be used for this purpose. Evidently, electricity generation cannot be solely dependent on wind power due to seasonality, the potential unreliability of wind as a power source, the need for different types of generation to match specific demand requirements, and the desirability of diversification in the energy mix.

13.8   Questions About the Role of Offsetting

The example above also raises critical questions about the role of offsetting. If a carbon-intensive business is sufficiently profitable, why not just offset in large volume? There are reasons why such an approach may not be desirable. For carbon-intensive activities, if an extremely high volume of carbon offsetting were to be required, it seems likely that there might be an insufficient offsetting capacity in environmental projects to meet demand, resulting in either a shortfall or the creation of substandard offsetting schemes, which may not yield the promised benefits (Carbon Trust 2006).[2] Another issue is that carbon emission estimates are prone to degrees of inaccuracy. Combined with the difficulties of accurately assessing the actual amount of carbon accounted for by offsetting schemes, it seems likely that although offsetting may be carried out in good faith, offsetting may then prove to fall short of what would be required to balance the climate need. However, neither of these points means that offsetting should not be used, simply that it might be wise to adopt a precautionary principle and use offsetting only as a temporary measure or last resort.

13.9   Offsetting Recommendations

Given these circumstances, the recommendation would be that carbon offsetting be used primarily in the two following circumstances.

- After low-carbon technologies have been used to reduce emissions to as low as possible. In this case, offsetting may be used to absorb any residual CO_2 emissions (C-Z.6).
- As a practical temporary measure to mitigate the worst effects of carbon emissions while a strategy for transitioning to lower carbon technologies is developed and implemented (C-Z.6, C-Z.4).

Investment policies must then define the requirements for credible, time-bound reduction strategies and the acceptable use of offsetting for companies. Given the longer time frame involved (typically a decade until 2030 or later), strategies should be realistic and use existing technologies to avoid the risk of failure at later stages. Policies should then seek to steer firms to operate within these requirements through a combination of engagement and selective investment policies, as used elsewhere in ethical investing (see, for example, Rayer 2017b).

13.10   Ethical Investing, Sustainability, and Carbon

For current purposes, little distinction is made between ethical, responsible, or sustainable investing; in this chapter these terms will be used interchangeably.[3] By adopting ethical investing as an overarching term, the author means to identify as morals-based any investment objective which does not solely maximize wealth (by legal means). Thus, an investment which maximizes wealth while considering sustainable, environmental, social, religious, or other factors is ultimately an ethical choice. Rather than defining an investor's utility function solely as a function of wealth, or $U(\$)$, ethical investment is defined as a function of both wealth and other ethical or moral considerations, or $U(\$, E)$. This follows Krosinsky and Robins' (2008) interpretation of Hudson (2006), which describes ethical investment as "an approach to investing driven by the value system of the key investment decision-maker." Broadly speaking, companies are encouraged to promote practices including environmental stewardship, consumer protection and human rights, and to support the social good (Krosinsky and Robins 2008; Krosinsky et al. 2012).

Ethical investors are already familiar with sustainable investing and its focus on ESG issues (Krosinsky and Robins 2008; Krosinsky et al. 2012). In sustainable investing, funds are directed toward companies with sustainable business practices, meaning their activities meet the needs of the present without compromising the ability of future generations to meet their own needs (Brundtland 1987).

ESG then identifies the following three key aspects for sustainable investing.

1) *Environmental*: CO_2 emissions or carbon intensity; forest and woodland degradation (which are important for absorption of atmospheric CO_2); air-borne, water-borne, or land-based pollution; usage of scarce resources, including water, living creatures, minerals, oil, and natural gas; mining activities which generate toxic by-products; overfishing, intensive agricultural methods, and so on.
2) *Social*: corporate social responsibility (CSR); child labor; modern-day slavery; payment of nonliving wages; hazardous, exploitative, or coercive working conditions; structures that reduce corporate taxation bills to levels incommensurate with the profits and activities taking place in those countries; antisocial working hours or conditions; displacement of indigenous peoples.
3) *Governance*: companies with weak internal controls whose management does not follow company policies, increasing risks of irresponsible behaviors, corruption, and bribery; nonexecutive directors (NEDs) who are unable to hold powerful executive directors in check, possibly damaging the company as well as its owners' (shareholders') interests and increasing the risk of excessive executive remuneration.

Once identified, issues of concern such as the above are addressed by the adoption of appropriate investment policies (Rayer 2017c, 2018b).

Within sustainable investing, environmental factors include addressing climate change and diminishing the effects of global warming caused by excessive build-up of atmospheric CO_2. While other environmental problems (as well as other social and governance issues) are of undoubted importance, few pose as existential a threat to human society as climate change does. Thus, the need for sustainability in relation to CO_2 sits squarely within an ESG investment framework.

13.11   The Need for Net Zero Carbon Emissions Investment Policies

Environmental policy requirements should thus include strategies for the achievement of NZCE or, for more carbon-intensive industries, the development of plans for the transition to low-carbon technologies. At present, the investment policies of UK-based ethical and sustainable fund managers with an emissions focus have considered reduction but have had no meaningful emphasis on neutrality. Indeed, it appears that there is confusion over definition of terms like "carbon neutral" and how emissions reduction differs from removal. In this respect, the British Standards Institute's PAS (Publicly Available Specification) 2060 standard can provide useful clarity (BSI 2010, 2014).

As a practical follow-up to the work reported here, the author is currently working with several UK-based ethical fund managers to explore how NZCE-related investment policies making use of PAS 2060 can be incorporated into their stock selection processes. If successful, this ongoing project should make it easier for underlying investors to select fund managers striving for NZCE strategies and start to promote some definitional rigor in this area.[4]

13.11.1   PAS 2060

Organizations that have followed a path of emissions reduction combined with carbon offsetting may wish to confirm or demonstrate that they are "carbon neutral." Launched in 2010, the British Standards Institute's PAS 2060 offers a common definition and method of validation for such claims (Carbon Clear n.d.).

Assessment of whether an organization meets the PAS 2060 standard includes all scope 1 and scope 2 emissions, as well as all scope 3 emissions that contribute to more than 1% of the entity's total footprint (Carbon Clear n.d.) (C-Z.8). The standard also sets requirements for the quality of carbon offsetting schemes used. According to certain definitions, an organization can be qualified as carbon neutral if its emissions are balanced by suitable offsetting, but the PAS 2060 standard does not allow for carbon neutrality to be claimed entirely through offsetting, except during early stages. Offsetting schemes must be high-quality certified credits, meeting criteria such as provenance from a PAS 2060-approved scheme, genuinely additional (not reductions that would have occurred without action being taken), verified by an independent third party to ensure they are permanent and not double counted, and retired after a maximum of 12 months to an appropriate registry (C-Z.5, C-Z.7).

13.12   Technology and the Internet of Things

Ethical investors can support change not only through encouraging firms to take up NZCE strategies, but also by directing funds away from emitters and toward firms developing solutions, including IoT technologies. The internet of things refers to the interaction between multiple technologies, including internet connectivity, computer devices, and the ability to transfer data over a network without human interaction. First mentioned by Kevin Ashton in 1999, IoT devices can employ real-time analysis, machine learning, sensors, and embedded systems (Wigmore 2014).

The IoT can contribute significantly to sustainability; however, its current use appears to focus on energy savings, including areas such as electricity distribution (smart metering), social and service industries (healthcare, education, and government) and transportation (route and traffic optimization) (Arias et al. 2018; Lueth 2018). These innovations are projected to help reduce GHGs by 15–20% by 2030 (#SMARTer2030, ICT Solutions for 21st Century Challenges 2015; Ericsson 2015) (C-I.1, C-I.3). While helpful, this is well short of the 40% required (Millar et al. 2018). However, IoT technologies can also help promote economic growth, which is useful if additional resources can be invested in the development of carbon-neutral strategies and techniques (Allen 2016).

Additionally, the IoT, which already uses temperature, precipitation, and humidity sensing, can help with the collection of data for climate models used to evaluate global warming which already make use of such data. It should also be able to extend this information to include improved data on ocean temperatures and sea levels (Technative 2017) (C-I.7, C-I.8).

13.12.1   IoT and Emissions Monitoring

Another crucial role the IoT can play, which appears to be overlooked by many commentators, is the monitoring of activities contributing to emissions, such as illegal logging and deforestation (Williams 2017). Significantly, the IoT can help monitor the use of fossil fuels, the emission of CO_2 and other GHGs, and the benefits of offsetting, feeding these data through to central databases and ultimately to computer climate models to determine their impacts (C-I.2, C-I.4). This could contribute to attribution analyses of emissions for specific countries and companies. Installing emission-monitoring devices in carbon-emitting technologies might even permit emissions levels to be determined at an individual level, thus supporting emissions transparency for investors and climate scientists alike (C-I.5, C-I.6).

Investment policies should cover all aspects. Companies and individuals developing IoT technologies should focus not just on efficiency gains but also how they can facilitate monitoring of GHG emissions and support strategies that will allow companies to achieve NZCE or carbon neutrality.

13.12.2   Why Should Firms Adopt IoT Emissions Monitoring?

Although the IoT can contribute to monitoring, it is unlikely that companies would adopt technologies that might make them liable for the cost of past, present, or future emissions, or force them to invest in carbon-neutral systems. Societal norms and social pressures may motivate some firms, while efficiency gains could spur others. Ethical investing, particularly as sustainability becomes the "investment norm," should provide a degree of pressure from financial markets via share-pricing and capital availability. Other financial market mechanisms could also be considered, such as carbon pricing. Adopting IoT technologies could then provide commercial gains in terms of appealing to customers, business partners, or financial markets.

Ultimately, however, if sufficiently rapid uptake of both emissions monitoring and carbon neutrality from the corporate sector is not forthcoming, voluntary or market forces approaches may not be enough and regulatory policies may be required. In this case, greenhouse gas emissions would need to be placed on a par with other pollutants and regulated accordingly.

Regarding emissions, the Paris Agreement makes little mention of loss and damage estimates associated with climate change as a basis for liability, even if a legal precedent for climate damage liability is established in future (Thornton and Covington 2016). This makes it hard to say how rapidly investors should react to the possibility of companies having (or deciding) to provide compensation for damages associated with climate change caused by their emissions. The barriers to a successful compensation case for climate damages remain substantial, but with the science developing, the possibility remains. For the major insurance companies or governments footing the bill, the prospect of multibillion dollar pay-outs may focus minds on whether the legal barriers could be overcome, to force firms to contribute to the costs incurred by climate damage, since this may allow them to pass on costs.

13.12.3 Data Protection Issues with IoT Technologies

Employing IoT technologies to collect and distribute data on energy, emissions, and climate for efficiency monitoring, greenhouse gas attribution, and enhanced climate analysis may raise discussion around data protection issues. In the UK, this is primarily covered by the Data Protection Act 2018 and the General Data Protection Regulation (GDPR) (UK Government 2018; Information Comissioner's Office n.d.).

Though the GDPR relates to the data of individuals, organizations must be aware that they are composed of collections of individuals, and be cautious so as to not collect any personally identifiable information. In this respect, if presented appropriately, anonymized data such as highest or lowest values, without names attached, should not present any risks. Firm-wide data on, say, aggregated CO_2 emissions should also not fall under GDPR.

While aggregated company or organization data on emissions and efficiency may not present data privacy issues from a GDPR perspective, there may be commercial sensitivities around the public release of such information, in case competitors could exploit it to their advantage. In the case of climate emissions data, it could be argued that global climate concerns should outweigh the commercial interests of any one organization or group. However, it should be noted that corporate emissions data are already being collected by organizations such as the CDP (2019) without undue difficulties arising; the implementation of IoT technologies would appear to be a natural extension of this sort of approach.

Finally, the ethical investing perspective also raises another aspect, though not directly related to NZCE; investors' policies should also require firms with IoT capabilities to respect individuals' privacy in relation to the data they collect. Apart from providing technology for efficiency, emissions, and other similar sensors, IoT firms may also be involved with activities where personal information can be collected – for example, smart domestic or personal appliances. Ethical investors should require the IoT firms they select to respect personal data and abide by legal requirements on data collection and use in relation to individuals (C-I.9).

13.13 Other Ethical Investing Policies Addressing the Climate Challenge

Ethical and sustainability-oriented investors have used several policies to support the reduction in industrial carbon emissions and hasten progress toward a carbon-neutral economy. Fossil divestment is one approach, although some investors argue that engaging with fossil companies is more effective in promoting change (for fossil divestment, see McKibben 2012). Technology can assist in emissions reduction and offsetting, as well as carbon capture or removal to help address climate challenges. This chapter has also emphasized adoption of NZCE strategies (an approach supported by the Oxford Martin Principles (Millar et al. 2018)). These investment policies can be supported by the IoT which may support multiple technologies for helping in emission reduction, enhanced data collection for climate analyses, and improved monitoring of GHG emissions by specific countries, industry sectors, and firms.

13.13.1 Fossil Engagement and Divestment

Fossil divestment involves severing ties with firms that extract fossil fuel reserves, and selling or refusing to own stock in fossil extractors and producers – a process backed by the UNFCCC in 2015 (Carrington 2015). Estimates from 2012 suggested that to keep global warming below $2\,°C$, no more than around 565 Gt of additional carbon dioxide could be released by mid-century, despite proven underground coal, oil, and gas reserves amounting to 2795 Gt – five times more than the climate can tolerate to stay below the recommended threshold (McKibben 2012). More recent estimates indicate that at least two-thirds of known fossil fuel reserves must remain unburned (McGlade and Ekins 2015). The solution is simple – most of this carbon needs to stay locked in fossil reserves underground.

Some investors fear that restricting investment may reduce diversification and impact performance, although many ethical investors disagree (see, for example, Grantham 2018; Rayer 2018b). Others accept the need to reduce CO_2 emissions but feel that engaging with – rather than ostracizing – fossil extractors and producers is more likely to achieve this goal (Howard 2015). They point out that a shareholding is needed to influence a firm, so divestment removes the possibility of encouraging companies to move away from fossil fuels toward renewables (McGee 2016). Critics suggest that engagement is most effective when backed up with a credible threat to divest (Operation Noah 2018; Rayer 2019). These investors have the same goal – a low-carbon or carbon-neutral future – but differ on whether engagement or divestment is the more effective tool.

If climate challenges are to be addressed, it is likely that all these approaches (and more) will be necessary.

13.14   Investment Policy Recommendations

Table 13.1 summarizes a number of policy recommendations. These recommendations relate to various groups: companies and organizations; ethical investors (including ethical fund managers), called "direct ethical investors," that invest in companies via both equities and bonds; and wealth managers and other parties who select ethical fund managers for investment (called "ethical fund selectors"). Recommendations consider both NZCE-related targets as well as how IoT technologies may be used to help support NZCE, in both relative and absolute terms.

Direct ethical investors may include portfolio managers who buy equities and bonds for portfolios, as well as those running ethical and sustainable funds for institutional and retail clients. Investors also include fund selectors, who use ethical funds and can include wealth management firms, charitable organizations, pension schemes, faith groups, trustees, and others. Direct ethical investors should adopt policies which support company-level recommendations (EI-Z.1 to EI-Z.5 and EI-I.1 to EI-I.5). Equally, ethical fund selectors should adopt policies which encourage ethical fund managers to follow investor recommendations (EF-Z.1 to EF-Z.3 and EF-I.1 to EF-I.3).

It should also be noted that both direct ethical investor organizations and ethical fund selectors are often companies or organizations. In this case, they should also seek to implement the policy recommendations for companies and organizations in Table 13.1. Although financial firms in this sector typically have modest carbon emissions, the nature of the policies proposed suggests that they should lead by example – and of course, all steps toward NZCE, however small, are important.

Table 13.1's recommendations also include the need for firms to terminate engagement with or support (financial or otherwise) for individuals and organizations that actively promote an anthropogenic global warming climate denial agenda. Such a position does not align with the established scientific consensus in the peer-reviewed literature (Oreskes 2004; Oreskes and Conway 2010) (C-Z.9). A final point is the promotion of the education of staff and stakeholders on the established scientific consensus on global warming (C-Z.10).

13.15   Conclusion

To meet climate change targets, emissions must be significantly reduced. Analyses suggest a reduction of 40% in carbon emissions by 2030 is needed to achieve the 1.5 °C Paris goal. It is unlikely that current reporting and reduction approaches will be enough. Therefore, to avoid a Malthusian trap, companies need to develop strategies for carbon neutrality. Carbon offsetting is helpful, but uncertainties in how effective it is mean it is preferable to use technology (or other methods) to reduce emissions as much as practically possible, using offsetting to address residual emissions. While strategies are being developed and implemented to achieve NZCE, offsetting should also be used temporarily.

Ethical investors need to start developing policies for judging companies based on their development and implementation of NZCE strategies, in addition to current ethical and sustainability criteria. For carbon-intensive industries, the same goal applies, although plans should include a preliminary step for transitioning to lower-carbon technologies on a realistic timescale. Current IoT technologies can contribute toward global warming containment, but their estimated impact of 15–20% emission reduction by 2030 falls short of the 40% required to meet the 1.5 °C Paris goal. Companies therefore need to develop business plans to achieve carbon-neutral operations; both ethical investors and IoT technologies can support progress in this area.

Apart from energy and efficiency savings, IoT technologies will need to contribute in other areas, including:

- enhancing data acquisition and collection for climate models used to evaluate global warming
- monitoring GHG emissions activities and offsetting benefits
- collecting data used to improve analyses of emissions attribution by specific countries and firms.

This chapter represents an initial step in highlighting the importance of targeting NZCE within climate policies. However, without a significant increase in profile, it remains easier for ethical fund providers to maintain the status quo, which includes offering carbon reporting and limited reduction strategies to investors, rather than including a clear, unambiguous emphasis on the need to target NZCE directly. Many ethically minded investors are not in a position to truly judge the approaches used by fund managers. This leaves fund management houses with the option to adopt minimal ethical criteria, and content themselves with a comfortable consensus on what is acceptable for ethical funds, while maintaining the profits and marketing advantages that ethically labeled funds provide them. Lacking expertise in this area and access to detailed information, many ethical investors are only weakly positioned to challenge this consensus. Climate science suggests that fund managers' failure to pursue NZCE directly may be an important opportunity lost.

For NZCE policies to be effectively adopted by the financial sector, they will need to be taken up by underlying investors, wealth managers, and fund providers. Fund providers must develop policies and the capability to judge and screen companies on their development and implementation of NZCE strategies. Interest from underlying clients will be required to generate the demand for products incorporating these strategies, while wealth managers can facilitate the direction of invested finances into the funds with the most robust NZCE policies.

Wealth management companies that use their expertise to select ethical and sustainable funds from those offered by fund management houses also have a pivotal role to play in this context. They often screen funds offered by management houses to determine which are most suitable for their clients. Providing they have the necessary ethical investment expertise and the capacity to complete the required screening and due diligence on fund managers, this puts wealth management companies in a strong position to select only those ethical funds that meet strict, progressive ethical and sustainability criteria. These ethically focused wealth management companies can also raise the importance of issues that the broader ethical fund management sector may not be ready to address yet (such as NZCE), and thus improve the policies of ethical investment funds on offer. By selectively directing client funds toward ethical funds that meet higher standards, they can promote the variety and quality of funds on offer.

However, to exert influence on fund management companies and improve their ethical products, wealth management companies, in turn, require the support of their clients. This entails the need to educate their clients and financial advisors on the crucial importance of the issues identified. These clients then ultimately provide them with the underlying monies to direct toward better-quality ethical funds.

As we have seen, to support progress toward containing global warming, it will be necessary for ethical investors and IoT technologists to work together at several levels; investors must support companies developing NZCE strategies (or IoT technologies promoting these strategies), and IoT developers must research and implement the novel technologies required. Table 13.1 presents a number of recommended policies to help support this goal. To maximize contributions in this area, it will be necessary for a wide range of ethical investors – from private individuals, charities, and trusts, to ethical wealth managers and ethical fund providers – to work together to raise the profile of this crucial aspect of sustainable environmental investment.

Acknowledgments

The author gratefully acknowledges helpful suggestions and input from Myles Allen, Professor of Geosystem Science at the Environmental Change Institute of the School of Geography and the Environment, and the Department of Physics, and Dr Pete Walton, Research Fellow with the UK Climate Impacts Programme, both at the University of Oxford, as well as helpful feedback from the anonymous reviewers.

Notes

1 In 2019 the G7 group of seven major advanced nations comprised Canada, France, Germany, Italy, Japan, the United Kingdom, and the United States.
2 For example, large-scale tree planting schemes to absorb atmospheric CO_2 might result in monoculture forests, which lacking biodiversity might be more than usually vulnerable to disease, pests, or wildfires.
3 For definitions of these terms see Krosinsky and Robins (2008). "Responsible investing" has been described as an approach used by institutional investors taking ESG factors into account (for the definition of ESG, see main text); "sustainable investing" is driven by long-term economic, environmental, and social risks. Related approaches include "clean technology" investment and "social investment" which seeks to generate social as well as financial returns.
4 In collaboration with practicing climate scientists, the author has created an NZCE investment target for fund managers, "NZC10" (Net-Zero Carbon 10). As at October 2019, this target has been adopted by four practicing UK-based fund managers, with over £2.2 billion assets following NZC10. For details see Rayer et al. (2019a, b) and Rayer and Walton (2019).

References

Allen, M.R. (2016). Drivers of peak warming in a consumption-maximizing world. *Nature Climate Change* 6: 684–686. https://doi.org/10.1038/NCLIMATE2977.

Arias, R., Lueth, K. L., & Rastogi, A. (2018). The effect of the internet of things on sustainability. www.weforum.org/agenda/2018/01/effect-technology-sustainability-sdgs-internet-things-iot

Brundtland, G. (1987). *Our Common Future, From One Earth to One World: An Overview by the World Commission on Environment and Development*. Oxford: World Commission on Environment and Development/Oxford University Press.

BSI. (2010). PAS 2060. www.bsigroup.com/en-GB/PAS-2060-Carbon-Neutrality

BSI. (2014) PAS 2060:2014 Specification for the demonstration of carbon neutrality. British Standards Institution, London. https://carbon-clear.com/what-we-do/carbon-offsetting/pas-2060-carbon-neutrality

Carbon Footprint. (2019). www.carbonfootprint.com

Carbon Trust. (n.d.). What are scope 3 emissions? www.carbontrust.com/resources/faqs/services/scope-3-indirect-carbon-emissions

Carbon Trust (2006). *The Carbon Trust Three Stage Approach to Developing a Robust Offsetting Strategy*. London: Carbon Trust.

Carrington, D. (2015, March 15). Climate change: UN backs fossil fuel divestment campaign. The Guardian.

CDP. (2019). www.cdp.net/en

Climate Action Tracker. (n.d.-a) Current policy projections. https://climateactiontracker.org/methodology/current-policy-projections

Climate Action Tracker (n.d.-b). Temperatures, 2100 warming projections. https://climateactiontracker.org/global/temperatures

Cook, J., Nuccitelli, D., Green, S.A. et al. (2013). Quantifying the consensus on anthropogenic global warming in the scientific literature. *Environmental Research Letters* 8 (024024): 7.

Energy Solutions. (2019). Historical electricity prices. www.energybrokers.co.uk/electricity/historic-price-data-graph.htm

Ericsson. (2015). ICT and the low carbon economy. Stockholm: Extract from the Ericsson Mobility Report, EAB-15:037849/1 Uen, Revision A.

Global e-Sustainability Initiative (GeSI) and AccentureStrategy (2015). #SMARTer2030, ICT Solutions for 21st Century Challenges. Brussels: Global e-Sustainability Initiative (GeSI) and AccentureStrategy.

Grantham, J. (2018). The mythical peril of divesting from fossil fuels. www.lse.ac.uk/GranthamInstitute/sustainablefinance

Greenhouse Gas Protocol. (n.d.). www.ghgprotocol.org/about-us

Griffin, P. (2017). The Carbon Majors Database, CDP Carbon Majors Report 2017. https://b8f65cb373b1b7b15feb-c70d8ead6ced550b4d987d7c03fcdd1d.ssl.cf3.rackcdn.com/cms/reports/documents/000/002/327/original/Carbon-Majors-Report-2017.pdf

Haustein, K., Allen, M.R., Forster, P.M. et al. (2017). A real-time global warming index. *Scientific Reports* 7: 15417. https://doi.org/10.1038/s41598-017-14828-5.

HM Government (2019). *Green Finance Strategy: Transforming Finance for a Greener Future*. London: HM Government.

Howard, E. (2015, June 23). A beginner's guide to fossil fuel divestment. The Guardian.

Hudson, J. (2006). *The Social Responsibility of the Investment Profession*. Charlottesville: CFA Institute.

Information Comissioner's Office. (n.d.). https://ico.org.uk/for-organisations/guide-to-data-protection

Investment Association. (2018). PDF Archive of Statistics. www.theinvestmentassociation.org/fund-statistics/full-figures

IPCC. (2018). Summary for Policymakers. In: Global warming of 1.5°C. An IPCC Special Report on the impacts of global warming of 1.5°C above pre-industrial levels and related global greenhouse gas emission pathways. Geneva: World Meteorological Organization.

Krosinsky, C. and Robins, N. (2008). *Sustainable Investing, the Art of Long-Term Performance*. London: Earthscan.

Krosinsky, C., Robins, N., and Viederman, S. (2012). *Evolutions in Sustainable Investing: Strategies, Funds and Thought Leadership*. Hoboken: Wiley.

Lueth, K. L. (2018). The effect of the internet of things on sustainability. https://dzone.com/articles/the-effect-of-the-internet-of-things-on-sustainabi

Malthus, T. R. (1798). An Essay on the Principle of Population. Oxford World's Classics reprint.

Masson-Delmotte, V., Schulz, M., Abe-Ouchi, A., et al. (2013). Information from Paleoclimate Archives. In: Climate Change 2013: The Physical Science Basis. Contribution of Working Group I to the Fifth Assessment Report of the Intergovernmental Panel on Climate Change. Cambridge and New York: IPCC.

Matthews, H.D. and Caldeira, K. (2008). Stabilizing climate requires near-zero emissions. *Geophysical Research Letters* 35: L04705. https://doi.org/10.1029/2007GL032388.

McGee, S. (2016, April 24). Is divesting from fossil fuels the best tactic for tackling climate change? The Guardian.

McGlade, C. and Ekins, P. (2015). The geographical distribution of fossil fuels unused when limiting global warming to 2°C. *Nature* 517: 187–190.

McGrath, M. (2018). Final call to save the world from 'climate catastrophe'. www.bbc.co.uk/news/science-environment-45775309

McKibben, B. (2012, August 2). Global warming's terrifying new math. Rolling Stone.

Meinshausen, M., Smith, S.J., Calvin, K. et al. (2011). The RCP greenhouse gas concentrations and their extensions from 1765 to 2300. *Climatic Change* 109: 213–241. https://doi.org/10.1007/s10584-011-0156-z.

Millar, R.J., Hepburn, C., Beddington, J., and Allen, M.R. (2018). Principles to guide investment towards a stable climate. *Nature Climate Change* 8: 2–4.

National Academies of Sciences, Engineering, and Medicine (2016). *Attribution of Extreme Weather Events in the Context of Climate Change*. Washington, DC: National Academies Press.

Operation Noah (2018). *Fossil Free Churches, Accelerating the Transition to a Brighter, Cleaner Future*. London: Operation Noah.

Oreskes, N. (2004). The scientific consensus on climate change. *Science* 306: 1686.

Oreskes, N. and Conway, E.M. (2010). *Merchants of Doubt*. New York: Bloomsbury Press.

Porritt, J. (2001). *The World in Context: Beyond the Business Case for Sustainable Development*. Cambridge: HRH The Prince of Wales' Business and the Environment Programme, Cambridge Programme for Industry.

Rayer, Q. G. (2017a). Should ethical investors target carbon-neutrality? Proceedings of the 12th Annual Green Economics Institute, Green Economics Conference, St Hugh's College, Oxford University, June, pp. 271–275.

Rayer, Q.G. (2017b). Exploring ethical and sustainable investing. *CISI, The Review of Financial Markets* 12: 4–10.

Rayer, Q.G. (2017c). Green dollars. *STEP (Society of Trust & Estate Practitioners) Journal* 25 (8): 52–53.

Rayer, Q. G. (2018a, December 20). What about zero?' asks Dr Quintin Rayer, Head of Ethical Investing at P1. The Review, p. 11.

Rayer, Q. G. (2018b, April 12). Why clean money matters. The Actuary, pp. 18–19.

Rayer, Q. G. (2019, March 14). Fossil engagement: if you won't divest, at least do this... https://p1-im.co.uk/research-articles/fossil-engagement-if-you-wont-divest-at-least-do-this/

Rayer, Q. G., & Millar, R. J. (2018a, February 8). Investing in extreme weather conditions. https://p1-im.co.uk/research-articles/investing-extreme-weather-conditions/

Rayer, Q.G. and Millar, R.J. (2018b). Investing in the climate. *Physics World* 31 (8): 17.

Rayer, Q. G., & Walton, P. (2019). Net Zero Carbon 10. Achieving Net Zero International Conference, 9–11 September, Wadham College, University of Oxford, poster.

Rayer, Q. G., Walton, P., Appleby, M., Beloe, S., Heaven, E., & Seery, A. (2019a). Sustainable investing: a target for progress towards carbon-neutrality. https://p1-im.co.uk/research-articles/sustainable-investing-a-target-for-progress-towards-carbon-neutrality/

Rayer, Q. G., Walton, P., Appleby, M., Beloe, S., Heaven, E., & Seery, A. (2019b). Zero tolerance. https://p1-im.co.uk/research-articles/zero-tolerance-net-zero-carbon/

Shragg, K.I. (2015). *Move Upstream, A Call to Solve Overpopulation*. Minneapolis-St. Paul: Freethought House.

Stanley, S.M. (1987). *Extinction*. New York: Scientific American Library.

Steffen, W., Rockström, J., Richardson, K. et al. (2018). Trajectories of the Earth system in the anthropocene. *PNAS* 115 (33): 8252–8259. https://doi.org/10.1073/pnas.1810141115.

Stern, N. (2006). *Stern Review Executive Summary*. London: New Economics Foundation.

Technative. (2017). Could IoT help combat climate change? www.technative.io/could-iot-help-combat-climate-change

Thomson, R. C., & Harrison, G. P. (2015). Life cycle costs and carbon emissions of offshore wind power. www.climatexchange.org.uk/files/4014/3325/2377/Main_Report_-_Life_Cycle_Costs_and_Carbon_Emissions_of_Offshore_Wind_Power.pdf

Thornton, J. and Covington, H. (2016). Climate change before the court. *Nature Geoscience* 9: 3–5.

UK Government. (2018). Data Protection Act 2018. www.legislation.gov.uk/ukpga/2018/12/contents/enacted

UN FCCC. (2015). Adoption of the Paris Agreement. United Nations Framework Convention on Climate Change. https://unfccc.int/resource/docs/2015/cop21/eng/l09r01.pdf

UN Principles for Responsible Investment. (n.d.). www.unpri.org

United Nations (1992). *United Nations Framework Convention on Climate Change. Chapter XXVII. Environment*. New York: United Nations.

University of Oxford Environmental Change Institute and University of Leeds Priestley International Centre for Climate. (2019). Current Global Warming Index. www.globalwarmingindex.org

Wigmore, I. (2014). Internet of Things (IoT). https://internetofthingsagenda.techtarget.com/definition/Internet-of-Things-IoT

Williams, K. (2017). How the Internet of Things can fight climate change. energypost.eu/internet-things-can-fight-climate-change/

Section IV

Financing the Environmental Transition

14

Building Sustainable Communities Through Market-Based Instruments

Ying Zhou[1], Amelia Clarke[1], and Stephanie Cairns[2]

[1] *School of Environment, Enterprise and Development (SEED), University of Waterloo, Waterloo, Ontario, Canada*
[2] *Smart Prosperity Institute, Ottawa, Ontario, Canada*

14.1 Introduction

Sustainable development and water management have become increasingly important at the local level. At the United Nations Conference on Environment and Development in 1992, national governments adopted Agenda 21, and one paragraph within that visionary docu-ment explicitly calls for critical action at the local scale (Clarke 2014). Emerging from Agenda 21, Local Agenda 21 (LA21) offers the opportunity for local governments to take a prominent role in sustainable community planning and development initiatives (Freeman 1996). More recently, the need for local sustainable development was reiterated through Sustainable Development Goal 11 – sustainable cities and communities (United Nations 2015).

LA21s integrate ecological, social, and economic topics in one strategic plan to help com-munities identify and document areas for sustainable improvements, and determine their long-term vision (anywhere from five to 100 years) (Clarke 2011, 2014). Some of the other standard terms for LA21 are Integrated Community Sustainability Plans (ICSPs), Sustainable Community Plan (SCP), Municipal Sustainability Plan, and Local Action Plans (Parenteau 1994; Clarke and MacDonald 2012). While many communities have adopted a form of LA21, only some are successful in implementing the objectives (Clarke and Fuller 2010; MacDonald et al. 2018). Thus, it becomes increasingly difficult to ignore the barriers between planning and implementing such plans (Gahin et al. 2010; Hendrickson et al. 2011).

Although municipalities have developed goals and targets to address their water con-cerns, one of the main barriers to implementation is the lack of sufficient and stable finan-cial resources (Cantin et al. 2005; Gahin et al. 2010; Hendrickson et al. 2011). Compared with the international community, Canadians have higher rates of water consumption but pay a small price for water use (Cantin et al. 2005). However, raising prices without consid-ering other alternatives, such as changing the pricing structure, may promote undesired

Environmental Policy: An Economic Perspective, First Edition. Edited by Thomas Walker, Northrop Sprung-Much, and Sherif Goubran.

and unsustainable water consumption patterns (Dinar et al. 1997). With over 1000 LA21s in Canada, most municipalities prioritize water near the top of their plans, with over 97.4% of the plans containing water-related goals and objectives (Clarke et al. 2019).

In order to advance water sustainability among Canadian communities, the "plan–implementation gap" of LA21s needs to be addressed (Hendrickson et al. 2011). The use of market-based instruments (MBIs) offers the potential to bridge the barriers associated with implementation. Pricing and market signals have the power to stimulate behavior changes and a paradigm shift through economic rationales (Hendrickson et al. 2011). MBIs serve as policy tools to mitigate the limitations of conventional regulatory and legislative approaches through combinations of pricing, taxes, charges, and subsidies (Bosquet 2000; Stavins 2003; Hendrickson et al. 2011). Thus, they could be complementary to the traditional regulatory approach of addressing water challenges. In fact, the use of pricing is deemed to be more cost-effective for demand management compared to regulations or other nonprice conservation methods (Olmstead and Stavins 2009).

This chapter will summarize the existing best practices at the intersection of sustainable development and water management. It will also examine the role of MBIs in the implementation of LA21s and their limitations at the local level. Finally, it will provide a list of MBIs relevant to water, wastewater, and stormwater management by local governments. The findings present over 15 MBIs across four different water subtopics and make a significant contribution to sustainable community development by providing an improved understanding of MBIs for implementing LA21s.

14.2 Literature Review

14.2.1 Sustainable Development and Sustainable Development Goals

Sustainable development was first defined in the Brundtland Report as "development that meets the needs of the present without compromising the ability of future generations to meet their own needs" (World Commission on Environment and Development 1987, p. 43). Governments, international agencies, and organizations undertook numerous initiatives across the world to address sustainability challenges (Roseland 2000). However, their impacts were mostly minimal (Mebratu 1998). In addition, these localized initiatives have led to many different interpretations of the concept of sustainable development (Mebratu 1998).

The most commonly accepted concept is the three pillars of sustainable development, which represent the environment, economy, and society. They are generally represented by a "Venn diagram model" or "concentric circles model" to illustrate the interaction and relationship between the three pillars (Campbell 1996; Lozano 2008). The two models present different perspectives on the connection between the three pillars. In the Venn diagram model, environment, society, and economy are equally important in achieving sustainable development (Campbell 1996). In contrast, the concentric circles model highlights a hierarchical relationship for sustainable development, where the environment is the most important (Lozano 2008). The three pillars of sustainable development can be further divided into those that need to be sustained: nature, life supports, and community; and those that need development: people, economy, and society (Robert et al. 2005). The concept of sustainable development has been continuously improved and refined ever since.

In September 2015, the UN member states adopted the 2030 Agenda for Sustainable Development (United Nations 2015). This 15-year framework builds on Agenda 21 and the Millennium Development Goals to ensure that all countries can take action in achieving sustainable development (United Nations 2015). It integrates and balances the initial three pillars of sustainable development with additional elements such as governance, peace, and justice. With over 17 goals, 169 targets, and 230 indicators, the new agenda for sustainable development envisions environmental prosperity, social and gender justice, eradication of poverty and hunger, and universal health and education (United Nations 2015). The Sustainable Development Goals (SDGs), which are part of the 2030 Agenda, have been highly instrumental in the field of sustainability management.

Canada has already begun actively participating in conferences, forums, mutual learning, and exchange of experience concerning the 2030 Agenda for Sustainable Development. Canada's Federal Sustainable Development Strategy for 2016–2019 is now linked to advancing many of the SDGs, and the federal government is developing an inclusive approach to implementation of the SDGs domestically and internationally (Government of Canada 2018).

14.2.2 Sustainable Community Plans/Local Agenda 21s

LA21s could be used to localize the SDGs. Currently, there are over 10 000 LA21 (or equivalent) initiatives around the world (ICLEI 2012). A sustainable community should continuously supply the social and economic needs of the residents, as well as maintain the environment's ability to sustain the demand (Roseland 2000). According to Roseland (2000), the six community capitals – natural, physical, economic, social, human, and cultural – need to be carefully managed to ensure that they will sustain the needs of future generations (Roseland 2000). A goal of sustainable community development and LA21s is for local governments to improve and strengthen all six forms of community capital through collaborative strategic planning and implementation (Clarke and Fuller 2010; Roseland 2012). The goals and objectives outlined in an LA21 are closely aligned with the six forms of community capital and integrate the three pillars of sustainability: society, economy, and environment (Clarke 2011; MacDonald et al. 2018).

In 1992, Hamilton established the first LA21 in Canada, Vision 2020 (Clarke 2012; Clarke and MacDonald 2012). Municipalities and communities who are actively involved in the planning of LA21s are also becoming interested in the implementation of LA21s to address environmental, social, and economic problems (Berke and Conroy 2000; Clarke and Erfan 2007; Hendrickson et al. 2011). Municipalities can continuously provide enormous opportunities to address sustainability challenges. They have the capability to manage their resources sustainably (Roseland 2000) and are important actors in reaching the new SDGs and targets. Local sustainable development and community planning can address environmental, social, and economic issues and generate possible additional revenue (Roseland 2000; Zokaei et al. 2017).

14.2.3 Sustainable Development and Water

Although Canada is fortunate to have an abundant supply of water resources, the available water resources per capita globally have dropped by more than half in the past 50 years

(Dinar and Saleth 2005). Furthermore, it is anticipated that by 2050, water availability per capita will drop by 10–20% of what it was in 1955 regardless of the initial water availability and state of development (Dinar and Saleth 2005).

With Canada ranked as the second highest in the rate of water consumption, one of the biggest challenges of providing water to communities is the ability to support the increased level of water consumption and the increasing urban population (Cantin et al. 2005). For Canada, urbanization presents a key issue since over 80% of Canadians currently live in urban communities and over 68% live in a census metropolitan area (CMA) (Statistic Canada 2008). Infrastructure maintenance, water quality, and demand management are also among the water concerns for municipalities (Cantin et al. 2005). However, most of the municipal water pricing structure continues to remain stagnant. If the misalignment between water pricing and consumption is not further addressed, municipal governments will likely be faced with numerous liability concerns (e.g., current subsidies for the price of water).

LA21s typically integrate key areas of municipal concerns. In terms of water management, most municipalities are responsible for drinking water and urban wastewater treatment and hold partial authority over water resource management for the local area (Environment Canada 2010). Thus, some of the critical concerns are the quality of water, consumption and treatment, wastewater and stormwater management, and protection of water resources and watershed ecosystems (Environment Canada 2010; United Nations 2015). Most Canadian municipalities have developed strategic goals and plans to address water, wastewater, and stormwater concerns in their community (Clarke et al. 2019). However, many Canadian communities are experiencing a significant "planning–implementation gap" along with the economic pressures associated with sustainable community governance and development (Hendrickson et al. 2011).

14.2.4   Market-Based Instruments and Water

Market-based instruments are policy tools that encourage behavioral change through market signals (Scoccimarro and Collins 2008). They are used to mitigate the limitations of conventional regulatory and legislative approaches (Hendrickson et al. 2011) and serve as implementation tools for LA21s.

The use of MBIs could help overcome the barriers associated with the SCP and implementation. Pricing signals and market power have the potential to stimulate behavior changes through economic rationales (Hendrickson et al. 2011). This research focuses on three types of MBIs: price-based instruments, rights-based instruments, and friction reduction instruments. Priced-based instruments address environmental impacts using pricing and economic signals (Sargent 2002; Whitten et al. 2003; Clarke and MacDonald 2012). This type of MBI can also be classified as a financial instrument that diversifies local revenue streams (Jacobs 1993; Roseland 2000). Rights-based instruments are those that control the type of goods and services produced (Whitten et al. 2003). Contrastingly, through price-based approaches, the government is able to establish limits on the quantity or quality of goods and services, while the price reflects the market's response (Whitten et al. 2003). Finally, friction reduction instruments aim to influence behavioral change through improving market functions, addressing market power (monopoly), externalities, and information failures (Hahn and Stavins 1991; Clarke and MacDonald 2012).

To move beyond the current limitations, there is a need for the implementation of innovative approaches to sustainability. The traditional approaches are often "command-and-control" regulations, where standards are uniform and environmental burdens are equally shared (Stavins 2003). These conventional approaches effectively limit environmental pollutants and distribute the costs equally (Stavins and Whitehead 1996). Thus, traditional approaches are inadequate in aligning economic drivers with sustainability objectives (Stavins 2003; Hendrickson et al. 2011). Moreover, they may also result in unacceptable expenses and high societal costs as individuals vary in their contribution to environmental problems (Stavins and Whitehead 1996). Furthermore, utilizing "command-and-control" regulations to achieve sustainable community development tends to result in nothing more than compliance (Stavins 2003). Little or no financial incentive exists for those who strive to achieve objectives beyond the minimal requirements and standards, while also discouraging changes in policies and governance structure (Roseland 2000; Stavins 2003).

By contrast, MBIs for sustainable community development offer greater flexibility, accountability, and transparency (Stavins 2003; Hendrickson et al. 2011). They also help to improve the allocation of environmental resources and the dissemination of information for individuals and society (Pirard and Lapeyre 2014). The financial incentives associated with MBIs motivate communities to better manage their community capitals, especially natural capitals (Roseland 2000; Henderson and Norris 2008). Additionally, MBIs are intended to be market-friendly and improve market efficiency if adequately designed (Hendrickson et al. 2011). MBIs can thus be used on their own or in conjunction with regulations.

Water pricing is often a mechanism to reduce water demands and consumption (Ruijs et al. 2008). Moreover, price structure (e.g., flat rates, unit pricing) has more influence on water demand and consumption compared to price level (Reynaud and Renzetti 2004). Cities in China with block pricing structures experienced a decrease in residential water demand by 3–5% compared with cities that had flat rates (Zhang et al. 2017). In fact, using pricing is considered to be more cost-effective for demand management than "command-and-control" regulations or other nonprice conservation methods (Olmstead and Stavins 2009). Aside from demand management, other well-known examples of MBIs for water include effluence charges and tradable permits (Stavins 2003, Cantin et al. 2005). Importantly, MBIs must not be regressive, and vulnerable populations also need to be considered. Thus, the proper design of MBIs remains crucial to alleviate these concerns.

Market-based instruments for water management have two key roles: (i) a financial role as a mechanism for generating municipal revenue, and (ii) an economic role for signaling the scarcity value and the real cost of water (Dinar and Saleth 2005). MBIs for water could also promote equity by identifying usage of individual users and points of pollution, thus accurately awarding beneficial behaviors and penalizing negative ones (Dinar and Saleth 2005). However, it is important to note that they also have their limitations. For example, there is no guarantee that one will gain advantages from using MBIs because two critical factors affect the use and effectiveness of the MBIs: (i) the nature of the environmental problem/objective; (ii) the state of the market and the government (Whitten et al. 2003; Broughton and Pirard 2011).

The success of MBIs is determined by the nature of the environmental problem/objective. To start, the gain from MBIs for environmental problems must exceed their cost to ensure success (Guerin 2003). Point sources and stationary environmental problems are more amenable to the use of market instruments compared to nonpoint sources and mobile environmental problems (National Center for Environmental Economics 2015). However, MBIs will be more cost-effective and beneficial if there is a higher degree of heterogeneity among the polluters (Stavins 2003). Since the degree of uncertainty regarding environmental problems affects effectiveness of MBIs, they tend to be more effective (Stavins 2003; National Center for Environmental Economics 2015). Lastly, clearly defining rights and responsibilities, as well as who pays and who will benefit, is necessary to ensure the effectiveness of an MBI (Whitten et al. 2003).

The market and the government have also played an influential role in the use and effectiveness of MBIs. Sufficient levels of political support are required to ensure the success of such instruments (Whitten et al. 2003). Moreover, transparency and information disclosure are critically important (National Center for Environmental Economics 2015). Lack of information is likely to discourage the proper design and use of MBIs (Kulsum 2012). Furthermore, market competitiveness also determines the design and price of MBIs (National Center for Environmental Economics 2015). Therefore, MBIs are by no means a replacement for the traditional command-and-control approach of implementation. In fact, they work to complement the traditional approach because each of the two could operate differently under different circumstances. The appropriate choice of MBIs will be essential in ensuring their successful implementation and practical results. This chapter explores the potential of MBIs as an alternative or complement to implementing water-related goals in the LA21s.

14.3   Research Design and Limitations

14.3.1   Research Design

This research utilized a multiphase qualitative approach to conduct an in-depth analysis of the use of MBIs in mid-sized municipalities from Ontario, Canada. Phase one of the research focused on the construction of the framework from academic and gray literature. In preparation for the case studies, a list of existing and emerging MBIs was created for the implementation of LA21s.

The case study approach was used in order to gather the amount of data necessary in the most effective manner (Yin 2009; Creswell 2014). Rather than using a single source, a multicase study analysis increased the level of accuracy and validity of the results (Creswell 2014). Thus, two municipalities were chosen for case studies, and a set of criteria were applied to identify potential municipalities for a comparative case study analysis.

- Due to funding restrictions, the communities must be within the Province of Ontario.
- The communities had LA21s (as determined by the Canadian Sustainability Plan Inventory) and were at least 2–3 years into the implementation phase of their LA21s (University of Alberta 2014).
- The communities had a population of over 100 000 (as determined by the population listed in the 2008 Census of Canada) and were similar in size.

- The communities had similar characteristics in their plans (e.g., age and time horizon of the plans).
- The communities were willing to engage in the research project and participate in focus groups.
- The communities had different governance structures. Each community must represent either a two-tier or a single-tier municipal structure.

A number of Canadian municipalities matched the above criteria. For this research, the City of Kingston was selected as the single-tier municipality, and the Region of Waterloo (which includes the cities of Kitchener, Waterloo, and Cambridge) was selected to represent the two-tier municipal structure. Both of the communities selected are located in southern Ontario, have LA21s (the lower-tier municipalities in the Waterloo Region all have their plans), and are considered leaders in community sustainability. The cities had already implemented numerous environmental initiatives, gained several awards, and received recognition for their efforts. They also had displayed a keen interest in and commitment to this research.

Moreover, both municipalities are also located within a major watershed. The City of Kingston is located within the Cataraqui watershed and the Region of Waterloo is part of the Grand River watershed. These major watersheds are essential parts of the water management for each municipality. The municipalities hold partial responsibility for their watershed and are active in the protection and management of the watersheds.

A half-day focus group was held in each community to gather data from the municipalities. The participants were the staff most familiar with the MBIs used for implementing the LA21 in their community. The objective of this focus group was to discuss the draft MBIs with the participants and gather feedback for further revisions. The participants were invited to provide feedback about the draft MBIs. The Chatham House Rule was enforced during the focus group, so the participants at the meeting were free to use information from the discussion but were not allowed to reveal the identity of the person commenting. This design enabled open discussion during the focus groups while ensuring that participants' specific comments remained anonymous.

14.3.2   Limitations

First, this chapter is focused on the use of MBIs for implementing water, wastewater, and stormwater goals in the LA21s. Thus, the MBIs outlined in this chapter only consider residential water concerns under municipal jurisdiction. For example, when considering water consumption, residential water uses account for only 5–10% of the total available water use, while irrigation accounts for 70–90% (Dinar and Saleth 2005). Therefore, the generalization of the MBIs in this chapter is limited to the implementation of water, wastewater, and stormwater goals within the municipal jurisdiction.

The second challenge of this research is attributable to funding constraints, which limited the case studies to Ontario communities Although focus group discussions and the resulting MBIs tool were both successful, more communities across a broader geographical boundary could have been involved in the research to help to ensure that broader generalizations could be made. The preference would have been to conduct multiple focus groups across Canada.

14.4   Results

14.4.1   Sustainability Alignment Methodology

The Sustainability Alignment Methodology (SAM) is used by practitioners to align the MBIs under the municipal jurisdiction with the environmental goals of the LA21s (Cairns et al. 2015). The Water section in the SAM identifies relevant MBIs for water, wastewater, and stormwater, as well as the associated municipal departments related to these MBIs. The listed MBIs are presented in Table 14.1.

Regarding the appropriateness of the subtopics, one participant pointed out that a critical subtopic for the water section is its source.

> Source really [influences] the types of programs you have, whether it is surface water or groundwater here. It changes your protection mechanisms ... and may affect the conservation bylaw.

Aside from the inclusion of the source of water, most participants agreed with the terms used for the subtopics. In addition to the modification of subtopics, the participants also suggested changing the topic names to water, wastewater, and stormwater to avoid confusion with the waste section.

Out of all the water-related MBIs identified from the literature, only water extraction charges are not within the municipal jurisdiction. Also, the structure of nitrogen and phosphorus levies may vary between communities. Surcharges or levies are available in communities with heavy loading bylaws; otherwise, fines are applied to the effluents. In addition, participants identified additional water-related MBIs under the municipal jurisdiction. One participant mentioned stormwater utility charges and associated water rebates.

> You pay for the stormwater you create, which is based on how impervious your site is. It was a new utility fee that was introduced a couple of years ago. As part of that, they also have [a] rebate program. If you have the infrastructure, such as a rain barrel, or a rain garden, you may pay less on the stormwater utility rate.

Other MBIs mentioned during the focus groups were water quality programs and education programs.

Regarding the location of information, the results obtained from both focus groups resonated with the publicly accessible information. The local utility company, water services departments, and environmental services departments are responsible for the implementation of water-related MBIs.

14.4.2   Policy Relevance

Canada has committed to over $59.8 million for programming to support the implementation of the SDGs. The information from this research offers a complementary approach toward achieving the global SDGs, particularly SDG 6: Ensure availability and sustainable

Table 14.1 Market-based instruments (MBIs) for water, wastewater, and stormwater

	Water quality	Water consumption and wastewater treatment	Water source (groundwater and surface sources)	Other
Price-based instruments	• Charges for biochemical oxygen demand (BOD) loading • Nitrogen levy • Phosphorus levy • Incentive for bio-swales	• Water rebates • Funds to support water, wastewater treatment infrastructure • Water pricing structure • Stormwater utility charges • Subsidies for rain barrels		• Other subsidies, funds, and grants
Rights-based instruments	• Water quality permit trading			
Market-friction reduction instruments	• Water quality program • Certification program (e.g., smart salt application) • Stormwater management	• Stormwater management	• Water source protection incentive programs or policy	• Green public procurement • Partnership approach • Education programs • Reporting requirements
Department/location	• Municipal utilities • Water services department • Environmental services department			

management of water and sanitation for all, and SDG 11: Make cities and human settlements inclusive, safe, resilient, and sustainable (United Nations 2015). To achieve the specific actions within these goals by 2030, communities must plan and take the approximate action to update and/or implement their LA21s (in Canada, often these are called ICSPs). The SAM helps the Canadian municipal government to accurately pinpoint the appropriate MBIs to implement individual water-related objectives in a LA21 and locates relevant municipal departments.

14.4.3 Limitations of Implementing Market-Based Instruments at the Local Level

The first lesson learned from the two focus groups is that currently, municipal governments only have limited authority in implementing MBIs. In many cases, municipalities are already stretching the limits of their power. It is crucial to identify and distinguish the types of MBIs within municipal jurisdictions. Many of the MBIs shared during the focus group were market friction reduction MBIs.

Another lesson is that the costs of implementing MBIs vary between MBIs. Some have high upfront costs, which may discourage implementation and municipal uptake. Thus, it is important to acknowledge both the environmental incentive and the financial burden for such an MBI. It is equally important to assess (i) the cost-effectiveness of various MBIs, especially with regard to subsidies and the free-rider effect; and (ii) the cost-effectiveness of a market-based approach, compared to alternative policies.

14.5 Conclusion

Although MBIs are becoming more prominent in sustainability, research on this topic remains scattered. The research conducted in this paper aimed to bridge the gap in the literature on the development of MBIs for water, wastewater, and stormwater, which was necessary in order to accelerate the implementation of LA21s and achieve the SDG targets. Over 20 MBIs are identified for the implementation of water-related goals and objectives in the LA21s. Furthermore, the market approach and the use of MBIs highlighted in this chapter are innovative methods of implementation of LA21s. Research on MBIs contributes positively to the understanding of market approaches for water management, as well as improving the understanding of MBIs for achieving the goals and targets of the 2030 Agenda for Sustainable Development.

However, research is still required to understand the different municipal jurisdictions and identify MBIs for water, wastewater, and stormwater that are distinct in other Canadian and international communities. Furthermore, additional research could also investigate the assessment criteria and scoring methodology for the SAM framework presented in this chapter. Although scoring seems to be useful to determine the performance of communities, further research is necessary to help assess the usefulness of scoring for the SAM framework and determine the best scoring methodology.

Overall, the chapter provided a list of MBIs that help to achieve the water-related goals in the LA21s and also established an essential foundation for future research in this direction.

Moreover, this chapter contributes to improving the understanding of MBIs that are applicable at the local level. Finally, the research helps provide alternative options and policy tools to implement LA21s, as these plans are an excellent mechanism for further implementation of SDGs at the local level.

References

Berke, P.R. and Conroy, M.M. (2000). Are we planning for sustainable development? *Journal of the American Planning Association* 66 (1): 21–33. https://doi.org/10.1080/01944360008976081.

Bosquet, B. (2000). Environmental tax reform: does it work? A survey of the empirical evidence. *Ecological Economics* 34 (1): 19–32. https://doi.org/10.1016/S0921-8009(00)00173-7.

Broughton, E. and Pirard, R. (2011). What's in a name? Market-based instruments for biodiversity. *Health and Environmental Report* 8: 1–44.

Cairns, S., Clarke, A., Zhou, Y., and Thivierge, V. (2015). *Sustainability Alignment Manual: Using Market-Based Instruments to Accelerate Sustainability Progress at the Local Level*. Ottawa: Sustainable Prosperity.

Campbell, S. (1996). Green cities, growing cities, just cities? Urban planning and the contradictions of sustainable development. *Journal of the American Planning Association* 62 (3): 296–312. https://doi.org/10.1080/01944369608975696.

Cantin, B., Shrubsole, D., and Aït-Ouyahia, M. (2005). Using economic instruments for water demand management: introduction. *Canadian Water Resources Journal* 30 (1): 1–10. https://doi.org/10.4296/cwrj30011.

Clarke, A. (2011). Key structural features for collaborative strategy implementation: a study of sustainable development/local agenda 21 collaborations. *Management & Avenir* 50 (10): 153. https://doi.org/10.3917/mav.050.0153.

Clarke, A. (2012). *Passing Go: Moving Beyond the Plan*. Ottawa: Federation of Canadian Municipalities.

Clarke, A. (2014). Designing social partnerships for local sustainability strategy implementation. In: *Social Partnership and Responsible Business: A Research Handbook* (eds. A. Crane and M. Seitanidi), 107–113. London: Routledge.

Clarke, A. and Erfan, A. (2007). Regional sustainability strategies: a comparison of eight Canadian approaches. *Plan Canada* 47: 15–18.

Clarke, A. and Fuller, M. (2010). Collaborative strategic management: strategy formulation and implementation by multi-organizational cross-sector social partnerships. *Journal of Business Ethics* 94 (85–101) https://doi.org/10.1007/s10551-011-0781-5.

Clarke, A., & MacDonald, A. (2012). Partner engagement for community sustainability: supporting sustainable development initiatives by reducing friction in the local economy. https://institute.smartprosperity.ca/sites/default/files/publications/files/Partner%20Engagement%20for%20Community%20Sustainability.pdf

Clarke, A., Huang, L., & Chen, H. (2019). Do collaborative planning processes lead to better outcomes? An examination of cross-sector social partnerships for community sustainability. Manuscript submitted for publication.

Creswell, J.W. (2014). *Research Design: Qualitative, Quantitative, and Mixed Methods Approaches*, 4e. Thousand Oaks: Sage Publications.

Dinar, A. and Saleth, R.M. (2005). Issues in water pricing reforms: from getting correct prices to setting appropriate institutions. In: *The International Yearbook of Environmental and Resource Economics* (eds. H. Folmer and T. Tietenberg), 1–51. Cheltenham: Edward Elgar Publishing.

Dinar, A., Rosegrant, M.W., and Meinzen-Dick, R. (1997). *Water Allocation Mechanisms: Principles and Examples*. Geneva: World Bank.

Environment Canada. (2010). Shared Responsibility. www.canada.ca/en/environment-climate-change/services/water-overview/governance-legislation/shared-responsibility.html

Freeman, C. (1996). Local government and emerging models of participation in the local agenda 21 process. *Journal of Environmental Planning and Management* 39 (1): 65–78. https://doi.org/10.1080/09640569612679.

Gahin, R., Veleva, V., and Hart, M. (2010). Do indicators help create sustainable communities? *Local Environment* 8 (6): 661–666. https://doi.org/10.1080/1354983032000152752.

Government of Canada. (2018). The 2030 Agenda for Sustainable Development. http://international.gc.ca/world-monde/issues_development-enjeux_developpement/priorities-priorites/agenda-programme.aspx?lang=eng

Guerin, K. (2003). Property Rights and Environmental Policy: A New Zealand Perspective. Treasury Working Paper Series. https://ideas.repec.org/p/nzt/nztwps/03-02.html

Hahn, R.W. and Stavins, R.N. (1991). Incentive-based environmental regulation: a new era from an old idea? *Ecology Law Quarterly* 18 (1): 1–42.

Henderson, B. and Norris, K. (2008). Experiences with market-based instruments for environmental management. *Journal of Environmental Management* 15 (2): 113–120. https://doi.org/10.1080/14486563.2008.10648738.

Hendrickson, D.J., Lindberg, C., and Connelly, S. (2011). Pushing the envelope: market mechanisms for sustainable community development. *Journal of Urbanism: International Research on Placemaking and Urban Sustainability* 4: 153–173. https://doi.org/10.1080/17549175.2011.596263.

ICLEI (2012). *Local Sustainability 2012: Taking Stock and Moving Forward*. Bonn: ICLEI.

Jacobs, M. (1993). *The Green Economy: Environment, Sustainable Development and the Politics of the Future*. Vancouver: University of British Columbia Press.

Kulsum, A. (2012). *Getting to Green – A Sourcebook of Pollution Management Policy Tools for Growth and Competitiveness*. Washington, DC: World Bank Group.

Lozano, R. (2008). Envisioning sustainability three-dimensionally. *Journal of Cleaner Production* 16 (2008): 1838–1846. https://doi.org/10.1016/j.jclepro.2008.02.008.

MacDonald, A., Clarke, A., Huang, L. et al. (2018). Cross-sector partnerships (SDG #17) as a means of achieving sustainable communities and cities (SDG #11). In: *Handbook of Sustainability Science and Research*, World Sustainability Series (ed. W. Leal), 193–209. New York: Springer https://doi.org/10.1007/978-3-319-63007-6_12.

Mebratu, D. (1998). Sustainability and sustainable development: historical and conceptual review. *Environmental Impact Assessment Review* 18 (6): 493–520. https://doi.org/10.1016/S0195-9255(98)00019-5.

National Center for Environmental Economics. (2015). Economic incentives. http://yosemite.epa.gov/EE/epa/eed.nsf/webpages/EconomicIncentives.html#market

Olmstead, S.M. and Stavins, R.N. (2009). Comparing price and nonprice approaches to urban water conservation. *Water Resources Research* 45 (4) https://doi.org/10.1029/2008WR007227.

Parenteau, R. (1994). Local action plans for sustainable communities. *Environment and Urbanization* 6: 183–199.

Pirard, R. and Lapeyre, R. (2014). Classifying market-based instruments for ecosystem services: a guide to the literature jungle. *Ecosystem Services* 9: 106–114. https://doi.org/10.1016/j.ecoser.2014.06.005.

Reynaud, A. and Renzetti, S. (2004). *Micro-Economic Analysis of the Impact of Pricing Structures on Residential Water Demand in Canada*. Vancouver: Environment Canada.

Robert, K.W., Parris, T.M., and Leiserowitz, A.A. (2005). What is sustainable development? Goals, indicators, values, and practice. *Environment: Science and Policy for Sustainable Development* 47 (3): 8–21. https://doi.org/10.1080/00139157.2005.10524444.

Roseland, M. (2000). Sustainable community development: integrating environmental, economic, and social objectives. *Progress in Planning* 54 (2): 73–132. https://doi.org/10.1016/S0305-9006(00)00003-9.

Roseland, M. (2012). *Toward Sustainable Communities: Solutions for Citizens and Their Governments*. Gabriola Island: New Society Publishers.

Ruijs, A., Zimmermann, A., and van den Berg, M. (2008). Demand and distributional effects of water pricing policies. *Ecological Economics* 66 (2–3): 506–516. https://doi.org/10.1016/j.ecolecon.2007.10.015.

Sargent, J.H. (2002). Economics of energy and the environment: the potential role of market-based instruments. *Canada – United States Law Journal* 28: 499–510.

Scoccimarro, M. and Collins, D. (2008). *Market Based Instruments Decision Support Tool*. Brisbane: Australian Government, Department of Natural Resource Management and Water.

Statistics Canada. (2019). Goal 6 – Clean water and sanitation. www144.statcan.gc.ca/sdg-odd/goal-objectif06-eng.htm

Stavins, R.N. (2003). Experience with market-based environmental policy instruments. *Handbook of Environmental Economics* 1: 355–435. https://doi.org/10.1016/S1574-0099(03)01014-3.

Stavins, R. N., & Whitehead, B. W. (1996). The next generation of market-based environmental policies. www.rff.org/files/sharepoint/WorkImages/Download/RFF-DP-97-10.pdf

United Nations. (2015). Sustainable Development Goals. https://sustainabledevelopment.un.org/?menu=1300.

University of Alberta. (2014). Canadian Sustainability Plan Inventory. https://wagner.augustana.ualberta.ca/cspi

Whitten, S., van Bueren, M., & Collins, D. (2003). An overview of market based instrument and environmental policy in Australia. Presented at the Annual National Australian Agricultural and Resource Economics Society Symposium, Canberra.

World Commission on Environment and Development (1987). *Our Common Future*. Oxford: Oxford University Press.

Yin, R.K. (2009). *Case Study Research: Design and Methods*. London: Sage Publications.

Zhang, B., Fang, K.H., and Baerenklau, K.A. (2017). Have Chinese water pricing reforms reduced urban residential water demand? *Water Resources Research* 53 (6): 5057–5069. https://doi.org/10.1002/2017WR020463.

Zokaei, K., Lovins, H., Wood, A., and Hines, P. (2017). *Creating a Lean and Green Business System: Techniques for Improving Profits and Sustainability*. Abingdon: Productivity Press.

15

Climate Justice and Food Security

Experience from Climate Finance in Bangladesh

Muhammad Abdur Rahaman[1] *and Mohammad Mahbubur Rahman*[2]

[1] *Climate Change Adaptation, Mitigation, Experiment & Training (CAMET) Park, Noakhali, Bangladesh*
[2] *Network on Climate Change in Bangladesh (NCCB), Dhaka, Bangladesh*

15.1   Introduction

Bangladesh is one of the most climate-vulnerable countries in the world and will become even more so because of climate-induced disasters (Rahaman et al. 2019a). Bangladesh is globally identified as the most vulnerable to tropical cyclones, the third most vulnerable to sea-level rise in terms of the number of people affected, and the sixth most vulnerable to floods (Francis and Maguire 2016; Rahaman et al. 2019a). Researchers have confirmed that the unpredictable nature of precipitation and temperature is steadily increasing in Bangladesh. Rainfall is becoming less anticipated and the monsoon is now characterized by higher volumes of rainfall within shorter intervals (Islam et al. 2014). Temperatures are becoming more extreme, with regional variations and an overall annual increase. Tropical cyclones are also predicted to increase in intensity (Rahaman et al. 2019a).

Both climate and weather patterns perform an important role in freshwater availability, agriculture, economic growth, and livelihoods (NAPA 2005; Rahaman et al. 2019a). The most devastating effects arise from flooding, drought, and heat stress (World Bank 2013; Rahaman et al. 2019a). The adverse effects of these on agricultural yield and the availability of fresh water are now apparent in several areas of Bangladesh. Harvest has decreased with drought, and perennial trees and livestock are damaged and lost due to floods each year (Rahaman et al. 2019a). Moreover, the rise in cyclonic storms, mixed with sea-level rise (SLR), increases the depth and risks of floods and storm surges and reduces the area of cultivable land (World Bank 2013; Rahaman et al. 2019a).

Tackling issues such as food insecurity that result from the impacts of environmental change is urgent. A recent report estimates that every seventh person in Bangladesh is underfed and does not receive sufficient food to meet dietary energy demands. The country has 24.4 million undernourished people, which means that 15.1% of its population is not receiving adequate food to meet the minimum daily requirements (FAO et al. 2017). Households which are desperately trying to combat malnutrition and food insecurity have been affected by the added burden of global warming. Global environmental changes are one of the key factors affecting the food system (Vermeulen et al. 2012). Increased frequency and severity of extreme weather events are the key risks of food insecurity for marginalized climate-vulnerable people (NAPA 2005). Government organizations, along with many development partners and nongovernment organizations, are working to combat climate crisis in Bangladesh and to ensure climate justice for climate-vulnerable people.

Climate justice means ensuring the human rights of climate-vulnerable people, safeguarding the rights of the most vulnerable, and equitably and fairly sharing the burdens and benefits of the climate crisis and its resolution. Climate change and food security for the most vulnerable people are the climate justice questions of this century (Noiret 2016; Robinson and Shine 2018). The right to food is one of the most basic rights of humankind, which is not only one of the fundamental rights enshrined in the Universal Declaration of Human Rights but also has been reflected in a series of UN conventions ranging from the Rights of the Child to the Convention on the Elimination of Discrimination against Women (CESCR 1999).

Though the government and its development partners have initiated several projects across the country to enhance food security in Bangladesh, the concern for justice has received less attention. Moreover, regional, racial, ethnic, and social disparities and inequalities exist in fund allocation and project implementation. Several studies (Yu et al. 2010; Mainuddin and Kirby 2015; Amin et al. 2015) have been undertaken to discuss the impacts of climate change on food security in Bangladesh, but none of these studies link climate justice issues with a food security concern.

This chapter provides an overview of the climate change policies, strategies, action plans, and climate financing systems on food security in Bangladesh in line with the different paradigms in spatial, racial, ethnic, and social exclusion perspectives. We begin with a brief description of the vulnerability of Bangladesh due to different climate-induced extreme weather events and focus on showcasing historical damage in the agriculture sector caused by different climate extremes. Next, we discuss the nexus of climate change, agriculture, and food security, highlighting the case of Bangladesh. Then, we discuss policies, approaches, and action plans in climate financing on food security in Bangladesh. We address the issues of regional, district-based, thematic area-wise, vulnerable area base and racial disparity and injustice in fund allocation and project implementation. Finally, we suggest policy recommendations to ensure climate justice and food security in Bangladesh.

The data were collected through case studies, and an in-depth literature review was conducted and includes a variety of sources, such as the national policy framework, plan of actions and strategies, climate invest plan, implemented project reports, published and unpublished scientific articles and reports, etc. However, the study finds that the poor and marginalized groups, including women and ethnic communities in Bangladesh, are often deprived of equity and, in the broader sense, justice, particularly in the food security sector. Moreover, fund allocation and implementation of projects often do not follow the justice mechanism and thus injustice is on the rise.

15.2   Impact of Climate Change on Food Security of Bangladesh

15.2.1   Vulnerability of Bangladesh

In the geographical context, Bangladesh is one of the most disaster-prone countries in the world (Rahaman et al. 2019a, b). It is a low-lying country in the tropics and has the largest delta in the world formed by three mighty rivers, the Ganges, the Brahmaputra, and the Meghna (Rahaman et al. 2019a). It has the Himalayan range to the north, the Bay of Bengal to the south with its funneling toward the Meghna estuary and the vast stretch of Indian land to the west. These special geographical features make a significant contribution to the climate system of Bangladesh (Islam 2010). As mentioned previously, impacts of climate change in Bangladesh include excessive flooding, severe cyclone and storm surges, increased salinity and drought, declining of agricultural productivity, lack of drinking water, and waterlogging due to the SLR (Rabbani et al. 2014). Considering the area, it is a small country with a huge population and a predominantly agrarian economy. According to the Maplecroft study (Maplecroft 2014), Bangladesh is the most climate-vulnerable country in the world. The Global Climate Risk Index (GCRI) 2010, covering the period 1990–2008, estimates that, on average, 8241 people died each year in Bangladesh while the cost of damage was around US $1.2 billion per year and the loss of gross domestic product (GDP) was 1.81% during the period (Harmeling 2010; Rahaman et al. 2019b).

Bangladesh, as a delta country, has been affected by devastating tropical cyclones. In their wake, the cyclones have left behind complete chaos, destruction, and despair. Most of the damage is caused by water in the form of storm surges. Bangladesh is particularly vulnerable to storm surge floodings due to the geography of the region. The Bay of Bengal is an extension of the Indian Ocean lying between India and Southeast Asia, which has an environmental setting that generates climatological disasters, thus aggravating the overall vulnerabilities of the communities living near the Bay of Bengal (Bangladesh.com n.d.). The southwest coastal region of Bangladesh is an area where the poverty rate is high (25–34%) (World Bank 2016a) and highly vulnerable to impacts associated with climatic variations. The bay narrows toward its northern shore where it touches the southern coast of Bangladesh. This narrowing can act as a funnel, directing cyclones toward Bangladesh's coast and intensifying them. The force of the storm surges that come with these cyclones is very dangerous, owing to the depressed, flat terrain that makes up most of Bangladesh (Bangladesh.com n.d.).

The economy of Bangladesh is heavily dependent on agriculture and a large portion of the population of the country is, directly and indirectly, engaged in agro-based activities (Golder et al. 2013). Therefore, the impacts of global warming on agricultural production in Bangladesh are widespread and devastating (Rahaman et al. 2019b). Such adverse impacts are responsible for declining crop production and fragile food security. However, agriculture is one of the most sensitive sectors to climatic variability and change (Cline 2007). In particular, extreme temperatures, rainfall patterns, droughts, floods, waterlogging and salinity intrusion, etc. have negatively impacted agricultural production in Bangladesh (Karim 2012).

Such extreme weather events have resulted in shortfalls in production, for example 0.8 and 1.0 metric tons in 1974 and 1987 respectively (Rahman and Alam 2003). The flash flood in 2017 affected about 4.67 million people from 1.03 million households, which is about one-fourth of the total population of the six affected districts (CPD 2017). The Centre for Policy Dialogue (CPD) (2017) estimated the loss of Boro rice production due to this flood to be about 1.58 million metric tons, which is equivalent to 8.3% of the national average yield. In monetary terms, the estimated loss was about 53 billion BDT, which is equivalent to 3.7% of the GDP in the agriculture crop sector (CPD 2017). In terms of monsoon flood, about 8.2 million people were affected in 32 districts and about 9% of the cultivated cropland was damaged, with a gross value of about 27 billion BDT. The CPD also states that the estimated loss of rice production was about 7–18 billion BDT in monetary value (CPD 2017). In the period 1962–1988, Bangladesh lost approximately 12.7 million metric tons of rice production, with an average annual loss of about 0.5 million tons yearly because of floods. This forced Bangladesh to import almost 30% of the average annual food grain demand of the country during the same period, which amounted to 1.6 million metric tons (Paul and Rasid 1993).

Sea-level rise, extreme temperatures, and changes in rainfall pattern also affect crop production in many parts of the country as arable land is decreasing (Hossain and Majumder 2018). At the same time, the shortened winter season is resulting in declining production of winter crops. In addition, the coastal people of Bangladesh experience salinity intrusion, with serious crop production restriction and food insecurity (Pachauri 2010). Ironically, this alluvial floodplain is very fertile and attracts farmers who grow crops in its rich soil (Bangladesh.com n.d.).

Around 32% of the total land and 25.7% of the total population of the country are in the coastal regions (Dasgupta et al. 2014). These areas are highly susceptible to various hydrological disasters and are experiencing increased severity and frequency of the disasters (Bronkhorst 2012).

Different studies have revealed that the impacts of climate change in the coastal regions of Bangladesh are evident (Hossain and Majumder 2018; Rahaman et al. 2019a, b; Saroar et al. 2019). These are impacting coastal communities since livelihoods rely on agricultural activities, related work, and use of natural resources (Islam 2010). The northern part of Bangladesh (Rajshahi and Rangpur divisions, the area lying west of the Jamuna river and north of the Padma river, and including the Barind tract) is characterized by climate-induced drought and flood. Drought typically affects Bangladesh in premonsoon and postmonsoon seasons (Rahaman et al. 2016). Regarding the emerging and ever-increasing concern about the impacts of climate change in the coastal and northern regions of Bangladesh, it has been realized that very little has been done collectively to enable vulnerable communities to adapt to the climate change impacts.

15.2.2 Climate Change, Agriculture, and Food Security in Bangladesh

Climate change-induced extreme events directly and indirectly threaten the food security of climate-vulnerable people all around the country (FAO 2008). They also threaten rights to life, water and sanitation, health, housing, self-determination, culture, and development (Pachauri et al. 2014). Concerning these issues, the Human Rights Council (HRC) and the Office of the High Commissioner for Human Rights have brought attention to human rights and climate change through a series of resolutions, reports, and activities and have advocated for a human rights-based approach to climate change. The basis of a rights-based approach to assure sufficient food is empowering poor people and those who are food insecure. Empowerment is essential to any plan that moves away from the benevolence model of food aid and rather emphasizes enabling conditions that support people in feeding themselves. Empowerment also eliminates the burden of other countries providing food. Nonetheless, as stated earlier, in the event that people are unable to feed themselves (because of household shocks or extreme weather events), the nation must take the pledge to help, whether through social safety nets or other programs and policies that guard vulnerable people against hunger (McClain-Nhlapo 2004).

During the last 50 years, Bangladesh has experienced about 20 drought conditions. During 1981 and 1982, droughts affected the production of the monsoon crops only. In the 1990s, drought in northwestern Bangladesh caused a shortfall of rice production by 3.5 million tons, which led to food insecurity and malnutrition in the affected districts (NAPA 2005). If losses to all crops (all Rabi crops, sugarcane, tobacco, wheat, etc.) as well as to perennial agricultural resources, such as bamboo, betel nut, fruits like litchi, mango, jackfruit, banana, etc., are considered, the losses would be substantially higher.

However, floods have the most deleterious effect on the crop production of Bangladesh. The 1988 flood caused a reduction of agricultural production by 45%. Under climate change scenarios, higher discharge and low drainage capacity, in combination with increased backwater effects, will increase the frequency of such devastating floods. Prolonged floods would tend to delay Aman planting, resulting in significant loss of potential Aman production, as observed during the floods of 1998 (DANIDA 2012). Loss of Boro rice crop from flash floods has also become a regular phenomenon in the riverine areas over the recent years.

Considering all the direct and induced adverse effects of climate change on agriculture, one may conclude that crop yield and agriculture in Bangladesh would be even more vulnerable in a warmer world. Natural calamities intensified by climate change and damaged field crops every year result in a lack of food. Living in a poverty cycle, the people of climate-vulnerable areas are trying to adapt to the impacts through community-based efforts and indigenous knowledge and skills. The government has also intensified its efforts by adopting different policies and programs, introducing new technologies and management practices to tackle the impacts.

15.3 Toward Climate-Resilient Food Security in Bangladesh

There have been laudable signs of progress in formulating policies and strategies to address climate change, disaster risks, and vulnerabilities in Bangladesh. The government has prepared the National Adaptation Programme of Action (NAPA) (2005), the Bangladesh Climate Change Strategy and Action Plan (BCCSAP) (2009), the Disaster Management Act (2012), the National Disaster Management Policy (2015), the National Plan for Disaster Management (2016–2020), and the Seventh Five-Year Plan (2016–2020), among others. The policies and plans have recognized the climate-induced problems like droughts, declining groundwater level, land degradation, and flooding and riverbank erosion. However, most of the programs that intend to tackle climate change issues focus on climatic concerns in the southern and southeastern coastal regions. Further, there are limited practical programs and actions with the vulnerable communities at the regional and local levels by the government and other actors.

Concerning the adaptation to climate change and the future well-being of the people, the Government of Bangladesh (GoB) has identified climate change as the burning threat to national development. Therefore, to overcome this worsening situation, the government has formulated NAPA in 2005, the BCCSAP in 2009 and established the Bangladesh Climate Change Trust Fund (BCCTF) in the same year to carry forward the BCCSAP. Within the framework of the BCCSAP (2009), the BCCTF has approved some projects for government and nongovernment organizations with a specific focus on the following thematic areas.

- Food security, social protection, and health
- Comprehensive disaster management
- Infrastructure
- Research and knowledge management
- Mitigation and low-carbon development
- Capacity building and institutional strengthening

The government has allocated approximately 435 million USD to the BCCTF over the seven fiscal years since the inception of the fund. As of June 2015, 368 projects had been undertaken with an estimated cost of 2319.70 crore taka (USD 27.42 million approximately) (BCCTF 2017). Different ministries, departments, and agencies of the government are implementing these projects.

Along with the BCCTF, there are four mechanisms by which the GoB has allocated funding and implemented projects to promote climate-resilient development all over the country based on the thematic areas of the BCCSAP.

- Bangladesh Climate Change Resilience Fund (BCCRF)
- Strategic Program for Climate Resilience (SPCR) –Bangladesh
- BCCTF
- Nongovernmental efforts

15.3.1 Bangladesh Climate Change Resilience Fund

The BCCRF is a partnership between the GoB, development partners, and the World Bank (WB), which is a trust sponsored by the international community to fund climate change-related activities in Bangladesh (GoB 2014). Of the total activities funded by the BCCRF, 84.6% was allocated to government institutions, 10% to nongovernmental organizations (NGOs), around 3% for civil society organizations, and 2% to the WB to provide analytical work, technical assistance, and fiduciary risk management (GoB 2014). These projects are expected to augment food security and improve the capacity to cope with disasters (Table 15.1).

According to the BCCRF annual report 2016 (World Bank 2016b), the BCCRF has supported the construction of 61 new cyclone shelters in the districts of Barguna, Patuakhali, Pirojpur, Satkhira, and Khulna under the multipurpose cyclone shelter project. As of December 31, 2015, the construction of the 61 new multipurpose disaster shelters was completed, and they have now been handed over to the school management committees. The

Table 15.1 Projects funded under the Bangladesh Climate Change Resilience Fund

Project name	(Million US$)	Percentage of the total fund
Multipurpose cyclone shelter project	25	16.0
Climate resilient participatory afforestation and reforestation project	33.8	21.6
Community climate change project	12.5	8.0
Agriculture adaptation in climate risk-prone areas of Bangladesh	22.8	14.6
Modern food storage facility project	25	16.0
Solar irrigation project	35	22.3
Establishment of the BCCRF secretariat/capacity building project	0.2	0.1
Analytical and advisory assistance	2.3	1.5
Total	156.6	100

Source: Authors (based on data from the BCCRF).

construction of three roads totalling 11.5 km in Barguna District was also completed. Under the afforestation and reforestation project, climate resilience activities were implemented in the 10 targeted forest divisions. As of the end of 2016, over 17 500 ha had been restored or reforested and over 2000 km of strip plantations had been completed, with over 60 000 direct project beneficiaries. Moreover, the Community Climate Change project (CCCP) has committed the full funds available to 41 NGOs to implement 41 subprojects.

By the end of December 2016, all planned field activities were completed and project development objectives had been achieved or exceeded. Disbursements for the solar irrigation project began in 2014 and by the end of 2016, 489 pumps had been installed covering 35 062 acres of land and serving 11 453 farmers. The allocation of $10 million was expected to be fully utilized within the grace period for disbursement by March 31, 2017. The project has made an influential and innovatory contribution to climate change mitigation efforts in Bangladesh by promoting a sustainable model for replacing diesel-powered pumps with solar irrigation pumps. In addition, disbursement for the establishment and capacity building of the BCCRF secretariat has been completed, and the project closed on December 31, 2014 (World Bank 2016b).

Although the principal aim of the BCCRF was to augment food security in Bangladesh, only three of its project activities were termed as a food security intervention project. But food storage establishment and solar irrigation are solely related to infrastructure and clean development mechanisms. Thus, it may be concluded that though the BCCRF is established to increase climate-resilient food security, the financial allocation does not fit with the justice aspect concerning climate-resilient food security to climate-vulnerable people.

15.3.2 Strategic Program for Climate Resilience – Bangladesh

Another mechanism that was developed to ensure climate-resilient development in Bangladesh is the SPCR – Bangladesh. This is the Pilot Program for Climate Resilience (PPCR) for supporting adaptation and resilience activities in the country. In October 2010,

Table 15.2 Allocated funds from the Strategic Program for Climate Resilience mechanism

Programs	Amount (million US$)	Percentage
Promoting climate-resilient agriculture and food security	325	34.86
Coastal embankments improvement and afforestation	325	34.86
Coastal climate-resilient water supply, sanitation, and infrastructure improvement	281.4	30.18
Climate change capacity building and knowledge management	0.5	0.05
Feasibility study for a pilot program of climate resilient housing in the coastal region	0.4	0.04
Total	932.3	100.00

Source: Authors (based on data from the SPCR).

US $110 million, $50 million in the form of grants and $60 million in concessional loans, were approved for Bangladesh by the Asian Development Bank (ADB), the WB, and International Finance Corporation (IFC) (GoB 2014). The GoB chooses SPCR components from among the 44 priority themes detailed in the BCCSAP and NAPA. The SPCR primarily focused on integrating climate-resilient interventions into specific sectors such as agriculture, food security, water and sanitation, and "climate-proof" coastal infrastructure (water, sanitation, roads, and embankments). The technical assistance component was designed to address specific technical needs, including capacity building and knowledge management for the Ministry of Environment and Forest (MoEF) (GoB 2014) with the allocated funds (Table 15.2).

15.3.3 Bangladesh Climate Change Trust Fund

The GoB assessed the improvement of the national capacity to cope with climate change-induced risks as a top priority and established the BCCTF in 2009 to promote the implementation of the BCCSAP. It enacted the Climate Change Trust Act of 2010. From fiscal year 2009–2010 to fiscal year 2013–2014, a total of 2700 crore BDT (about US$319.73 million) (Table 15.3) was allocated to the BCCTF (BCCTF 2015).

According to the climate-induced hazard risk map (Figure 15.1), the Rangpur and Rajshahi divisions of Bangladesh are highly vulnerable to drought and flood, the Khulna division is the division most vulnerable to SLR and salinity intrusion, and the Sylhet division is highly vulnerable to flash flood. Meanwhile, the Barisal and Chittagong divisions are less vulnerable than other divisions in the country (BCCSAP 2009). Under the BCCTF, the Barisal division received the highest amount (4981.52 lakh[1] BDT) of the fund allocated for 2009–2014, and the Chittagong division received the second highest amount (3814.78 lakh BDT) allocated within the same period. However, Rajshahi, a drought-prone and ethnically dominated division, received less priority (1619.73 lakh BDT) (Figure 15.2).

Rangpur, Rajshahi, and Khulna are also economically stricken areas and most of the districts of the divisions are below the poverty line (Figure 15.3) (World Bank 2010). However,

Table 15.3 Budget allocation for the Bangladesh Climate Change Trust Fund

Fiscal year	Allocated amount (crore BDT)
2009–2010	700.00
2010–2011	700.00
2011–2012	700.00
2012–2013	400.00
2013–2014	200.00
Total	2700.00

Source: Authors (based on data from the BCCTF).

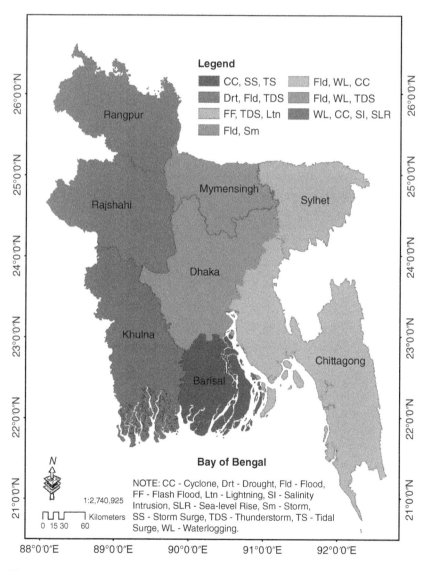

Figure 15.1 Multihazard map of Bangladesh. *Source:* Authors (based on data from the DDM 2016).

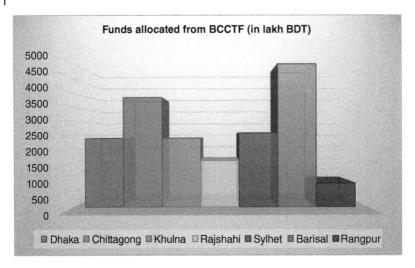

Figure 15.2 Fund allocation from the BCCTF for different divisions of Bangladesh during the period 2009–2014. *Source:* Authors (based on data from the BCCTF).

those divisions received less priority in terms of climate finance from the BCCTF. Similarly, the Chittagong Hill Tracts is also an economically and ecologically fragile area, and the BCCTF gave it less priority to regenerate and reconstruct its economy and ecology.

In the period 2009–2017, the BCCTF allocated 16 210.46 million BDT to manage climate-induced risks all over the country. With this fund, 15.4 km of coastal sea dike have been constructed and about 6760 cyclone-resilient houses have been erected. At the same time, 142 km of embankments have been built, 122 km of river bank protective work have been completed, 535 km of canals have been excavated or reexcavated, 44 elements of water control infrastructure including regulators/sluice gates have been constructed, and 166 km of drainage have been constructed in urban areas to reduce waterlogging. Moreover, 500 water sources and 550 rainwater reservoirs have been established, agro-met stations for early forecasting have been set up in 4 Upazilas, 143.35 million trees have been planted, and 4971 ha of land have been afforested. Additionally, 7800 biogas plants have been installed, 528 000 improved cook-stoves have been distributed, and 12 872 solar home systems have been installed in remote off-grid areas. In addition to these, stress-tolerant crop varieties such as BINA Rice 7, BINA Ground Nut 1 and 2, and BRRI Rice 40, 41, and 47 have been developed. Furthermore, 4500 metric tons of stress-tolerant seeds have been produced and distributed (GoB 2014; BCCTF 2017).

Figure 15.4 reveals that the BCCTF has given less priority to ensure climate-resilient food security for climate-vulnerable people in different climate-vulnerable hotspots. It has allocated only 9% of the total funds, which represents 14 957.21 lakh BDT, to food security. Unfortunately, the BCCTF has given the highest priority to infrastructure development activities and disbursed 64% (103 713.63 lakh BDT) of the total fund under this thematic area, which is disadvantageous for ensuring justice for the most climate-vulnerable people.

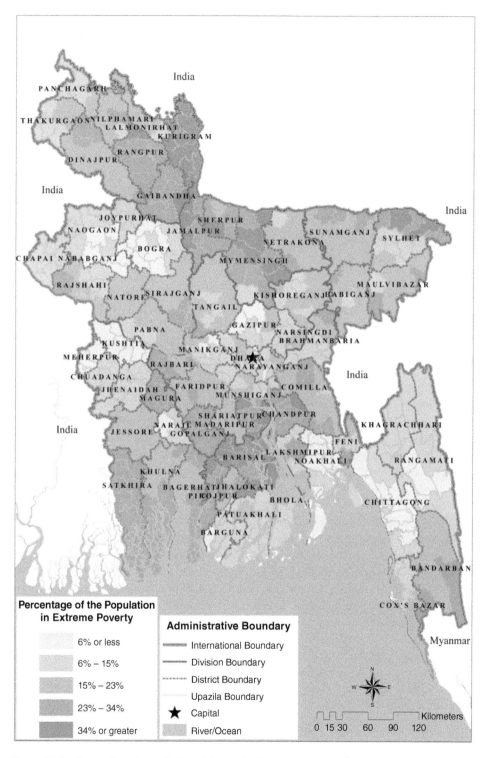

Figure 15.3 Proportion of population of Bangladesh living below the "lower poverty line" in 2010. *Source:* Authors (based on data from the World Bank).

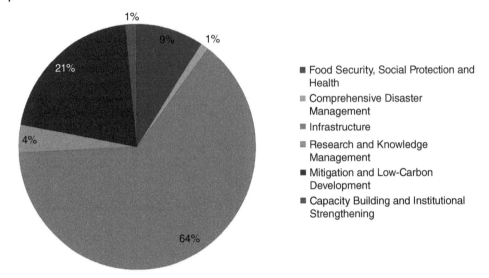

1%

9%   1%

21%

4%

64%

- Food Security, Social Protection and Health
- Comprehensive Disaster Management
- Infrastructure
- Research and Knowledge Management
- Mitigation and Low-Carbon Development
- Capacity Building and Institutional Strengthening

Figure 15.4 Allocation by thematic areas from the BCCTF during 2009–2017. *Source:* Authors (based on data from the BCCTF).

15.3.4 Nongovernmental Efforts: The Case of the Palli Karma-Sahayak Foundation

As a multi-donor trust fund, the BCCRF was established for implementing the BCCSAP. Around 90% of the available BCCRF fund has been allocated to public sector projects while only 10% was channeled through NGOs for community-level climate actions through a separate project titled the CCCP. The Governing Council of BCCRF designated the Palli Karma-Sahayak Foundation (PKSF) to implement the community-level climate change adaptation activities through the CCCP. The project focuses on three climate risks prevalent in Bangladesh: salinity, drought, and flood. Based on the severity of vulnerability and poverty, the CCCP has identified climate risk areas where 41 projects were granted in 15 districts (Figure 15.5).

The target communities of the project are poor and extremely poor populations in hotspots that are the most vulnerable to the adverse impacts of climate change. These areas cover the Kurigram, Nilphamari, Khulna, Bagerhat, Cox's Bazar, Mymensingh, Jessore, Jamalpur, Satkhira, Barguna, Patuakhali, Chuadanga, Rajshahi, Natore, and Naogaon districts. Figure 15.6 illustrates that Satkhira and Kurigram receive the highest priority as 14% of the total CCCP grant was allocated to these districts. Bagerhat receives the second-highest priority at 11%. Khulna, another multihazard risk district, receives third priority and is allocated 8% of the fund.

Although the CCCP gives attention to food security issues and allocates more than 80% (9335.45 million BDT) of the funds to food security and social protection (Figure 15.7), it allocates 38.58% (3501.6 million BDT) for flood-prone areas, 38.16% (3562.46 million BDT) for salinity ingression areas, and only 23.39% (2171.39 million BDT) for drought-prone areas, which does not ensure justice for drought-vulnerable people (Figure 15.8). Moreover, the major field-level activities of the CCCP include raising plinths, courtyards, and

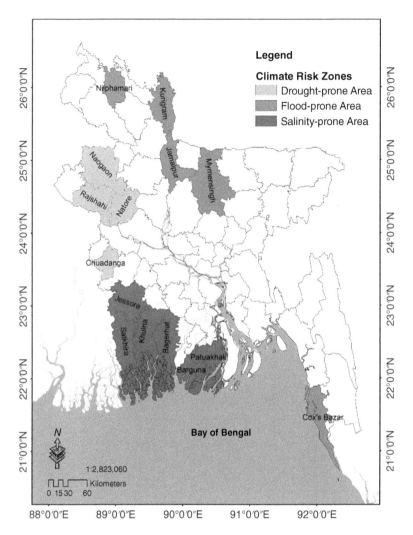

Figure 15.5 Geographical coverage of the Community Climate Change Project. *Source:* Authors (based on data from the PKSF).

community grounds through earth filling to make them climate resilient. To make safe water available, the CCCP undertook activities such as the installation of shallow and deep/ semi-deep tube wells according to local climatic risks, and pond and canal reexcavation to ensure drinking water, irrigation, and water for domestic purposes, water purification systems for safe drinking water in saline areas (pond sand filter and desalinization plants), and rainwater harvesting systems for individuals and communities. Moreover, this program installed improved latrines, distributed environment-friendly cooking stoves, and repaired damaged roads and embankments (PKSF 2016). A small portion of the fund goes to support climate-resilient crops, pumpkin cultivation in sandbars, crab fattening, goat, and sheep rearing, poultry, and duck rearing using the semi-scavenger method, homestead gardening, cage fishing, and vermicomposting.

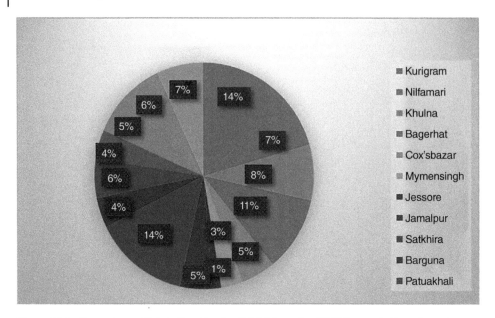

Figure 15.6 Percentage of total allocation by district from the CCCP. *Source:* Authors (based on data from the PKSF).

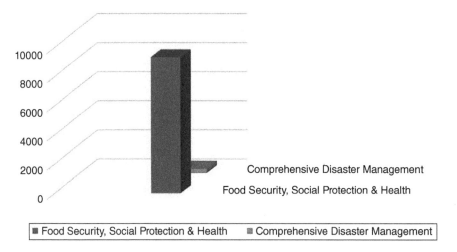

Figure 15.7 Grant allocation by thematic areas under the CCPP. *Source:* Authors (based on data collected from the PKSF).

The CCCP aims to ensure the climate-resilient capacity of poor and marginal people. Ethnic people are more vulnerable than any other communities due to economic status, social disintegration, social exclusion, ecological constraints, and religious barriers. The ethnic minorities demand the highest priority in terms of climate-resilient food security interventions. However, the CCCP only allocates 11% of the total fund for ethnic minority people as well as to the ethnic minority-dominant areas of north and southwest Bangladesh (Figure 15.9). Although the Chittagong Hill Tracts are dominated by ethnic minorities, the CCCP did not implement any project in this area.

Figure 15.8 Grant allocation by climate-induced disaster-prone areas. *Source:* Authors (based on data collected from the PKSF).

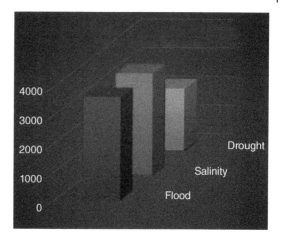

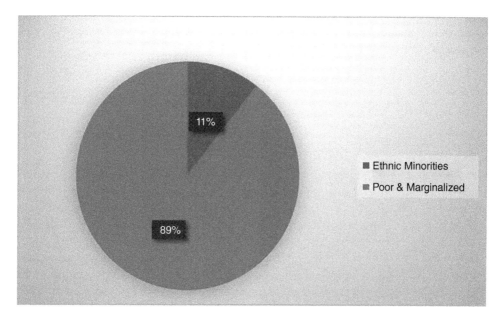

Figure 15.9 Grant allocation by racial class. *Source:* Authors (based on data from the PKSF).

15.4 Policy Recommendations and Way Forward

Bangladesh has made significant strides in developing and enhancing productivity in the agriculture sector, which has led to improvements in the overall food security situation in the country. However, climate change events have severely affected agriculture and crop productivity mostly in climate change hotspots, such as drought-prone areas, flood-prone areas, saline-prone areas, etc. As a result, food security for poor and vulnerable people is facing an increasing threat where marginalized people, women, and children are the victims. These groups should be taken into account while planning, designing, and executing national adaptation policies and strategies for climate change and food security.

Assessment of Bangladesh's existing sector-specific policies shows that the National Agriculture Policy (1999), the BCCSAP (2009), and the Master Plan for Agricultural Development in the Southern Region of Bangladesh (2013) have addressed the issue of food security. Yet, it should be noted that policy goals associated with climate change and food security are mostly addressed as mutually exclusive issues, and hence justice at all levels of the policy implementation process is not always ensured. Regional, racial, and gender-specific disparities in the distribution of resource allocations are observed, which need to be addressed. Otherwise, justice in the food security of Bangladesh will not be achievable. Therefore, the following recommendations are suggested to improve the situation.

- Revise the current policies and strategies regarding food security by considering gender-specific, racial, and regional needs and demands.
- Ensure justice and equity at all levels of the implementation process of the various government and nongovernment development projects.
- Make sure that the climate financing process is more transparent and accountable, thus recognizing the demands of the most vulnerable people.

Pathways for ensuring climate justice and food security in Bangladesh are suggested in Table 15.4.

15.5 Conclusion

In Bangladesh, the frequency and intensity of extreme weather-related events have increased significantly in recent decades due to global environmental changes. These extreme climatic events have greatly affected gross agricultural production. As a result, the food security of the country is increasingly threatened. However, justice is the key tool for ensuring sustainable food consumption and supply for all citizens. In Bangladesh, poor and marginalized groups, including women and ethnic communities, are often deprived of equity and, in the broader sense, justice. As a result, equal needs and rights based on the distribution of resources are often not ensured, which ultimately increase social and cultural inequalities and raise food insecurity, especially in climate hotspots.

Analysis of the four principal climate financing sources of Bangladesh, the BCCRF, SPCR – Bangladesh, BCCTF, and nongovernmental efforts, shows significant shortcomings in fund distribution and implementation of various projects. Moreover, fund disbursement and project implementation often avoid vulnerability and social and economic risk factors while selecting project beneficiaries. Such activities result in food insecurity among the most vulnerable communities and particularly in climate hotspots across the country.

To overcome this situation, review is needed of existing policies and strategies to enhance justice for securing food security across the country, especially among the most vulnerable people. Thus, a bottom-up and inclusive approach, as well as region- and community-specific interventions, must be taken to reduce regional, racial, and sectoral disparities and inequalities, and hence ensure climate justice in achieving food security. The policy recommendations of this study could be repeated in other developing countries.

Table 15.4 Pathways (policy, institution, finance, and technology response) of climate justice and food security in Bangladesh

Particulars	Gaps/issues	Recommendation
Policy	• Agricultural policies, plans, and strategies are not perfect for climate-resilient agriculture and climate justice for regional, racial, and gender groups. • Government does not subsidize or provide any incentive for climate-resilient agricultural practices for smallholders.	• It is very important to reshape agricultural policies, plans, and strategies to ensure climate justice and food security for climate vulnerable and poor people, ethnic minorities, and smallholders. Government should introduce incentives for practicing climate-resilient agriculture for smallholders.
Institution	• There is no responsible government institution to promote climate justice, climate-resilient food security, and resilient agriculture. • Capacity-building mechanisms for climate-vulnerable communities are not available in terms of resilient food security and agricultural practice.	• Enhanced coordination among different departments responsible for climate justice and food security led by the Ministry of Environment, Forest, and Climate Change. • Inclusion of resilient food security, climate justice, and capacity building in resilient agriculture practice.
Financial	• Government provides a social safety net, but it is not adequately targeted to climate justice and food security issues. Social safety nets cover only social and sudden-onset disasters. Provision of the social safety net is irrespective of gender, location, and economic status. It often does not address the needs of diverse farmers.	• Crop insurance should be incorporated. Climate justice safety net should be introduced. Financial allocation should be ensured to reduce discrimination toward women, disadvantaged groups, and farmers in remote locations.
Technology	• Farmers use traditional agricultural technology haphazardly. More research is needed for the fine-tuning of resilient agriculture and food security for smallholders.	• Provide training to farmers before offering support on resilient farming and food security, which would improve quality and profitability.

Source: Authors.

Note

1 A lakh or lacs is a unit in the Bangladeshi/Indian numbering system equal to one hundred thousand (100 000).

References

Amin, M., Zhang, J., and Yang, M. (2015). Effects of climate change on the yield and cropping area of major food crops: a case of Bangladesh. *Sustainability* 7 (1): 898–915.

Bangladesh.com. (n.d.). Bangladesh – cyclones and floods. www.bangladesh.com/blog/bangladesh-cyclones-and-floods

BCCSAP (2009). *Bangladesh Climate Change Strategy and Action Plan*. Dhaka: Government of Bangladesh.

BCCTF (2009). *Bangladesh Climate Change Trust Fund*. Dhaka: Government of Bangladesh.

BCCTF (2015). *Bangladesh Climate Change Trust Fund*. Dhaka: Government of Bangladesh.

BCCTF (2017). *Bangladesh Climate Change Trust Fund*. Dhaka: Government of Bangladesh.

Bronkhorst, V.B. (2012). *Disaster Risk Management in South Asia: Regional Overview*. Washington, DC: World Bank.

CESCR. (Committee on Economic, Social and Cultural Rights). (1999). Comment 12: The right to adequate food (Art. 11). www.nichibenren.or.jp/library/ja/kokusai/humanrights_library/treaty/data/CESCR_GC_12e.pdf

Cline, W.R. (2007). *Global Warming and Agriculture: Impact Estimates by Country*. Washington DC: Peterson Institute.

CPD (2017). *Flood 2017: Assessing Damage and Post-Flood Management*. Dhaka: Centre for Policy Dialogue.

DANIDA. (2012). Preliminary assessment of ASPS-II components vulnerability to climate change. Danish International Development Agency. https://um.dk/en/danida-en/

Dasgupta, S., Kamal, F. A., Khan, Z. H., Choudhury, S., & Nishat, A. (2014). River salinity and climate change: evidence from coastal Bangladesh. http://documents.worldbank.org/curated/en/522091468209055387/pdf/WPS6817.pdf

DDM (Department of Disaster Management) (2016). *Multi-Hazards Risk and Vulnerability Assessment, Modeling and Mapping*, vol. 3. Dhaka: Department of Disaster Management, Ministry of Disaster Management and Relief.

FAO. (2016). 925 million in chronic hunger worldwide. www.fao.org/news/story/en/item/45210/icode

FAO, IFAD, UNICEF, WFP, & WHO (2017). *The State of Food Security and Nutrition in the World 2017. Building Resilience for Peace and Food Security*. Rome: FAO.

Francis, A. and Maguire, R. (2016). *Protection of Refugees and Displaced Persons in the Asia Pacific Region*. Abingdon: Routledge.

GoB (Government of the People's Republic of Bangladesh). (2014). Bangladesh Climate Fiscal Framework. Government of Bangladesh, Dhaka.

Golder, P.C., Sastry, R.K., and Srinivas, K. (2013). Research priorities in Bangladesh: analysis of crop production trends. *SAARC Journal of Agriculture* 11 (1): 53–70.

Harmeling, S. (2010). *Global Climate Risk Index. Who Suffers Most from Extreme Weather Events? Weather-Related Loss Events in 2009 and 1990 to 2009*. Bonn.: Germanwatch.

Hossain, M.S. and Majumder, A.K. (2018). Impact of climate change on agricultural production and food security: a review on coastal regions of Bangladesh. *International Journal of Agricultural Research, Innovation and Technology* 8 (1): 62–69.

Islam, A. K. M. S., Murshed, S. B., Khan, M. S. A., & Hasan, M. A. (2014). Impact of climate change on rainfall intensity in Bangladesh. Bangladesh University of Engineering and Technology, Dhaka.

Islam, M.R. (2010). Vulnerability and coping strategies of women in disaster: a study on coastal areas of Bangladesh. *Arts Faculty Journal* 4: 147–169.

Karim, A. (2012). Climate Change & its Impacts on Bangladesh. www.ncdo.nl/artikel/climate-change-its-impacts-bangladesh

Mainuddin, M. and Kirby, M. (2015). National food security in Bangladesh to 2050. *Food Security* 7 (3): 633–646.

Maplecroft. (2014). Climate change and lack of food security multiply risks of conflict and civil unrest in 32 countries. https://reliefweb.int/report/world/climate-change-and-food-insecurity-multiplying-risks-conflict-and-civil-unrest-32

McClain-Nhlapo, C. (2004). Implementing a human rights approach to food security. 2020 Africa Conference. IFPRI Policy Brief 13. http://ebrary.ifpri.org/utils/getfile/collection/p15738coll2/id/64619/filename/64620.pdf

NAPA. (2005). National Adaptation Programme of Action. Government of Bangladesh, Dhaka.

Noiret, B. (2016). Food security in a changing climate: a plea for ambitious action and inclusive development. *Development* 59 (3–4): 237–242.

Pachauri, R.K. (ed.) (2010). *Dealing with Climate Change: Setting a Global Agenda for Mitigation and Adaptation*. New Delhi: Energy and Resources Institute.

Pachauri, R. K., Allen, M. R., Barros, V. R., et al. (2014). Climate Change 2014: Synthesis Report. Contribution of Working Groups I, II and III to the Fifth Assessment Report of the Intergovernmental Panel on Climate Change. Geneva: IPCC.

Paul, B.K. and Rasid, H. (1993). Flood damage to rice crop in Bangladesh. *Geographical Review* 83: 150–159.

PKSF (Palli Karma-Sahayak Foundation). (2016). Community Climate Change Project (CCCP). Government of Bangladesh, Dhaka.

Rabbani, M., Huq, S., and Rahman, S.H. (2014). Impacts of climate change on water resources and human health: empirical evidences from a coastal district (Satkhira) in Bangladesh. In: *Impact of Climate Change on Water and Health* (ed. V. Grover), 272–285. Abingdon: Routledge in association with GSE Research.

Rahaman, K.M., Ahmed, F.R.S., and Islam, M.N. (2016). Modeling on climate induced drought of north-western region, Bangladesh. *Modeling Earth Systems and Environment* 2 (1): 45.

Rahaman, M.A., Rahman, M.M., and Rahman, S.H. (2019a). Pathways of climate-resilient health systems in Bangladesh. In: *Confronting Climate Change in Bangladesh* (eds. S. Huq, J. Chow, A. Fenton, et al.), 119–143. Cham: Springer.

Rahaman, M.A., Rahman, M.M., and Hossain, M.S. (2019b). Climate-resilient agricultural practices in different agro-ecological zones of Bangladesh. In: *Handbook of Climate Change Resilience* (ed. W. Leal Filho), 1–27. Cham: Springer.

Rahman, A. and Alam, M. (2003). *Mainstreaming Adaptation to Climate Change in Least Developed Countries (LDC)*. London.: IIED.

Robinson, M. and Shine, T. (2018). Achieving a climate justice pathway to 1.5 C. *Nature Climate Change* 8 (7): 564.

Saroar, M.M., Rahman, M.M., Bahauddin, K.M., and Rahaman, M.A. (2019). Ecosystem-based adaptation: opportunities and challenges in coastal Bangladesh. In: *Confronting Climate Change in Bangladesh* (eds. S. Huq, J. Chow, A. Fenton, et al.), 51–63. Cham: Springer.

Vermeulen, S.J., Campbell, B.M., and Ingram, J.S. (2012). Climate change and food systems. *Annual Review of Environment and Resources* 37: 195–222.

World Bank (2010). *Updating Poverty Maps of Bangladesh*. Washington, DC: World Bank.

World Bank. (2013). Turn down the heat: climate extremes, regional impacts, and the case for resilience. A report for the World Bank by the Potsdam Institute for Climate Impact Research and Climate Analytics. World Bank, Washington, DC.

World Bank. (2016a). Climate Risk and Adaptation Country Profile. World Bank, Washington, DC.

World Bank. (2016b). Bangladesh Climate Change Resilience Fund (BCCRF) Annual Report 2016 (English). Washington, DC: World Bank.

Yu, W., Alam, M., Hassan, A. et al. (2010). *Climate Change Risks and Food Security in Bangladesh*. Abingdon: Routledge.

16

A Survey of UK-Based Ethical and Sustainable Fund Managers' Investment Processes Addressing Plastics in the Environment

Quintin G. Rayer

P1 Investment Management, Exeter, UK

16.1 Introduction

Sustainable investment includes environmental issues as part of the three dimensions of ESG (environmental, social, and governance) (UN PRI 2006; United Nations 2006). Current practice in ethical and sustainable investing tends to focus on exclusions based around the so-called "sextet of sin" (tobacco, alcohol, gambling, porn, arms, nuclear) (Knoll 2002), animal testing (Rayer 2017a), and climate change via fossil divestment. On one hand, there is much less emphasis placed on pollution and other problems caused by plastic waste. On the other hand, this issue has been extensively raised by the media and has drawn the attention of the public to the environmental damage caused by plastic waste (BBC One 2017).

As a wealth manager constructing ethical and sustainable portfolios, P1 Investment Management Ltd. (P1) conducted research to gain insight into the investment policies used by fund managers to address plastics. After considering the background and motivation behind plastic concerns, the chapter outlines the questionnaire design philosophy and method. The results are analyzed, and the various investment policies used are discussed, which include overall investment policy; screening and exclusions; impact investing; best-in-class investing; engagement; focus on reduction, reuse and recycling of plastics; disposal; and corporate standards relating to plastics. Apart from understanding current approaches, the research is intended to identify best practices and develop concrete recommendations to effectively address plastic waste by investment policies.

To the best of the author's knowledge, this is the first survey intended to capture the investment policies used by fund managers to address plastics. It is hoped that the survey forms a useful starting point for more comprehensive analyses of investment practice concerning plastics, as well as raising awareness. Following discussion of the results, specific recommendations for ethical fund managers' investment policies on plastics are indexed back to the main text using the numbers in Table 16.4 (e.g. [R1], etc.) and broader policy recommendations in Table 16.5 (indexed as [BR1], etc.).

16.2   Background and Motivation

Many current practices in ethical investing have focused on exclusions. Climate change concerns have resulted in fossil fuel exclusions, with emphasis on extraction and production, commonly known as fossil divestment. Sustainable investment has focused on factors primarily defined by the three dimensions of ESG and endorsed by the UN PRI (2006) (see also Rayer 2017a; Amel-Zadeh and Serafeim 2018, for overviews). Generally, practice has focused on the so-called "sextet of sin" (tobacco, alcohol, gambling, porn, arms, nuclear) (Knoll 2002) and more recently animal testing (cosmetic or medical) (Kreander et al. 2005; Rayer 2017a), environmental and pollution risk (Derwall et al. 2005; Guenster et al. 2010), environmental and social issues (Brammer et al. 2006; Eccles et al. 2012; Investment Leaders Group 2014), corporate governance (Gompers et al. 2003; Bebchuk et al. 2008; Trojanowski and Shaukat 2017), corporate social responsibility (CSR) (Porter and Kramer 2006), social issues (Edmans 2012), general ethical sector funds and socially responsible investing (Bauer et al. 2007; Kempf and Osthoff 2007; Stenström and Thorell 2007), and sustainability (Grewal et al. 2017). However, the literature indicates little focus on the issue of plastic pollution directly, an area which has become of significant concern. Programs such as the BBC TV series Blue Planet II (BBC One 2017) and others have pointed out the alarming global reach of plastic detritus, its chemical toxicity and appallingly harmful effects on wildlife and sea creatures.

A 2018 WWF report (Elliott and Elliott 2018) indicates that total plastic production in the 28 EU member states was 58 million tonnes in 2015, and 4.9 million tonnes of plastic waste was generated in the UK in 2014, with packaging accounting for 67% of this. The problem has been deemed sufficiently severe that the UK Government has pledged to eradicate avoidable plastic waste as part of a 25-year plan (GOV.UK 2018a), while the European Commission has created a plastic strategy as part of a transition toward a more circular economy (European Commission 2018). National and other initiatives are also being developed (see the section on standards below). Although supermarket chains appear to be making some efforts with plastic packaging, a recent "Which?" report indicates that considerable progress is still required (Simmonds 2018).

Investors can support action on this issue by actively directing resources to companies providing solutions, while avoiding firms that contribute to plastic pollution (Rayer 2018a). The trouble is that plastics are extremely useful for both manufacturers and consumers; they are cheap, strong, lightweight, waterproof, chemically inert, and durable, which make them ideal for many purposes including storage and food packaging. As a result, even when aware of the issues, consumers struggle to excise them from their lives (Rayer 2018a). Although some fund managers are taking plastics into account, many approaches appear weak. Thus, further action in this area is likely to be useful in tackling the problem.

16.3   Philosophy of Approach

The fund management sector is intensely competitive, with fund management groups seeking to convince client (and potential client) investors that each has a superior offering. In practice, distinctions between fund management approaches can be slender; thus, from a marketing perspective, each company feels obliged to emphasize differences between their products and those of their competitors as much as possible. Since investment processes can seem arcane in a sector replete with jargon, this can make it difficult to distinguish between funds that genuinely attempt to address ethical and sustainability issues (including plastics) and those that merely seek to provide a sophisticated marketing gloss to try to gain a competitive edge (Rayer 2017a).

As a result, financial advisers and other sector professionals may struggle to distinguish between genuinely committed fund managers and imitators seeking to appear ethical for marketing purposes. Financial advisers (and clients) can be swayed by compelling examples or case studies presented by fund managers. Advisers pass these on to clients as examples to motivate them to invest and to ensure that their services are used. Finding themselves unable to evaluate competing fund offerings objectively, advisers and clients often resort to heuristics (mental shortcuts or rules of thumb) including reliance on investment examples of engagement or activism. Individuals confronted with uncertainty typically use heuristics to guide decision making (Kahnman and Tversky 1979).

Consequently, the fear is that fund managers can become adept at presenting *ad hoc* examples of their activities in the form of interesting or exciting investment "stories" given to advisers, typically involving engagement. Manager engagements can vary widely in quality from "fire and forget" letters requesting action with no follow-up, to carefully designed programs with time-bound targets and predetermined investor actions to be taken in the event of inaction, including ultimately the threat of divestment. While many fund managers' engagement programmes are well designed, marketing pressures represent a real concern for clients and advisers.

Even with the best of intentions, initiatives by investee firms may also prove less enduring than hoped. For example, in 2008 Starbucks pledged that by 2015 it would serve 25% of beverages in reusable containers and recycle all cups in North American stores. Ten years later, less than 2% of beverages are served in reusable cups and only 60% of North American stores have a recycling bin. As a result, "As You Sow" (a shareholder advocacy nonprofit organization) presented a shareholder proposal to reinvigorate the previous commitment and extend it geographically, gaining a 44.5% vote in favor (As You Sow 2019). Thus, even with successful engagement, there is a significant risk that good intentions may not materialize or result in permanent progress.

16.3.1   Behavioral Finance Traps

By using heuristics, advisers and clients may be falling into behavioral finance traps (for overviews see Elton et al. 2017; Rayer 2018b), and these include representativeness and availability heuristics as well as confirmation bias (cognitive dissonance). The representativeness heuristic assumes that an example presented is widely representative of activities, while it may have been selected as a unique case. The availability heuristic presumes that conveniently available information is appropriate for the decision in hand, although such examples must be readily available if a fund manager provides them. Cognitive dissonance may apply after an investment selection has been made, with the tendency to seek examples supporting the choice made, rather than facing uncomfortable evidence suggesting a poor decision (Festinger 1957). As more emphasis is placed on information confirming prior decisions, this may also be known as "confirmation bias."

These heuristics question whether such activities might be a deliberate exploitation of clients, potential clients and advisers, response to demand, or merely delivering what works from experience. The Financial Conduct Authority (FCA) has published occasional papers on aspects of behavioral finance, exploring how people make financial decisions (Erta et al. 2013; Iscenko et al. 2014; Adams et al. 2016; Lukacs et al. 2016). They seek to understand the mistakes made by consumers and how financial providers respond to these mistakes, how this affects competition and any interventions they might consider. One concern is that financial providers' product design and sales processes may accentuate rather than ameliorate the effects of consumer biases. Questions also arise as to the extent to which financial advisers are themselves victims of such strategies or whether they are content to play along (knowingly or unknowingly) to propagate these strategies to their clients, the underlying investors.

In this environment, many ethical fund managers provide specific examples of good works they have carried out on plastics (e.g., see Ast 2018), typically engagements with particular companies on recycling (and use of recycled content), discussions around the role of government (e.g., plastic bag taxes), and the need for technological innovation. Regrettably, specific examples of individual engagements may be short-lived and may also be one-off examples used to address an individual client's enquiry. Nevertheless, these one-off examples are still contributions toward solving an issue and thus have some value. When clients enquire about how a specific issue is being tackled, they would prefer to know the permanent underlying investment policies that are used to address it. Implementation of these long-term policies should, of course, result in examples that can be presented as investment case studies or "stories." While the existence of a long-term investment policy addressing an issue implies that examples of implementation and outcomes would develop, the reverse is not necessarily true; examples of case studies on an issue do not necessarily imply the existence of an enduring policy addressing the issue in question.

16.3.2 Investment Policies Versus *Ad Hoc* Examples

Fund managers may be able to generate *ad hoc* examples of work in a specific area. Short-term projects can be developed to meet perceived client demand and dropped when the attention has moved elsewhere. A significant concern, therefore, was that questioning fund managers could yield specific examples, with little insight gained into long-term investment policies. The questionnaire was designed to mitigate this, and questions related to fundamental investment policies used to address plastics were asked. Although investment policies can be changed, they are more likely to be enduring and genuinely representative of the activities of a given fund manager than a few specific examples, no matter how compelling the story.

This approach evidently discomforted some fund managers who would have preferred to present *ad hoc* case studies (investment "stories") demonstrating superior credentials in this area. A few of the questionnaire responses were accompanied by (often extensive) examples of case studies; in one case the fund manager felt unable to respond directly to questions, and instead provided long, verbose answers to "yes/no" questions.

Conversely, a few fund managers were disarmingly honest, admitting weaknesses in policies addressing the issue. This honesty is commendable, as the author made clear in dealings with these managers, seeking to reassure and encourage this level of openness. Transparency can be interpreted as evidence of a willingness to improve, which is to be supported.

16.4   Method

P1 approached 12 fund management firms running 20 ethical or sustainable UK-based retail funds with total assets under management of £6.3 billion as of January 2018. Due to P1's strong working relationships with the firms involved, all firms and fund managers responded to the questionnaire and they also consented for their provided data to be combined with results from other fund managers and published, provided this was done in an anonymized way. Accordingly, no mention is made of the names of the fund management houses or individual funds involved. The funds represented 40.4% of the UK ethical funds universe by assets under management (Investment Association 2018). Further details are presented in Table 16.1. Given the range in fund sizes from £56 million to £1123 million, assets under management was considered a better measure of the proportion of the sector covered than the number of funds.

The fund managers completed a questionnaire about their investment policies relating to plastics (see Appendix A). The primary emphasis was on investment policies rather than recent *ad hoc* engagements. While *ad hoc* engagement initiatives have value, it was felt that these run the risk of being short-lived. The emphasis on ongoing processes was intended to provide more enduring insights and better reflect fund managers' commitment to the issue.

The fund managers approached had passed P1's proprietary selection process, which involved the submission of requested documentation explaining the fund managers' ethical investment policies and followed up with interviews and further submissions to clarify

Table 16.1   Size of survey sample relative to UK investors' fund sectors, as of January 2018 (Investment Association 2018)

Sector	Assets under management	Proportion of all funds	Proportion of ethical funds sector
Questionnaire sample	£6.3 bn	0.5%	40.4%
Ethical funds sector	£15.6 bn	1.3%	100.0%
All funds	£1223.0 bn	100.0%	

The "Ethical funds sector" includes all those funds classified as "ethical" by the Investment Association.

any issues identified. The process is designed to gauge the quality of the fund's ethical investment policies and determine the level of commitment of both the fund managers and their organization to ethical investing. P1's due diligence process is audited by its external ethical oversight committee and subject to continuous improvement. Only those funds judged to have superior ethical investment policies are included in P1's ethical model portfolios.

Because of the above, the survey sample is not claimed to be statistically meaningful and was not based on specific ethical styles, nor were those chosen necessarily representative of practice across the UK ethical funds sector. However, the sample can be claimed to avoid those funds with an only superficial commitment to ethical or sustainable investment.

16.5 Analysis

The collective responses of the fund managers were analyzed using Likert (1932) scoring. Likert scoring allows questionnaire responses to be converted into numerical values permitting further analysis or statistical testing of the results. Here, higher questionnaire scores indicated more actions taken to address plastics. The total scores of the funds over the questionnaire covered the range 30–56, with a mean of 48.25 (median 50, mode 51) and standard deviation of 6.33. Results ranged from 2.88 standard deviations below the mean to 1.22 standard deviations above, indicating that a meaningful variation had been obtained. The results are presented in Figure 16.1 as a bar chart with a Gaussian distribution based on the above mean and standard deviation overlaid. Following Likert (1932), the results are assumed to be normally distributed for the analysis that follows. Due to the relatively small sample size (20 funds), it was not deemed worthwhile to analyze the higher moments of

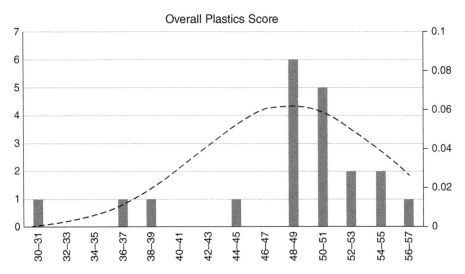

Figure 16.1 Distribution of overall scores of fund managers on the investment processes they used in relation to plastics. Normal distribution overlaid with mean of 48.25 and standard deviation of 6.33.

the distribution. The negative tail is due to the inclusion of funds with no investment policies on plastics, which had low scores.

16.5.1   Fund Manager Claims Versus Practice

The initial questions asked about the fund managers' commitment to address plastics.

- *Question 1*: Does your investment policy seek to assess the involvement or activities of companies you invest in, in relation to plastics pollution issues? (Yes / No)
- *Question 1a*: If your investment policy does seek to address plastics issues (i.e., you answered "yes" to question 1), how important are plastics issues in relation to your other ethical or sustainability criteria on a scale of 1 (least important) to 5 (most important)?

Later questions explored the specific investment policy actions taken to address plastics. It was therefore possible to compare the claimed commitment (questions 1 and 1a) with investment practice, that is, whether there was evidence that fund managers were following up claims with genuine investment policies.

The scores of questions 1 (investment policy "yes/no"), 1a (how important on scale 1–5) and "claimed importance" (defined as the sum of scores from questions 1 and 1a) were compared with an overall "plastics score" representing the sum of the scores from the remaining questions, as a measure of the range of investment policies used to address plastics. Fund managers who did not respond to questions 1 or 1a (including because they answered "no" to question 1) were scored "0" for question 1a. One might expect the "claimed importance" (Q1 + Q1a) of plastics from a fund manager to be a reasonable predictor of the investment policies they used; however, as seen in Figure 16.2, there does not appear to be a very strong relationship. While there is a relationship at lower "claimed importance" (2–4), there is no improvement for the values above 4. For managers claiming the highest importance regarding plastics, the score possibly decreased slightly.

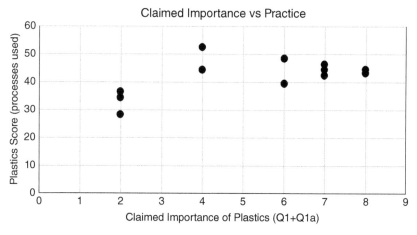

Figure 16.2   Fund manager investment processes on plastics ("plastics score") against claimed importance of plastics (sum of score from Q1 and Q1a).

Further examination (Figure 16.3) indicates that the responses to question 1a appeared to have no bearing on the investment policies ("plastics score"). The level of the "plastics score" does not change, although its dispersion appears to decrease for fund managers who state that plastics are of increasing importance.

The better indicator of a fund manager's commitment to plastic investment policies was the simple "yes/no" question (Q1), while an attempt to elicit the degree of commitment (Q1a) added no further value in terms of the investment response. In Figure 16.4, the plastics scores are presented as "box and whisker" plots with a visible distinction between fund managers who stated they had no investment policy on plastics and those who did (Q1).

The results presented in Figure 16.4 are reassuring, as they indicate that the fund managers who claimed a policy on plastics were following this up with genuine investment process responses. However, it was somewhat disappointing that the degree of importance that managers claimed to assign to plastics (Q1a) appeared to have no relation to their

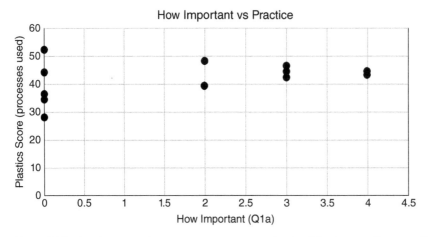

Figure 16.3 Fund manager investment processes on plastics ("plastics score") against "how important are plastics" (score from Q1a).

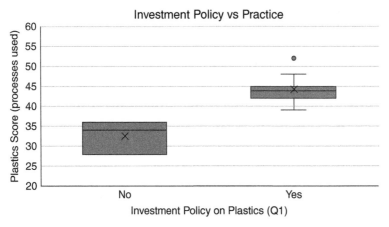

Figure 16.4 Fund manager investment processes on plastics ("plastics score") against existence of investment policy on plastics (Q1).

Table 16.2 Statistics on plastics scores of fund managers that either did not have ("No") or had ("Yes") an investment policy on plastics (Q1)

Statistic	Q1 response: "No"	Q1 response: "Yes"
Number of funds	3	17
Minimum plastics score	28	39
Maximum plastics score	36	52
Range of plastics scores	8	13
Mean plastics score	32.67	44.24
Standard deviation of plastics scores	3.40	2.88
Median plastics score	34	44
Mode of plastics scores	N/A	44

investment policies. The statistics of the two population responses to Q1 (investment policy on plastics) are summarized in Table 16.2.

If the plastics scores in Figure 16.4 are interpreted as normally distributed (following Likert 1932), the distribution of the difference between the two means can be used to explore whether they can be regarded as being drawn from independent populations (Duncan 1974; Hogg and Craig 1989). The difference in the sample means is $\bar{x}_{Yes} - \bar{x}_{No} = 44.24 - 32.67 = 11.57$. By hypothesis, the two populations are assumed to be drawn from the same parent population, thus $\mu_{Yes} - \mu_{No} = 0$. The standard error of the difference between the two means is:

$$\sqrt{\frac{\sigma^2_{Yes}}{n_{Yes}} + \frac{\sigma^2_{No}}{n_{No}}} = \sqrt{\frac{2.88^2}{17} + \frac{3.40^2}{3}} = 2.08$$

The two populations have means $11.57/2.08 = 5.56$ standard errors apart, with standard normal z-score given by:

$$z = \frac{\left(\bar{x}_{Yes} - \bar{x}_{No}\right) - \left(\mu_{Yes} - \mu_{No}\right)}{\sqrt{\frac{\sigma^2_{Yes}}{n_{Yes}} + \frac{\sigma^2_{No}}{n_{No}}}} = \frac{11.57 - 0}{2.08} = 5.56$$

The associated probability is $P(z > 5.56) = 1.4 \times 10^{-8}$. To a very high degree of confidence, it appears that the investment responses of the fund managers in relation to plastics genuinely differ based on their claim to have an investment policy in this area.

16.6 Discussion of Results

Question results are presented individually with a discussion, including relevant ethical investment approaches and other implications. This is supported by the Likert (1932) scoring analysis to investigate how well fund managers' claims about their investment processes in relation to plastics appeared to be supported by detailed policy actions. The Likert scoring used response formats which collected data on a 1–5 (or 0–4) scale and "yes/no" questions for information capture. A question requesting information permitting a range of views was scored on a 1–5 scale, while another relating to the number of standards revealing responses between 0 and 4 was scored directly. Questions calling for "yes/no" answers were scored 4 points for "yes," 2 points for "no" and 3 points for a blank (which was interpreted as "don't know"), which follows the values used by Likert (1932).

As indicated above, the questions focused on investment policies used by the managers to address plastics rather than specific examples of, say, engagement actions.

The ethical funds analyzed included the following asset classes: UK equity, European equity, global equity, and sterling-denominated bonds. No differences in the questionnaire were made for different asset classes. The funds analyzed had average assets under management of £315 million, covering a range from £56 million to £1123 million around the time the questionnaires were being completed (January 2018).

16.6.1 Overall Investment Policy on Plastics

Of the respondents, 85% (17 funds) felt their investment policy sought to assess the involvement or activities of companies concerning plastic pollution. For the 15% who felt their investment policy did not cover plastics, this did not necessarily mean that they did nothing in this area. Two funds from different management companies engaged with companies to improve the environmental management of plastics and had a policy of voting shares against companies with harmful plastic practices. One of these was also aware of corporate standards relating to plastics. However, one fund had no plastic policies.

The 85% of funds with an investment policy on plastics were further asked to indicate how important plastics were relative to other ethical or sustainability criteria on a scale of 1 (least important) to 5 (most important) using a Likert (1932) scale. Most respondents judged that plastics were rated "3" (medium importance) compared with other ethical or sustainability issues, with smaller numbers opting for higher or lower ratings. Three funds gave a "2" (slightly important) rating and two a "4" rating (rather important). The responses had both median and mode answers of "3" (medium importance). Figure 16.5 shows the results, including that 17.6% (three funds) indicated that the level of importance given would depend on the circumstances.

The largest proportion selected a medium rating of "3," which seems unsurprising. Plastics must be balanced against other environmental issues (including climate change, pollution, and biodiversity) as well as social and governance issues within a sustainability ESG framework, so it would seem unlikely that plastics could have the highest priority. Indeed, no replies rated plastics as "5" (most important).

Since all the respondents had already indicated that they had an investment policy in place on plastics, it seems reasonable that none reported that plastics issues were rated "1" (least important). For those who replied that the importance depended on circumstances ("it depends"), they stated that it would depend on their assessment of the degree of risk or exposure of the company and industry in question.

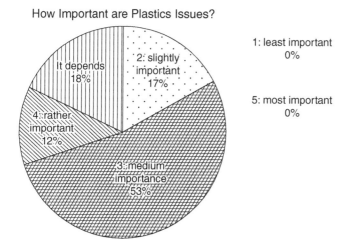

How Important are Plastics Issues?

1: least important
0%

5: most important
0%

Figure 16.5 The importance of plastics issues for fund providers with an investment policy in relation to plastics. Scale of 1 (least important) to 5 (most important). Three respondents indicated the importance would depend on circumstances ("it depends").

16.6.2 Screening and Exclusions

Screening is commonly employed by ethical investors. Investments are tested against requirements aligned with positive and negative impacts, or other criteria. Companies' impacts are identified as positive, negative, or "ethically neutral" (broadly doing neither good nor harm) (Rayer 2017a). Decisions about which firms to invest in are made based on these criteria. An investor must decide whether to avoid ethically neutral companies (see Figure 16.6). Negative screening avoids unethical companies but invests in ethically neutral companies, while positive screening only invests in ethically beneficial companies, avoiding both ethically neutral and unethical companies. Ethically beneficial companies identified under positive screening would be those that actively contribute to providing solutions. Regarding plastics, these might include firms that offer alternative products,

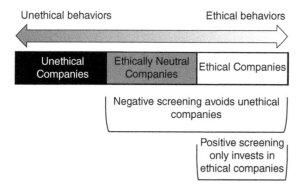

Figure 16.6 Negative and positive screening. *Source:* Reproduced from (Rayer 2017a).

allow or facilitate the reuse of plastics, help improve recycling methods, or provide for environmentally safe disposal.

On plastics, the fund managers were asked whether they used either exclusions (negative screening) to avoid companies, or positive screening selectively to invest in companies that address problems. Two funds (10%) used negative screening to avoid companies with harmful practices, while 60% said they used positive screening to invest preferentially in companies that help address plastic problems.

While investments may be excluded based on negative screening, opportunities can be identified among the companies that offer solutions to plastics issues (positive screening). Alternatively, firms that screened as "neutral" on plastics may be worthwhile investment opportunities on other grounds, including other ethical issues, or on performance grounds. Hence, while negative or positive screening could be applied to the whole fund, often positive screening would be used to identify selected investment opportunities rather than being applied to the entire portfolio.

Overall, while only a small proportion of funds (10%) applied exclusions based on plastics, more (60%) sought the investment opportunities that offer solutions to plastic problems. For stronger impact, it would be necessary for a much higher proportion than 10% of ethical and sustainable fund managers to entirely refuse to invest in firms that generate significant plastic pollution (R1).

16.6.3 Impact Investing

Impact investors focus on companies providing solutions to problems with a positive impact on issues of concern. Superficially, impact investing appears similar to positive screening. The difference is that less emphasis is placed on the investment return. Impact investors might not expect an investment to generate a loss, but they are prepared to accept part of their "investment return" as the utility of the benefit (or impact) it creates. In economic terms, the utility function of an impact investor includes contributions not only regarding the money made, but also the value of environmental, social, or other contributions. Typically for ethical funds, an impact investor would require some return. However, it would mean that the requirement for investment gains would be placed at a slightly lower level than would otherwise be the case while keeping anticipated returns within an acceptable envelope.

Only 10% (two funds) described themselves as using impact investing to address plastics. Consistent with the replies on positive screening, both funds using impact investing also used positive screening on plastics. Of the two, one described itself as using negative screening and one not.

16.6.4 The Relationship Between Screening and Impact Investing

To discuss the relationship between negative and positive screening and impact investing, imagine a fund (here denoted "NPI"; "N" negative, "P" positive, "I" impact) that negatively screens all its investments and applies positive screening to a subset of these, with a further subset of assets selected based on impact investing. Such a fund would hold no "unethical" stocks, some "ethically neutral" stocks, some "ethical" stocks and some "impact" stocks.

Another approach would be for a fund (denoted "PI") to positively screen all its investments, with a subset selected as impact investments. This fund would hold no "unethical" stocks and no "ethically neutral" stocks, with the portfolio being composed of "ethical" and "impact" stocks. The difference between the NPI and PI funds is that the NPI fund holds some "ethically neutral" firms, while the PI fund does not. There are also implications for the proportions of "ethical" stocks and "impact" stocks in each. Superficially, one might expect the NPI fund to hold a lower proportion of "impact" stocks than the PI fund. A further complication is that a fund will generally also have ethical objectives in areas other than plastics.

Since impact investing places a lower priority on returns, this could raise concerns that the PI fund might generate weaker performance than the NPI fund. However, this argument should be treated with caution. The above offers no detail on the proportions of "ethical" and "impact" stocks in the notional funds above, nor does it make any comment on the relative performance merits of "ethically neutral," "ethical," and "impact" investments. Although "impact" investments might surrender some performance, there is no indication of how much return is given up, and from which initial level.

Underlying investors can balance whether they would be content to merely exclude firms with harmful practices against their requirements for investment performance. If they have higher return expectations or are content only to avoid companies with harmful practices, they may prefer the NPI fund. If they wish to emphasize investment in firms that are actively contributing to providing solutions to plastics and are either less concerned about performance or do not accept the underperformance argument above, they may feel that the PI fund is most suitable.

Of the 12 funds that used negative screening, positive screening, or impact investing, 75% (nine funds) used only positive screening (P), and one fund each used either negative and positive screening (NP); positive screening and impact investing (PI); or all three (NPI). When addressing plastics issues, positive screening in isolation was the most popular approach employed by fund managers. As indicated above, for greater impact on plastic pollution, it would appear necessary for more fund managers to include negative screening on plastics issues as an investment policy (R1), while the ones not using positive screening should also consider it (R2).

The topic of ethical fund performance relative to conventional investing is an extensive subject. While there is a "common wisdom" that ethical funds must underperform (see, e.g., Posner and Langbein 1980), there are arguments as to why ethical investments could be expected to outperform (Rayer 2017a), and some academic studies for specific ethical approaches over a range of historical periods support this (Gompers et al. 2003; Derwall et al. 2005; Kempf and Osthoff 2007; Bebchuk et al. 2008; Edmans 2012). These are not further discussed here, but for an overview see Rayer (2017b, 2018c).

16.6.5   Best-in-Class Investing

Best-in-class investing focuses on the best operators within the class considered, that is, the best companies within a specific sector. For sectors with harmful practices, it can imply investment in the "least bad" companies (Rayer 2017a). The goal is to motivate companies in ethically challenging sectors to improve. However, investors need to feel sufficiently confident in this approach to accept challenges from those who may question why they have any involvement with the sector in question. For some investors, depending on their motivations and profile, any such investment might be unacceptable.

These investments might be in plastic manufacturers or retailers that use significant amounts of plastic packaging. A best-in-class approach might mean investing in companies that are most active in developing biodegradable (or compostable) plastics or make their products easier to recycle. At present, they may still produce significant volumes of harmful plastics, but rather than avoiding these firms, investment might support progress toward producing less harmful products (R4). However, while plastic biodegradability may seem straightforward, there are differences between biodegradable, compostable, degradable, and oxo-degradable, which can lead to confusion and misuse (Eenee 2019). These terms are explored in more detail below.

Food retailers may use significant amounts of plastic packaging, due to its convenient properties. Best-in-class investment might mean selecting firms that are seeking to replace packaging with non-plastic equivalents where possible, preferentially selecting biodegradable forms of plastic, and helping their customers by making recycling easier and so on. They may still currently be using unacceptable volumes of harmful plastic packaging, with insufficient thought given to disposal, but if they are making genuine efforts to improve in these areas, progress may be hastened by inward investment rather than avoidance (R4).

Investors should be aware that they could be criticized for supporting a plastic manufacturer or a firm using plastic packaging in volume. However, if investors are confident that they have selected the "least bad" firms in this area because they are striving to improve, they would help ensure that resources are directed toward progressive firms rather than firms in similar activities with no intention of changing their ways. "Best-in-class" depends on being confident that the selected companies are among the least bad in their area, and it can often be effectively combined with engagement, an approach discussed below.

Of respondents, 50% (10 funds) stated that they used a "best-in-class" approach. All but one of the funds using this approach also used positive screening. For the one that did not use positive screening, best-in-class was only used with engagement.

Overlap between the positive screening and best-in-class approaches may reflect an interpretation issue. If positive screening is regarded as "selectively investing in companies that help address problems," while best-in-class is interpreted as "investing in companies that do the least harm or are better than peers," there can be an overlap. Depending on whether "least harm" is also interpreted as "doing good," and how "peers" are defined, they could be regarded as asking the same question. Therefore, a recommendation for future questionnaires would be to ensure that best-in-class firms comprise the companies that would not be suitable for inclusion based on negative (or positive) screening, or impact investing. This recommendation assumes the precondition that it only relates to firms engaged in undesirable practices to assess whether some are better or worse on a scale of harm.

16.6.6 Engagement

Investors engage with companies to encourage them to improve their practices. This encouragement can take different forms. For example, fund managers should actively use their shareholder votes, that is, their rights as part-owners in a company, to transmit their views to management (R5). Traditionally, many conventional fund managers have been passive in this area, taking the view that if they did not like a company's management approach, they could sell their shareholding. A passive approach has advantages, such as the ability to earn fee income by stock lending to counterparties who can sell the shares "short" for a fee. The stock lending agreement would ensure that fund managers receive equivalent shares back at the end of the loan period and are compensated for any dividend payments due, so economically speaking, stock lending would have no direct consequence, except for earning a fee for loaning the stock, which would contribute to the overall portfolio value. However, during this period the fund managers would be unable to exercise share voting rights since they would not own the shares.

The UK Stewardship Code encourages fund managers to act responsibly and exercise their ownership rights appropriately, pointing out that stewardship involves more than just shareholder voting and that investors should monitor and engage with companies on matters including strategy and risk (Financial Reporting Council 2012). One would expect that ethical fund managers would make a point of exercising their voting rights as a matter of course.

Bond fund managers do not own shares and therefore have no voting rights. Yet, an ethical bond fund manager may be able to exert voting power through their colleagues' equity portfolios if part of a larger fund management group that also manages equity funds.

Regarding share voting, 25% of the funds responding to the questionnaire stated that they had a policy of voting against companies with harmful plastic practices. A substantial proportion, 95%, stated that they engaged with company management to improve their environmental management of plastics. Given the remarks about *ad hoc* initiatives above, it may be that short-term initiatives are caught in this question, making a positive response easier.

Another aspect of engagement is "policy engagement" with governments or regulators to enhance standards or regulations in areas of concern. Few funds (15%) engaged with governments or regulators to improve the environmental management of plastics (R6).

The most popular engagement style was with individual companies (95%). Little use has been made of share voting to encourage firms to change behaviors (25%), and less use of policy engagement with governments and regulators has been recorded (15%). Considering plastic bag taxes (GOV.UK 2018b), potential bans on plastic earbuds and straws (Perkins 2018), possible taxes (BBC 2018), and other government initiatives (Dalton 2018), the UK government is demonstrating more proactive leadership than fund managers on policy.

16.6.7 Reduction, Reuse, and Recycling

The degree to which fund managers were selecting firms addressing plastics was further explored using the "reduce – reuse – recycle" paradigm. Emphasis is placed on products or services that support reduction, reuse, or recycling of plastics, preferentially in that order (i.e., reduction is preferable to reuse of plastics, with recycling as a last resort).

From the questionnaire, 85% (17 funds) had investment policies that considered reduction in plastics use, 80% had policies that explored reuse, and 75% of investment policies included recycling. See Figure 16.7 (disposal is covered in the next section).

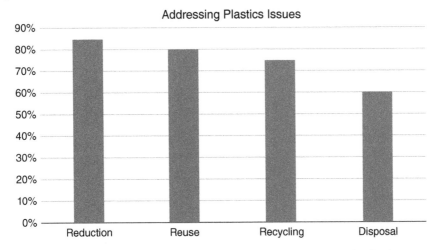

Figure 16.7 Proportion of funds with investment policies that explored reduction, reuse, recycling, and disposal in relation to plastics.

Most funds considered all three – reduction, reuse, and recycling – which seems appropriate given the scale of the problem. More emphasis was placed on the "upstream" side, with greater interest in firms that reduced plastic use. This appears positive as, over time, such investment would be expected to cut off excessive plastic use "at the root," rendering reuse and recycling less necessary. On reuse and recycling, funds that explored investment in reuse were a subset of those that included reduction in plastics, and all those that invested in recycling also considered reduction and reuse. In each case, the decrease over each step of "reduction – reuse – recycling" was by a single fund.

The three funds (15%) that did not consider selection based on reduction, reuse, or recycling of plastics also did not use negative screening, positive screening, impact investing or best-in-class approaches. Of these three, two used voting and company engagement, while the other did not.

16.6.8 Disposal of Plastics

For plastics, the waste management paradigm of "reduce – reuse – recycle" may usefully be extended to include disposal. Considering the amount of plastic waste that has accumulated in the waste processing sector (Laville 2017; Harrabin 2018) and in ecosystems (Le Guern Lytle 2017), it is natural to explore technologies that can dispose of plastics in an environmentally safe way (Singh and Sharma 2016; Elliott and Elliott 2018). Equally, a significant amount of plastic material is still in use, being produced, and likely to be produced for some time to come. Thus, methods of safe disposal for a wide variety of plastics, particularly nonbiodegradable and noncompostable, are likely to be required.

Fund managers were asked whether their investment policy on plastics explored environmentally safe disposal. The responses showed that 60% (12 funds) used safe disposal as a criterion for investment concerning plastics issues. As indicated in Figure 16.7, this number is lower than the one for recycling (R7).

16.6.9 Corporate Standards

External parties can be more confident that companies and fund managers are responsibly addressing plastics by looking at the degree to which they subscribe to external corporate or environmental standards. Although standards can be helpful, caution is required since some standards are purely aspirational, while stronger standards may be externally audited.

For fund managers selectively investing in companies that are addressing plastics or which have superior standards, one might expect an awareness and use of verifiable corporate plastic standards. These standards would be valuable tools and could confirm the company's understanding of the plastic exposures of the investee firms they are considering. For this reason, the questionnaire included questions around plastics standards. Managers were asked whether they were aware of any corporate standards that addressed environmental issues surrounding the creation, use, reuse, recycling, and disposal of plastics, whether run as formal national or international standards, voluntary initiatives, or otherwise.

The standards that fund managers identified ranged from European initiatives such as the Circular Economy Package (European Commission 2019) and guidelines (GOV.UK 2017; European Commission 2018), UK packaging waste regulations (The Environment Exchange 2005), sustainable textiles (Blue Sign 2019), and the wide-ranging initiatives led by the Ellen MacArthur Foundation (Ellen MacArthur Foundation 2017; New Plastics Economy 2017). Other standards had a focus on sustainability and waste (WWF 2017; WRAP 2019). With the possible exception of the New Plastics Economy (2017), these documents appear to be primarily aimed at businesses or government. Of the standards identified, only one was directly aimed at investors, the Plastics Solutions Investor Alliance (PSIA) (As You Sow 2018).

Many of the initiatives are focused on sustainability or circular economies (Ellen MacArthur Foundation 2017; WWF 2017; Blue Sign 2019; European Commission 2019; WRAP 2019) or a general waste management focus (The Environment Exchange 2005; GOV.UK 2017), and thus capture plastic concerns as part of a wider remit. Several have a specific plastic focus (New Plastics Economy 2017; European Commission 2018; As You Sow 2018). As stated above, only the PSIA specifically aims to use investor influence to address plastics.

Of fund responses, 70% (14 funds) indicated that they were aware of plastics standards. However, only 5% (one fund) used such standards as a selection criterion for investment (R8). No fund managers used any of these standards inside their own organizations (R9). Despite 70% of managers being aware that formal standards existed, only 45% were able to identify specific standards, implying that 25% (70–45%) of fund managers were aware that plastics standards exist but were unable to identify any by name. The plastics standards identified by fund managers are summarized in Table 16.3.

Figure 16.8 shows the proportions of plastics standards that were identified by the 45% of fund managers able to do so. The most commonly identified standard was WRAP (WRAP 2019), which works with governments, businesses, and communities to deliver practical solutions to improve resource efficiency. Its mission is to accelerate the move to a sustainable, resource-efficient economy by reinventing design, production and sale of products, rethinking consumption, and redefining what is possible through reuse and recycling.

Table 16.3 Summary descriptions of plastics standards identified by fund managers

Standard	Short Name	Description
Blue Sign	Blue	Sustainable textile production standard. Aims to eliminate harmful substances from the beginning of the manufacturing process and sets and controls standards for an environmentally friendly and safe production (Blue Sign 2019).
Circular Economy Package	CEP	European Commission initiative for a Europe-wide EU Strategy for Plastics in the Circular Economy to transform the way plastics and plastics products are designed, produced, used, and recycled. By 2030, all plastics packaging should be recyclable (European Commission 2019).
European Commission and UK government guidelines	EU-UK	European strategy on plastics as part of a transition toward a more circular economy with objectives to make recycling profitable, curb plastic waste, stop littering at sea, drive investment and innovation, and spur global change (European Commission 2018). The UK Government provides guidance on producer responsibilities on packaging waste (GOV.UK 2017).
New Plastics Economy	NPE	The New Plastics Economy is a three-year initiative to build momentum toward a plastics system that works. Led by the Ellen MacArthur Foundation, it is applying the principles of the circular economy and brings together key stakeholders to rethink and redesign the future of plastics, starting with packaging (New Plastics Economy 2017).
Packaging Recovery Note system	PRN	PRNs (packaging recovery notes) are the evidence required by producers of packaging waste to comply with the Producer Responsibility (Packaging Waste) Regulations 2005 in the UK. PRNs are issued by accredited reprocessors and act as an incentive to recycle and as a means for businesses to offset the amount of packaging that they place into the UK market (The Environment Exchange 2005).
Plastics Solutions Investor Alliance	PSIA	An international coalition of investors that will engage publicly traded consumer goods companies on the threat posed by plastic waste and pollution, supported by 25 institutional investors from four countries with a combined $1 trillion of assets under management (As You Sow 2018).
Ellen MacArthur Foundation	EMcA	The Ellen MacArthur Foundation works with business, government, and academia to build a framework for an economy that is restorative and regenerative by design (Ellen MacArthur Foundation 2017).
Waste and Resources Action Programme	WRAP	WRAP works with governments, businesses, and communities to deliver practical solutions to improve resource efficiency. Its mission is to accelerate the move to a sustainable, resource-efficient economy by reinventing how we design, produce, and sell products, rethinking how we use and consume products, and redefining what is possible through reuse and recycling (WRAP 2019).
World Wide Fund for Nature	WWF	To protect nature, it is necessary to take on the big issues facing our planet. These include facing up to challenges such as the urgent threat of climate change, promoting sustainable use of resources, and helping us all change the way we live and working with business and government to protect our planet for generations to come. Only then can we develop a world where people and wildlife thrive (WWF 2017).

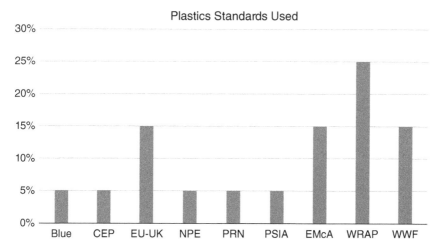

Figure 16.8 Proportion of plastics standards identified by fund managers. For definitions of the standards' descriptions, see Table 16.3.

16.6.10 Recyclable, Degradable, Biodegradable, and Compostable

The differences between biodegradable, compostable, degradable, and oxo-degradable plastics can be confusing and are often misused (Eenee 2019). Consequently, well-intentioned schemes to replace harmful plastics with seemingly more benign equivalents may not yield hoped-for benefits.

Degradable or oxo-degradable are traditional plastics treated with additives which cause them to disintegrate over several years. However, they do not truly decompose (compost) (Eenee 2019) and instead break down into many small pieces. As a result, they contaminate compost, reducing its value or limiting its use, and any recyclable plastic feedstock. Compostable is a superior standard: broadly, a biodegradable material is not necessarily compostable, but a compostable material is always biodegradable (Bag to Earth 2019). The compostable standard requires that materials conform to certain standards such as the European Standard EN 13432 and US Standard ASTM D6400, which lay down criteria for what can be described as compostable or biodegradable. These standards are intended to ensure that the materials will break down in industrial composting conditions. Further details and definitions are available in Eenee (2019) and Bag to Earth (2019).

16.6.11 Other Plastics Standards

The discussion on degradable and compostable plastics suggests that other standards may be relevant. Although these standards might be more appropriate for use by companies directly, rather than investors, they may help investors select firms. These include the following.

- European standard for compostable materials: EN 13432 (European Bioplastics 2000).
- USA standard for compostable materials: ASTM D6400 (ASTM International 2019).
- ISO: ISO/TC 61 working group and associated standards including ISO 17088:2012 (International Organization for Standardization 2012) for compostable plastics.
- Australia for compostable materials: AS 4736-2006 (Australian Standard® 2006), which is similar to EN 13432 but with the addition of worm toxicity test.

16.7 Recommendations

Most investment policies for funds addressed plastics, regarding them as of medium importance. As plastics must be balanced against other ethical and sustainability issues, this is unsurprising.

For significant issues, negative screening or exclusions appear to be perceived as a strong ethical investment response. However, they do not appear to be widely adopted for plastics, which could be due to multiple factors, such as a reluctance to extend current screens. A more significant difficulty may be defining which activities should be considered for plastics exclusion criteria. Production might be straightforward, but given the widespread use of plastics in packaging, transportation, food sales, and distribution, establishing exclusion criteria may not be easy. Distinction would be required between firms using plastics and those not, perhaps based on a metric representing the weight of plastic used per unit of turnover (or sales). Using sector data, *de minimis* levels could be determined to define a cut-off point, with firms using more than a specific weight of plastic per unit of turnover (or sales) excluded on a plastic negative screening criterion. Additional data would be required by fund managers and other investors, with the associated data collection challenges. Further research would be required to determine whether the metric suggested above would be the most suitable in this context. Care appears necessary to ensure that whichever metric was adopted could not be easily manipulated by firms, to "game" outcomes (R3).

The questionnaire did not ask fund managers to define how they negatively screened for plastics, but given the small number of funds involved, revealing the details about their exclusion process might also reveal their identities, counter to the promise of anonymity. It seems logical that ethical fund managers would develop and adopt negative screens (exclusions) based on the production of harmful plastics or excessive production, use, and waste generation for incorporation into investment policies, particularly for single-use plastics. This would place plastics on a par with the other important ethical issues (R1).

Overall, although positive screening was more widely used than exclusions, not all fund managers used it, and wider adoption would be beneficial (R2).

Half of the funds used best-in-class approaches to address plastics. Even if absolute standards may be difficult to determine, relative estimates of which firms are more (or less) proactive in addressing issues may be more straightforward and thus easier for fund managers to implement in investment strategies. Since half of fund managers did not use this approach, it could be more widely adopted (R4).

Most funds engaged on plastics. All engagements should help, but a cautious interpretation is that engagement could be evidence of the *ad hoc* approaches to plastic investment described at the outset. The worry is that such engagements may be short-lived, eye-catching initiatives designed for marketing benefit which can be quietly dropped once media focus moves elsewhere. Most likely the truth lies between the two extremes. It appeared surprising that most fund managers did not systematically vote on company resolutions involving plastics. This basic form of shareholder engagement should be more widely employed (R5).

Very few managers appeared to use policy engagement with governments or regulators. It appears that governments are providing stronger leadership than fund managers.

It would be helpful for ethical fund managers to be more proactive, since a combined approach involving investors as well as governments would be more likely to promote effective and rapid change in company behaviors than by governments alone (R6).

Investment policies mostly made use of the "reduce – reuse – recycle" paradigm, with primary emphasis placed on upstream reduction and smaller emphasis downstream (reuse, recycling). This focus appears sensible as cutting off the supply of plastics at the source seems to be the most effective means of addressing issues in the long term. Furthermore, fewer funds explored safe disposal. Given the longevity of plastics in the environment, the large amounts currently being used, and the time necessary to significantly reduce use, disposal would be an area where fund managers could support progress by investment in environmentally safe removal (R7).

The use of corporate standards with plastics appeared disappointingly low. It can be challenging to assess genuine commitment when addressing such issues within an organization, making independent standards a useful tool. Many fund managers (70%) were aware of corporate plastics standards but only 45% could name specific standards, with only 5% using them in investment policies (R8). No fund managers used plastics standards within their own organizations to demonstrate their commitment in this area. It is disappointing that so few fund managers make use of independent corporate plastics standards, which would likely improve the robustness of their investment policies and provide valuable support to these standards, giving additional momentum for their uptake by companies (R9). Initiatives include those such as the PSIA (As You Sow 2018), which fund managers could adopt. The lack of uptake suggests weak commitment, consistent with suspicions that engagement is an *ad hoc* activity, as mentioned above. The adoption of appropriate standards would help reassure client investors using ethical funds that the commitment to address plastics is genuine, enduring, and not merely a marketing exercise (R8, R9).

A Likert score analysis of the questionnaire responses ("plastics scores") indicates that a simple "yes/no" question was effective at determining whether fund managers' investment policies meaningfully addressed plastics. Fund managers who answered "yes" to "do you have a policy on plastics?" had plastics scores significantly higher than those answering "no," at the 5-sigma level. No further insight into the commitment of the fund managers on plastics appeared to be gained from asking "how important?". This result is disappointing. Although plastic may not be a primary issue for ethical fund managers, it was hoped that the question might elicit some further information. There is a clear message for investors seeking funds that address plastic pollution: unless detailed information about investment policies is to be collected from fund managers, the simple "yes/no" format is more effective at indicating commitment. It appears that the fund managers' own judgments of the importance they assign to plastics add no further value.

A summary of key recommendations is provided in Table 16.4 and has been keyed to the discussion above and earlier text by the recommendations numbers (R1,..., R9).

16.7.1 Broader Policy Recommendations

The focus of the chapter has been on investment policies that could be adopted by ethical investors to address plastic problems. However, several of these permit extensions to broader policy recommendations, external to investment practice.

Table 16.4 Investment policy recommendations for best practice by ethical and sustainable fund managers committed to addressing plastics-related issues

No	Area	Recommendation
R1	Exclusions	Adopting negative screening (exclusions) criteria based around plastics production, use and waste.
R2	Positive screening	More extensive use of positive screening to select companies that are helping address plastics issues in a range of areas including the development of alternatives (supporting reduction), reuse of plastics, recycling, and environmentally safe disposal.
R3	Metrics for plastics usage	Researching suitable metrics to determine levels of plastics usage. These might be based on the weight of plastic used per unit of turnover (or sales) or some other measure. Such metrics should be carefully designed to ensure that they cannot easily be manipulated by firms, to avoid the likelihood of companies "gaming" such a measure to their advantage.
R4	Best-in-class approaches	Considering wider use of best-in-class approaches to help ensure that investment is directed toward best practice in this area, and away from firms with minimal commitment.
R5	Share voting	Systematic use of share voting on plastics issues.
R6	Policy engagement	More proactive policy engagement to discuss plastics-related issues with governments and regulators seeking to address problems. This will provide useful commercial insight to government and regulators and provide additional support for policy initiatives from the investment community.
R7	Positive investment in disposal	Consider positive investment into companies developing commercial propositions for the environmentally safe disposal of plastics and plastic waste. This should support the removal of plastics whether in the form of detritus or once no longer required, whether at the end of life or due to replacement by alternatives.
R8	Standards for investment selection	Use of independent or corporate standards on plastics-related issues when selecting companies for investment. This helps provide confidence in investments selected and supports more widespread adoption of plastics standards by firms and corporate organizations.
R9	Investor standards	Adoption of independent investor standards on plastics. This will reassure client investors and attract potential investors. It will also help ensure that fund managers will maintain focus on the problems that plastics create so that they are less likely to reallocate attention elsewhere if plastics drop out of media and investor focus.

Table 16.5 Broad policy recommendations to help address plastics

No	Area	Recommendation
BR1	Critical plastics factors	Governments, local governments, and environmental bodies better define factors that measure the environmental and social costs of different forms of plastic waste, including longevity in the environment, toxicity, ability to recycle or dispose of, for use by firms and investors.
BR2	Encourage investor input on policy engagement	Governments and regulators to actively reach out to investors for input on policy engagement.
BR3	Research on safe disposal	Governments to support research into environmentally safe disposal of plastics, both academic and applied.
BR4	Wider adoption of plastics standards	Governments and industry bodies to encourage wider adoption of plastics standards by industry.
BR5	Reassessment of plastics degradability and compostability standards	Consider withdrawal of degradability standards that still result in plastic contamination. Consider development of standards for compostability that can be used under domestic composting conditions.

Governments, local governments, and environmental bodies could develop guidelines to better define critical factors in plastic usage (supporting R3) (Table 16.5, BR1). These institutions would be better positioned to determine the environmental and social costs of various forms of plastic waste and determine factors that matter (e.g., types of plastic, specific toxicities, longevity in the environment, ability to recycle or dispose of), and provide feedback to help firms and investors determine which types of plastic should be targeted as a priority.

Governments and regulators should actively reach out to encourage investors' input on policy engagement (supporting R6) in order to strengthen standards and regulations (BR2). Governments can also support additional research into environmentally safe disposal of plastics, in terms of both academic and applied industrial research, which investors can support as it nears commercial viability (supporting R7) (BR3).

Governments and industry bodies should also encourage firms to adopt plastics standards (supporting R8) (BR4). Ideally, regulators could also encourage ethical and sustainable investors to widely adopt plastics standards (supporting R9); however, the current lack of consistent terminology and taxonomy for ethical and sustainable investing styles would probably hinder this adoption at present.

Given the potential confusion between degradable and compostable plastics, the reassessment of the benefits these standards bring might be worthwhile. If degradable plastics still result in environmental contamination, perhaps they are of little value and provide only a cosmetic solution. In this case, consideration should be given to whether standards in relation to degradability should be withdrawn, with emphasis placed on compostability standards instead. Compostability standards could also be enhanced, as many plastics can only be broken down under industrial composting conditions. Thus, the development of standards to identify plastics that break down under domestic composting conditions would likely be beneficial (BR5).

These broader policy recommendations are summarized in Table 16.5 and indexed to the discussion above using the keys (BR1,..., BR5).

16.8 Conclusion

Ethical and sustainable investing offers a route for investors to encourage company actions to address the environmental damage caused by plastic waste. Ethical fund managers are aware of this, with some promoting their activities in this area. However, ethical fund managers have many other issues to deal with, and they may be more concerned with eye-catching initiatives than with the implementation of fundamental ethical investment policies intended to address plastics.

The author designed and submitted a questionnaire on ethical investment policies on plastics to selected fund managers. The analysis suggests that although some efforts are being made, the commitment by fund managers to address plastics may be somewhat weak. It appears that plastic may not be taken as seriously as other traditional, ethical, and sustainability issues (the "sextet of sin," animal rights testing, fossil divestment). Most activities appeared to be around company engagement, which runs a risk of being *ad hoc*, and may not follow through when the focus of media and the underlying investors' attention move elsewhere. To address plastics more strongly, ethical fund managers should develop specific investment policies, including those in Table 16.4.

For investors seeking funds that address plastic pollution issues, an analysis of the results suggests that unless they can question fund managers in detail about the processes they use, a simple "yes/no" question as to whether a fund manager's investment policies address plastics is the most effective indicator of their commitment. Attempts to refine this by enquiring about the level of importance they assign to plastics appeared to add no further value.

The results are discussed in detail above with specific recommendations, which the author believes would strengthen ethical and sustainable fund managers' investment policies in this area. Broader policy recommendations are also included in Table 16.5. The adoption of these by governments, regulators, and others would support the investment policy recommendations.

It is hoped that by considering the above recommendations, fund managers, company management, and others will be able to strengthen policies to support broader efforts in society to address the damaging global effects of plastic waste, threats posed by its chemical toxicity, and its appallingly harmful effects on wildlife and sea creatures.

Acknowledgments

The author gratefully thanks all the managers, ethical and sustainable analysis teams, and client support staff of the fund management firms contributing to the questionnaire. All contributors agreed for their questionnaire results to be used on condition of anonymity.

The author appreciates that ethical and sustainable investment decision making can be challenging and that even taking small steps adds value. While some criticism has been necessary, it is hoped that fund managers will take this as a genuine opportunity to explore how they can improve their investment processes to match best practice in this area. The author would also like to thank Professor Panagiotis Andrikopoulos of the Centre for Financial and Corporate Integrity (CFCI) at Coventry University and Professor Grzegorz Trojanowski of the Xfi Centre for Finance and Investment at the University of Exeter for their helpful feedback and constructive suggestions as well as the feedback from the anonymous referees.

A. Appendix: Questionnaire

#	Question	Response	
1	Does your investment policy seek to assess the involvement or activities of companies you invest in, in relation to plastics pollution issues?	Yes	No
1a	If your investment policy does seek to address plastics issues (i.e., you answered "yes" to question 1), how important are plastics issues in relation to your other ethical or sustainability criteria on a scale of 1 to 5 (1 = least important / 5 = most important)?	1–5	
2	Do you address investment in companies involved in plastics issues by avoiding them (negative screening or exclusions)?	Yes	No
3	Do you address investment in companies involved in plastics issues by selectively investing in companies that help address the problems caused (positive screening or impact investing)?	Yes	No
4	Do you address investment in companies involved in plastics issues by selectively investing in companies that help address the problems caused with investments where the returns or performance are secondary to the importance of the environmental benefits obtained (impact investing)?	Yes	No
5	Do you address investment in companies involved in plastics by investing in those companies that do least harm in this area, or are at least better than broadly equivalent peers?	Yes	No
6	Do you have a policy of voting against company issues that relate to harmful practices in relations to plastics?	Yes	No
7	Do you engage with company management to seek to improve their environmental management of plastics?	Yes	No
8	Do you engagement with governments or regulators to seek to improve the environmental management of plastics?	Yes	No
9–12	Does your investment policy in relation to companies with plastics exposure explore:		
	9. reduction in use of plastics by use of alternative materials or packaging, or replacement with more sustainable materials?	Yes	No
	10. reuse of plastics, for example bottle deposit schemes?	Yes	No
	11. recycling of plastics, use of plastics capable of being recycled, or use of plastics in ways that can be separated for recycling?	Yes	No

(Continued)

#	Question	Response	
	12. environmentally safe disposal, ensuring that plastic waste is disposed of responsibly so that it does not enter oceans or ecosystems and cause harm?	Yes	No
13	Are you aware of any corporate standards that address environmental issues surrounding the creation, use, reuse, recycling and disposal of plastics, whether run as formal national or international standards, voluntary initiatives or otherwise?	Yes	No
13a	If you are aware of any standards above (i.e., you answered "yes" to question 13), do you use any of these standards as selection criteria for investing in companies?	Yes	No
13b	If you are aware of any standards above (i.e., you answered "yes" to question 13), do you use any of these standards in relation to your own organization?	Yes	No
13c	If you are aware of any standards above (i.e., you answered "yes" to question 13), please list the standards you are aware of.	List	
14	So long as the results of this survey are used in an anonymized way, are you happy that the results of this questionnaire should be combined with the results from other fund managers we contact and published in publicly available media?	Yes	No

References

Adams, P., Aquilina, M., Baker, R., et al. (2016). Full disclosure: a round-up of FCA experimental research into giving information. Occasional Paper 23. www.fca.org.uk

Amel-Zadeh, A. and Serafeim, G. (2018). Why and how investors use ESG information: evidence from a global survey. *Financial Analysts Journal* 74 (3): 17.

As You Sow. (2018). As You Sow launches investor alliance to engage companies on plastic pollution. www.asyousow.org/blog/2018/6/14/as-you-sow-launches-investor-alliance-to-engage-companies-on-plastic-pollution

As You Sow. (2019). Nearly half of Starbucks shareholders support As You Sow proposal on reusable drinking containers and recycling goals. www.asyousow.org/press-releases/starbucks-shareholder-vote-plastic-recycling-goals

Ast, M. (2018). How advisers can help tackle the plastics scourge. https://citywire.co.uk/new-model-adviser/news/how-advisers-can-help-tackle-the-plastic-scourge/a1138538?ref=new-model-adviser-features-list

ASTM International. (2019). Standard Specification for Labeling of Plastics Designed to be Aerobically Composted in Municipal or Industrial Facilities. www.astm.org/Standards/D6400.htm

Australian Standard®. (2006). Biodegradable plastics – Biodegradable plastics suitable for composting and other microbial treatment. www.saiglobal.com/pdftemp/previews/osh/as/as4000/4700/4736-2006.pdf

Bag to Earth. (2019). We create paper bags that return to the earth. https://bagtoearth.com/about#sustainability

Bauer, R., Derwall, J., and Otten, R. (2007). The ethical mutual fund performance debate: new evidence from Canada. *Journal of Business Ethics* 70: 111–124.

BBC. (2018). Public 'back' taxes to tackle single-use plastic waste. www.bbc.co.uk/news/uk-45232167

BBC One. (2017). Blue Planet II. www.bbc.co.uk/programmes/p04tjbtx

Bebchuk, L., Cohen, A., and Ferrell, A. (2008). What matters in corporate governance? *Review of Financial Studies* 22 (2): 783–827.

Blue Sign. (2019). Follow the Blue Way. www.bluesign.com

Brammer, S., Brooks, C., and Pavelin, S. (2006). Corporate social performance and stock returns: UK evidence from disaggregate measures. *Financial Management* 35 (3): 97–116.

Dalton, J. (2018). Plastic cutlery and plates may be banned in UK in drive to halt oceans pollution. Independent. www.independent.co.uk/news/uk/home-news/plastic-cutlery-plates-banned-uk-oceans-sea-pollution-marine-life-economic-defra-a8425626.html

Derwall, J., Guenster, N., Bauer, R., and Koedijk, K. (2005). The eco-efficiency premium puzzle. *Financial Analysts Journal* 61 (2): 51–63.

Duncan, A.J. (1974). *Quality Control and Industrial Statistics*, 4e. Homewood: Richard D. Irwin Inc.

Eccles, R. G., Ioannou, I., & Serafeim, G. (2012). The impact of a corporate culture of sustainability on corporate behavior and performance. Harvard Business School Working Paper 12-035. www.eticanews.it/wp-content/uploads/2012/10/Studio-Harvard.pdf

Edmans, A. (2012). The link between job satisfaction and firm value, with implications for corporate social responsibility. Academy of Management Perspectives. http://faculty.london.edu/aedmans/RoweAMP.pdf

Eenee. (2019). Compostable vs biodegradable. www.biodegradable.bags.tas.eenee.com/contents/en-us/d7_compostable_biodegradable.html

Ellen MacArthur Foundation. (2017). www.ellenmacarthurfoundation.org

Elliott, T., & Elliott, L. (2018). Plastics consumption and waste management. www.wwf.org.uk/sites/default/files/2018-03/WWF_Plastics_Consumption_Report_Final.pdf

Elton, E.J., Gruber, M.J., Brown, S.J., and Goetzmann, W.N. (2017). *Modern Portfolio Theory and Investment Analysis*, 9e. Chichester: Wiley.

Erta, K., Hunt, S., Iscenko, Z., & Brambley, W. (2013). Applying behavioural economics at the Financial Conduct Authority. Occasional Paper 1. www.fca.org.uk

European Bioplastics. (2000). Harmonised standards for bioplastics. www.european-bioplastics.org/bioplastics/standards

European Commission. (2018). Plastic waste: a European strategy to protect the planet, defend our citizens and empower our industries. http://europa.eu/rapid/press-release_IP-18-5_en.htm

European Commission. (2019). Implementation of the Circular Economy Action Plan. http://ec.europa.eu/environment/circular-economy/index_en.htm

Festinger, L. (1957). *A Theory of Cognitive Dissonance*. Stanford: Stanford University Press.

Financial Reporting Council. (2012). UK Stewardship Code. www.frc.org.uk/investors/uk-stewardship-code

Gompers, P., Ishii, J., and Metrick, A. (2003). Corporate governance and equity prices. *Quarterly Journal of Economics* 118 (1): 107–155.

GOV.UK. (2017). Guidance – packaging waste: producer responsibilities. www.gov.uk/guidance/packaging-producer-responsibilities

GOV.UK. (2018a). Policy paper – at a glance: summary of targets in our 25 year environment plan. www.gov.uk/government/publications/25-year-environment-plan/25-year-environment-plan-our-targets-at-a-glance

GOV.UK. (2018b). Policy paper – carrier bags: why there's a 5p charge. www.gov.uk/government/publications/single-use-plastic-carrier-bags-why-were-introducing-the-charge

Grewal, J., Hauptmann, C., & Serafeim, G. (2017). Stock price synchronicity and material sustainability information. Harvard Business School: Working Paper 17-098. https://dash.harvard.edu/bitstream/handle/1/33110114/17-098.pdf?sequence=1&isAllowed=y

Guenster, N., Bauer, R., Derwall, J., and Koedijk, K. (2010). The economic value of corporate eco-efficiency. *European Financial Management* 17: 679–704.

Harrabin, R. (2018). UK faces build-up of plastic waste. www.bbc.co.uk/news/business-42455378

Hogg, R.V. and Craig, A.T. (1989). *Introduction to Mathematical Statistics*, 4e. New York: Macmillan.

International Organization for Standardization. (2012). ISO 17088:2012 Specifications for compostable plastics. www.iso.org/standard/57901.html

Investment Association. (2018). Fund statistics. www.theinvestmentassociation.org/fund-statistics

Investment Leaders Group (2014). *The Value of Responsible Investment*. Cambridge: University of Cambridge Institute for Sustainability Leadership.

Iscenko, Z., Duke, C., Huck, S., & Wallace, B. (2014). How does selling insurance as an add-on affect consumer decisions? Financial Conduct Authority. Occasional Paper 3. www.fca.org.uk

Kahnman, D. and Tversky, A. (1979). Prospect theory: an analysis of decision under risk. *Econometrica* 47 (2): 263–292.

Kempf, A. and Osthoff, P. (2007). The effect of socially responsible investing on portfolio performance. *European Financial Management* 13: 908–922.

Knoll, M.S. (2002). Ethical screening in modern financial markets: the conflicting claims underlying socially responsible investment. *Business Lawyer* 57: 681–726.

Kreander, N., Gray, R.H., Power, D.M., and Sinclair, C.D. (2005). *Evaluating the Performance of Ethical and Non-Ethical Funds: A Matched Pair Analysis*. Scotland: Centre for Social and Environmental Accounting Research, School of Management, University of St Andrews.

Laville, S. (2017). Chinese ban on plastic waste imports could see UK pollution rise. The Guardian. www.theguardian.com/environment/2017/dec/07/chinese-ban-on-plastic-waste-imports-could-see-uk-pollution-rise

Le Guern Lytle, C. (2017). When the mermaids cry: the great plastic tide. http://plastic-pollution.org

Likert, R. (1932). A technique for the measurement of attitudes. *Archives of Psychology* 22: 5–55.

Lukacs, P., Neubecker, L., & Rowan, P. (2016). Price discrimination and cross-subsidy in financial services. Occasional Paper 22. www.fca.org.uk

New Plastics Economy. (2017). https://newplasticseconomy.org

Perkins, A. (2018). Cotton buds and plastic straws could be banned in England next year. The Guardian. www.theguardian.com/environment/2018/apr/18/single-use-plastics-could-be-banned-in-england-next-year

Porter, M.E. and Kramer, M.R. (2006). Strategy and society: the link between competitive advantage and corporate social responsibility. *Harvard Business Review* December: 78–91.

Posner, R. and Langbein, J. (1980). Social investing and the law of trusts. *Michigan Law Review* 79: 72–112.

Rayer, Q.G. (2017a). Exploring ethical and sustainable investing. *CISI, Review of Financial Markets* 12: 4–10.

Rayer, Q. G. (2017b). Why ethical investing matters. https://p1-im.co.uk/wp-content/uploads/2018/12/2018-11-30-JP-Scot-why-ethical-matters-AS-PUBLISHED.pdf

Rayer, Q. G. (2018a). Saving our seas. Citywire New Model Adviser®, 576, 19.

Rayer, Q.G. (2018b). *Portfolio Construction Theory*, 6e. London: Chartered Institute for Securities and Investment.

Rayer, Q. G. (2018c,). Ethical investing: the unlikely route to alpha? Citywire New Model Adviser®, 585, 25.

Simmonds, E. (2018). Up to 29% of supermarket packaging is not recyclable, Which? finds. www.which.co.uk/news/2018/07/up-to-29-of-supermarket-packaging-is-not-recyclable-which-finds

Singh, P. and Sharma, V. (2016). Integrated plastic waste management: environmental and improved health approaches. *Procedia Environmental Sciences* 35: 692–700.

Stenström, H.C. and Thorell, J.J. (2007). *Evaluating the Performance of Socially Responsible Investment Funds: A Holding Data Analysis*. Stockholm: School of Economics, Master's thesis within Finance.

The Environment Exchange. (2005). Packaging Recovery Notes (PRNs). www.t2e.co.uk/packaging-recovery-note.html

Trojanowski, G. and Shaukat, A. (2017). Board governance and corporate performance. *Journal of Business Finance and Accounting* 45: 184–208.

UN PRI. (2006). What is responsible investment? www.unpri.org/pri/an-introduction-to-responsible-investment/what-is-responsible-investment

United Nations. (2006). Secretary-General launches 'Principles for Responsible Investment' backed by world's largest investors. www.un.org/press/en/2006/sg2111.doc.htm

WRAP. (2019). Our vision: WRAP's vision is a world in which resources are used sustainably. www.wrap.org.uk/about-us/about

WWF. (2017). www.wwf.org.uk

Index

Page numbers in *italics* refer to figures.
Page numbers in **bold** refer to Tables.

Environmental Policy: An Economic Perspective, First Edition. Edited by Thomas Walker,
Northrop Sprung-Much, and Sherif Goubran.
© 2020 John Wiley & Sons Ltd. Published 2020 by John Wiley & Sons Ltd.